PROGRESS IN

Nucleic Acid Research and Molecular Biology

Volume 82

PROGRESS IN
Nucleic Acid Research and Molecular Biology

edited by

P. Michael Conn
Oregon National Primate Research Center
Oregon Health and Science University
Beaverton, Oregon

Volume 82

AMSTERDAM • BOSTON • HEIDELBERG • LONDON
NEW YORK • OXFORD • PARIS • SAN DIEGO
SAN FRANCISCO • SINGAPORE • SYDNEY • TOKYO
Academic Press is an imprint of Elsevier

Academic Press is an imprint of Elsevier
32 Jamestown Road, London, NW1 7BY, UK
Radarweg 29, PO Box 211, 1000 AE Amsterdam, The Netherlands
30 Corporate Drive, Suite 400, Burlington, MA 01803, USA
525 B Street, Suite 1900, San Diego, CA 92101-4495, USA

This book is printed on acid-free paper. ∞

Copyright © 2008, Elsevier Inc. All rights reserved

No part of this publication may be reproduced or transmitted in any form or by any means, electronic or mechanical, including photocopy, recording, or any information storage and retrieval system, without permission in writing from the publisher

Permissions may be sought directly from Elsevier's Science & Technology Rights Department in Oxford, UK: phone (+44) (0) 1865 843830; fax (+44) (0) 1865 853333; email: permissions@elsevier.com. Alternatively you can submit your request online by visiting the Elsevier web site at http://elsevier.com/locate/permissions, and selecting *Obtaining permission to use Elsevier material*

Notice
No responsibility is assumed by the publisher for any injury and/or damage to persons or property as a matter of products liability, negligence or otherwise, or from any use or operation of any methods, products, instructions or ideas contained in the material herein. Because of rapid advances in the medical sciences, in particular, independent verification of diagnoses and drug dosages should be made

British Library Cataloguing in Publication Data
A catalogue record for this book is available from the British Library

Library of Congress Cataloging-in-Publication Data
A catalog record for this book is available from the Library of Congress

ISBN: 978-0-12-374549-1
ISSN: 0079-6603

For information on all Academic Press publications
visit our website at elsevierdirect.com

Printed and bound in the USA
08 09 10 11 12   10 9 8 7 6 5 4 3 2 1

Working together to grow
libraries in developing countries

www.elsevier.com | www.bookaid.org | www.sabre.org

ELSEVIER   BOOK AID International   Sabre Foundation

Contents

Drosophila Orthologues to Human Disease Genes: An Update on Progress 1
Sergey Doronkin and Lawrence T. Reiter

I. Introduction	2
II. Neurological Disease	2
III. *Drosophila* in Cancer Research	5
IV. Tumorigenesis, Neuroprotection, and Fortitude: The Hypoxic Response in *Drosophila*	7
V. Blood, Immune Response, and Infectious Disease	10
VI. Future Candidates	24
References	24

Applications of Fluorescence Correlation Spectroscopy to the Study of Nucleic Acid Conformational Dynamics. 33
Kaushik Gurunathan and Marcia Levitus

I. Introduction: Fluorescence and Energy Transfer	34
II. Fluorescence Correlation Spectroscopy	37
III. Conformational Dynamics of Nucleic Acids	46
IV. Experimental Techniques	60
V. Concluding Remarks	65
References	66

RNA Structure and Modeling: Progress and Techniques. 71
Dinggeng Chai

I. Introduction	72
II. Chemical and Enzymatic Methods	78
III. Physical Approaches to Study RNA Folding and Structure	82
IV. A Molecular Dynamic View of RNA Molecules	87

	V. Computer-Assisted Modeling	89
	VI. Conclusion	93
	References	93

DNA Polymerase ε: A Polymerase of Unusual Size (and Complexity) 101

Zachary F. Pursell and Thomas A. Kunkel

I.	Introduction	102
II.	Pol ε Structure	104
III.	Physical and Functional Interactions of Pol ε	109
IV.	Biochemical Properties of Pol ε	111
V.	Pol ε in DNA Replication	114
VI.	The Role of Pol ε in Checkpoint Control	119
VII.	Pol ε Involvement in Regulating Chromatin States	121
VIII.	Pol ε Relationship with Chromatin Remodeling Complexes	123
IX.	The Roles of Pol ε in Excision Repair of DNA Damage	126
X.	Pol ε in Recombination	129
XI.	*Schizosaccharomyces pombe* Pol ε	131
XII.	Xenopus Pol ε	133
XIII.	Concluding Remarks	134
	References	134

Site-directed Spin Labeling Studies on Nucleic Acid Structure and Dynamics 147

Glenna Z. Sowa and Peter Z. Qin

I.	Overview	148
II.	Basic Physics Underlying SDSL	150
III.	Site-Specific Attachment of Nitroxides to Nucleic Acids	153
IV.	Distance Measurements Using SDSL	162
V.	Site-Specific Structural and Dynamic Information from a Single-Labeled Nitroxide	172
VI.	Beyond Distance Measurements and Nitroxide Dynamics Analysis	187
VII.	Future Directions	189
	References	190

Molecular Computing with Deoxyribozymes. 199

Milan N. Stojanovic

I.	Introduction to Molecular Computing by Deoxyribozymes	199
II.	Deoxyribozyme-Based Logic Gates	200

III. Deoxyribozyme-Based Circuits for Arithmetical Operations	204
IV. Deoxyribozyme-Based Automata: Circuits that Play Tic-Tac-Toe	206
V. Deoxyribozyme-Based Control of Downstream Elements	210
VI. Expanding Molecular Logic to Nanoparticles	213
VII. Other Approaches to Autonomous Computing with DNA	215
VIII. Conclusions and Future Visions	215
References	217

Molecular Colony Technique: A New Tool for Biomedical Research and Clinical Practice........ 219

Alexander B. Chetverin and Helena V. Chetverina

I. Introduction	220
II. Detection of Airborne RNAs	221
III. Monitoring Reactions Between Single Molecules	225
IV. Cell-Free Gene Cloning	228
V. Molecular Colonies as a Precellular Form of Life	233
VI. Molecular Colony Diagnostics	235
VII. Gene and Gene Expression Analysis	241
VIII. Opportunities Provided by the Molecular Colony Technology	247
References	249
Index	257

Drosophila Orthologues to Human Disease Genes: An Update on Progress

Sergey Doronkin and
Lawrence T. Reiter

Department of Neurology, Department of Anatomy and Neurobiology, University of Tennessee Health Science Center, Memphis, Tennessee 38163

I. Introduction.. 2
II. Neurological Disease.. 2
III. *Drosophila* in Cancer Research..................................... 5
IV. Tumorigenesis, Neuroprotection, and Fortitude: The Hypoxic Response in *Drosophila* 7
V. Blood, Immune Response, and Infectious Disease.............. 10
VI. Future Candidates... 24
References.. 24

Modeling human disease in flies is possible because many basic processes of cellular proliferation, motility, regulation and interaction are highly conserved among multicellular organisms. Despite years of extensive study, a clear understanding of the basic biology of many human illnesses still remains elusive. In part, this is due to a deficit in adequate genetic model systems to study pathogenesis of disease dynamics in a developing organism. *Drosophila melanogaster* is emerging as a model of choice to study the molecular genetic underpinnings of human disease. It should be noted as well that the selection of *Drosophila* to model human genetic disease is not only based on homology, but also on the wide variety of tools and a century of classic genetics that provide outstanding experimental capabilities. Recent advances in methodology have increased the value of this model system to study the basic science of human disease and opened up new opportunities. The elucidation in flies of the underlying regulated mechanisms of human disorders may eventually reveal new therapeutic targets for the treatment of diseases. Here we describe recent advances in the study of neurological disorders, blood diseases and even cancer. We also outline future directions in research on modeling many devastating diseases in the fruit fly *Drosophila melanogaster*.

I. Introduction

The goal of this chapter is to present to a broader view of the state of human disease modeling in *Drosophila melanogaster* and to outline new directions in the study of the genetic basis of human disorders in flies. The *Drosophila* classical genetics powerhouse in combination with rapidly developing genomic and postgenomic tools accelerates the identification and characterization of gene networks. Because the molecular mechanisms controlling a variety of physiological pathways are largely conserved between flies and humans, flies are quite useful in modeling a variety of human diseases. These include nervous system disorders, cancer, immune responses, elements of the cardiovascular system, and many more (1). In addition, *Drosophila* genetic tools can also be used to study systems that are not evolutionarily conserved or common between flies and humans. In fact, fly genetics has been applied to the dissection of certain basic metabolic pathways in human organs that are not even present or undeveloped in flies. Due to obvious anatomical differences, the humble fruit fly certainly will never compete with mammalian models in every aspect of human diseases research, but a century of fly genetics should not be underestimated. As a genetic model organism, *Drosophila* has much to offer human disease researchers in terms of genetic screening power, a wide variety of molecular tools, multiple stock centers packed with a variety of allele, transgene, and deficiency collections and at the same time, any fly geneticist will tell you that they offer an elegant simplicity that drives basic research discoveries even in inexperienced student investigators.

II. Neurological Disease

In terms of modeling human genetic disease in *Drosophila*, neurological diseases have been the most lucrative [reviewed in (2, 3)]. This is not surprising considering a significant level of sequence and function conservation of nervous system genes and pathways that are directly relevant to human neurological disease [links between human disease and fly genes can be found using the Homophila database at http://homophila.sdsc.edu; (1, 4)]. Although the basic processes of neurogenesis, neuronal pathfinding, and synaptogenesis have been studied in *Drosophila* for some time, recently, there has been a boost in *Drosophila* research focusing directly on models for neurodegenerative diseases such as Parkinson's disease, Alzheimer's disease, spinocerebellar ataxias, and Huntington's disease. These efforts have not only contributed to a better understanding of the underlying basic genetic and molecular mechanisms of these disorders but also opened new avenues for practical pharmacotherapy and potential drug screening.

Contemporary genetic studies imply misexpression of the gene for α-*synuclein* in familial forms of Parkinson's disease. The common symptoms of this locomotion disorder are the presence of pathological aggregates of α-synuclein into inclusions known as Lewy bodies accompanied by the loss of dopaminergic neurons in substantia nigra (5–7). Despite the fact that there is no endogenous α-*synuclein* in flies, *Drosophila* models of Parkinson's disease have been created through transgenic expression of wild type or mutant forms of the human α-*synuclein* gene in flies. Gain-of-function expression of α-*synuclein* in the fly brain leads to Parkinson's pathology, recapitulating several important aspects of the disease including degeneration of dopaminergic cells and formation of Lewy body-like inclusions (8, 9). These flies also show age-dependent loss of movement control. The *Drosophila* model of Parkinson's disease can also be treated by some of the same drugs including dopamine agonists with positive results (10, 11). Although flies have no homologues to α-*synuclein*, they do have their versions of another two genes that genetically cause Parkinson's disease—*parkin* and *pink1*. In flies, *parkin* appears to be downstream of *pink1* in the same pathway (12, 13). Mutations in these genes lead to mitochondrial defects, muscle and locomotor dysfunction, but do not damage dopaminergic neurons (14, 15). On the other hand, overexpression of *parkin* in flies can suppress the effect of human α-synuclein-dependent degeneration phenotype (16). Co-overexpression of HSP70 and α-synuclein in the fly brain can rescue dopaminergic neurons against α-synuclein-induced neurodegerative phenotype (17), revealing potential therapeutic targets.

Alzheimer's disease has also been modeled in flies. Unlike the Parkinson's disease model, there are homologues in *Drosophila* to the human Alzheimer's disease-associated genes—the *amyloid precursor protein (APP)* gene and *Presenilin-1*. Just as in humans, Presenilin (fly version) is responsible for the release of the Aβ peptide from APP via proteolytic cleavage. The hallmark lesion in Alzheimer's disease is characterized by the formation of Aβ peptide-containing amyloid plaques in brain [reviewed in (18, 19)]. The mechanism of APP processing has been investigated in *Drosophila* using a genetic screening approach that showed *Drosophila* Presenilin is involved in the cleavage of the Notch protein (20). *Drosophila* APP appears to participate in axonal transport and if misexpressed leads to axonal vesicular accumulation (21–23). Fly models of Alzheimer's disease have striking similarities to phenotypic defects resembling Alzheimer's disease, in particular age-dependent learning defects, progressive neurodegeneration, and protein aggregate formation (24–28).

Even more impressive is the finding that the ubiquilin protein (UBQLN1), which when mutated can cause AD, can suppress Psn overexpression-induced phenotypes in flies (29).

The second hallmark of Alzheimer's disease, tauopathy, has also been modeled in flies. tau is a microtubule-associated protein and the principal component of the neurofibrillary lesions that can be associated with some amyloid plaques. Currently, *tau* receives more attention in the emerging view that the synergistic action of Aβ and tau is causal in Alzheimer's disease (30, 31). *Drosophila* models have been particularly instrumental in understanding the relations between tau and the Aβ peptide-containing amyloid plaques. When wild type and mutant forms of human tau are expressed in *Drosophila*, flies recapitulate the major human disease phenotypes, including progressive neurodegeneration, accumulation of abnormal tau, and neurotoxicity. The Alzheimer's disease-like neurofibrillary pathology is also observed when expression of wild-type human tau is combined with its *Drosophila* GSK-3 homologue (32), suggesting that GSK-3 may be a potential drug target. Expression of wild-type human tau also causes impaired axonal transport with vesicle aggregation and loss of locomotor function (33). *Drosophila* models support a role for cell-cycle activation, leading to apoptosis of postmitotic neurons *in vivo*. As in Alzheimer's disease, target of rapamycin kinase (TOR) activity is increased in fly models and promotes neurodegeneration. TOR activation enhances tau-induced neurodegeneration in a cell cycle-dependent manner and, when ectopically activated, drives cell-cycle activation and apoptosis in postmitotic neurons (34).

Possibly the most successful area of human neurological disease modeling in *Drosophila* is the models of polyglutamine tract repeat disorders [reviewed in (35)]. The fly eye is an excellent readout for polyglutamine tract repeat disorders like Huntington's disease and the spinocerebellar ataxias. In both conditions, there is a critical threshold of polyglutamine repeats that must be reached before a clinical presentation is observed. In flies, expression of the human Huntingtin protein or the SCA3/MJD protein containing the clinically relevant number of repeats leads to degeneration of photoreceptor neurons (36, 37). As in humans, these defects become more severe as the flies age and can become quite extreme as the number of polyglutamines increases suggesting conserved mechanisms. Using the *Drosophila* eye for misexpression studies provides an easy and convenient opportunity for genetic screening approaches. In screens for mutations that modify polyglutamine repeat phenotypes, it has been found that heat-shock proteins HSP70, HSP40, and other proteins can ameliorate these defects and may even serve as neuroprotectors (38–40). In fact, there may even be shared pathways between the SCAs and HD which suggest that these fly models could produce therapeutic targets that will work for all polyglutamine expansion disorders (41). There is even hope that polygluatime repeat disorder fly models could be used to screen for small molecules that can suppress these phenotypes as well.

III. *Drosophila* in Cancer Research

The major functional components of the cell cycle and metabolic and signaling pathways leading to cancer are highly conserved between fruit flies and humans. *Drosophila* is quite useful in modeling cancer or at least simple morphological aspects of cancer such as cell division, apoptosis, or cell migration [reviewed in (42–47)]. It should also be noted that the fruit fly has a glorious past serving in studies on the delineation of signaling pathways involved in oncogenesis. For example, signaling by Wnt proteins (Wingless in *Drosophila*), Ras/MAPK, Notch, and Hedgehog have well-characterized roles during the fly's embryonic development and in adults. All of these signaling pathways are clearly implicated in mammalian tumorigenesis and metastasis [reviewed in (48–52)].

Modeling human disease in flies, especially cancer, may at first seem a bit overambitious, except for the fact that at the molecular level, these cell-cycle genes and the process of cell proliferation, cell division, and cell motility are highly similar among multicellular organisms. Therefore, the fly becomes a genetic model for the identification of pathway members and not a model that recapitulates the cell biological characteristics of cancer such as tumor growth, differentiation, and vascularization. For example, the mosaic technique in flies provides the ability to work with strong or lethal mutations during various stages of development or in the adult tissues. Animals homozygous for a strong cell cycle or growth mutation could be lethal at very early stages of development without obvious morphological defects making them hard to study. The mosaic technique makes it possible to generate homozygous clones of lethal mutations in an animal that is otherwise heterozygous (i.e., morphologically wild type) for the same mutation [reviewed in (53)]. Genetic mosaicism of these chimeric flies mimics the loss of heterozygosity observed in the somatic cells of cancer patients. Mosaic flies carrying mutations can display clones of cells with irregular growth and overproliferating phenotype. This technique allows for the examination of homozygous mutant phenotypes and the design of genetic screens to identify tumor suppressor genes.

Using this mosaic analysis method in a nonvital somatic tissue like the compound eye led to the identification of the tumor suppressor gene *archipelago (ago)*. This screen was designed so that mutants that display increased cell proliferation could be easily identified using clonal analysis in the *Drosophila* compound eye (54). The F-box protein Archipelago is involved in a mechanism that suppresses cell proliferation by promoting the degradation of Cyclin E, a protein required for entry into S phase, the DNA synthesis phase of the cell cycle (54). Its human orthologue, *hCDC4/hAGO* has a similar function and, perhaps not surprisingly, is mutated in some breast and ovarian cancer cell lines (54–56). In addition, up to 16% of endometrial carcinomas may be the result of *hCDC4/hAGO* mutations (57). Mutations in another gene identified in this screen, *erupted*, the *Drosophila* orthologue of mammalian *tumor susceptibility gene 101*, causes dramatic non-cell-autonomous overproliferation of adjacent wild-type tissue (58). In a different screen for *Drosophila* mutations that result in tissue overgrowth, *salvador* (*sav*), a gene that promotes both cell cycle exit and cell death, was identified. The human orthologue of *salvador* (*h*WW45) is also mutated in cancer cell lines (59).

In addition to cell overproliferation phenotypes, *Drosophila* models have been used to study developmental signaling pathways that regulate pattern formation and cell migration. In particular are the processes that model cell motility, a critical step in tumorigenesis and metastasis (60, 61). During normal development and also during tumorigenesis, cells change their position extensively. The basic mechanisms involved in cell locomotion have been studied primarily *ex vivo* in cultured cells. An obvious disadvantage of this approach is that these cultured cells are now isolated from the comprehensive signaling networks that underlay guided cell migration *in vivo*, not to mention tissue-specific cellular interactions. Recently, major advances have been made in the study of migrating cells in *Drosophila* and have shed light on the basic mechanisms of cell locomotion. These studies of cell migration take place in a number of cell types including hemocytes (embryonic blood cells), primordial germ cells, border cells in the ovary, and tracheal cells.

A number of elegant studies in *Drosophila* using a variety of genetic approaches have contributed to our current understanding of guided cell migration. *Drosophila* genetic screens have uncovered new genes that are relevant to human cell migration, including tumor invasion and metastasis [reviewed in (44, 62–65)]. For example, loss of function genetic screening in mosaic clones revealed that Taiman, the p160-type steroid hormone coactivator, is required for the border-cell migration and the proper distribution of adhesion molecules (66). The human homologue of *Taiman*, called *Amplified in Breast Cancer 1*, is upregulated in many ovarian and breast cancers (67). Taiman acts as a coactivator for the estrogen receptor. Blocking estrogen signaling in cancer patients can prevent metastasis and recurrence (68). Understanding more about function of Taiman may provide insight into its role in estrogen receptor signaling and mechanisms of metastasis. On the other hand, Taiman has been reported to have a role in cell migration independent of its role in estrogen receptor signaling. SRC-3, a homologue of Taiman, promotes cell migration of human ovarian cancer cells regardless of the estrogen receptor status of the cells (69).

Border cells in the *Drosophila* ovary are a group of 6–10 epithelial cells that become invasive and eventually migrate to the oocyte border. Studies of border-cell migration have revealed that transformation of nonmotile cells within follicular epithelium into invasive cells occurs via activation of the JAK/STAT signaling pathway through the Domeless receptor (Dome) (70–72). Similar JAK/STAT-dependent signaling mechanism is also applied in tracheal cell migration in response to activated expression of Trachealess and the FGF receptor (73, 74). Cancer cells appear to use similar mechanisms: some mammalian STATs are upregulated or activated in cancer cells, in fact, STAT3 can promote cell-cycle progression and protect against apoptosis (75).

An interesting connection between cell motility and programmed cell death was recently revealed in *Drosophila*. DIAP1, the *Drosophila* inhibitor of apoptosis protein (IAP), is required for border-cell migration (76). IAPs are evolutionarily conserved proteins that bind to caspase proteases blocking their activity and thereby inhibiting apoptosis. IAPs also control cell growth during carcinogenesis [reviewed in (77, 78)]. One human IAP, XIAP, is a key determinant of sensitivity to cisplatin in ovarian cancer cells, and failure of cisplatin to downregulate XIAP is a hallmark of chemoresistance (79, 80). In border cells, DIAP1 was identified in a genetic overexpression screen to rescue the migration defect caused by expression of a dominant-negative form of the small GTPase RAC1 (76). Although the mechanistic explanation of these phenomena has yet to be uncovered, given the functional similarities of border cells and ovarian cancer cells, it seems reasonable to predict that XIAP or another human IAP could contribute to cell motility in ovarian cancer cells.

Useful cancer-suppressing therapeutic agents may also be developed as a result of studies in *Drosophila*. Genetic screening in the Hedgehog pathway first clarified that to initiate a signaling cascade that regulates early tissue differentiation, the *hedgehog* gene interacts with the transmembrane protein *patched* and its partner *smoothened*. Cyclopamine, a compound in the corn lily, *Veratrum californicum*, was found to act as an inhibitor of *smoothened* and, as a consequence, suppressor of *hedgehog* signaling (81, 82). Mutations in the human homologue of *patched* have been reported in basal cell nevus syndrome (OMIM #109400), also known as Gorlin syndrome. *Drosophila* studies suggest the possibility that topical cyclopamine could be potentially beneficial in the treatment of skin cancer in humans (83, 84).

IV. Tumorigenesis, Neuroprotection, and Fortitude: The Hypoxic Response in *Drosophila*

Oxygen deprivation, or hypoxia, and the cellular mechanisms that can regulate hypoxia are key factors in the pathogenesis of cancer, stroke, and familial inherited disorders such as those that occur in Von Hippel-Lindau syndrome (OMIM #193300) [reviewed in (85–87)]. For example, localized hypoxic effects play a central role in limiting tumor growth and may also be involved in blunting the actions of important chemotherapies [reviewed in (88, 89)]. Oxygen deprivation causes devastating effects during acute ischemic injury and the accompanying cardiac infarct and stroke often lead to brain injury. In fact, some neurons in the brain are particularly vulnerable to hypoxia causing rapid irreversible damage. While complete interruption of blood flow can kill cardiac myocytes or kidney cells in 20–40 min, it takes less than 5 min to start trigger the death of neurons in brain [reviewed in (90, 91)]. Using simple genetic model organisms like *Drosophila* to delineate the mechanisms of hypoxic response and adaptations to low oxygenation these animals have acquired may provide new tools in therapeutic interventions to preserve the status quo in hypoxia-vulnerable cells, increase their tolerance for lower oxygen, and promote survival.

Many nonmammalian organisms can tolerate extended hypoxic episodes (92, 93). Studies in yeast and in zebrafish embryos had demonstrated that development in these organisms can be arrested reversibly in response to hypoxia (92–95). The ability to temporary shut down metabolism is not limited to embryos or early stages of development. Cold-blooded animals like turtles are known to hibernate in wintertime in essentially complete anoxia (96). Even some small mammals like mice can survive several hours of hypoxia with little or no neurological damage. Animation state in mice can be suspended by H_2S-induced strong suppression of oxygen use (97). In these conditions, the body temperature is dropping but the mouse can survive hypoxia for a few hours (98).

It is not surprising that *Drosophila* also exhibits a protective response to hypoxia because flies spend their embryonic and larval life stages submerged in rotting fruit, where they must compete for limited oxygen supplies. Early work on oxygen deprivation in flies showed that the initial mitotic cycles of fly embryos can be temporary arrested by hypoxia and embryos remain viable despite prolonged periods of hypoxia (99, 100). As one might expect, *Drosophila* tolerates much longer exposures to hypoxic conditions than mammals (101–104). In fact, even after a week in the near absence of oxygen, arrested embryos recover and develop when oxygen is restored.

Recent studies have revealed a sophisticated signal transduction system that has evolved to ensure animals survival during hypoxia. Activation of this signaling system alters the behavior of the organism and stalls cell proliferation. Although little is known about the mechanisms that elicit the rapid metabolic turnover upon sudden reduction of oxygen supply, some evidence suggests the contribution of processes that trigger a rapid switch to energy conservation upon abrupt imposing of severe hypoxia [reviewed in (105, 106)]. The powerful genetic manipulations available in *Drosophila* may prove to be the key tools required to answer the profoundly difficult question of identifying ways to enhance the survival of hypoxia-sensitive cells.

Considerable progress has been made in understanding the aspects of hypoxic response under control of a transcription factor called the hypoxia-inducible factor 1 (HIF-1). HIF-1 plays a pivotal role in cellular adaptation to oxygen availability and is directly regulated by ubiquitin-dependent machinery [reviewed in (*107*)]. HIF-1 is composed of two subunits, the oxygen-sensitive HIF-1α and constitutively expressed HIF-1β. When oxygen levels are normal, the HIF-1α subunit of this factor is targeted for ubiquitination and subsequent rapid degradation by the ubiquitin proteasome system (*108, 109*). This process involves the modification of HIF-1α by prolyl hydroxylases, or PHDs (*108–111*). Hydroxylated HIF-1α is then recognized by the E3 ubiquitin ligase that contains the product of the *von Hippel-Lindau* (or *VHL*) gene, a tumor-suppressor gene, and targets it for degradation (*112–114*). Hypoxic conditions stabilize the HIF-1α protein and allow it to accumulate in the cell. HIF-1-induced transcription contributes to hypoxia response that allows cells to accommodate to at least mild reductions of oxygen (*112–116*).

A homologous hypoxia-responsive system has been described in flies. *Drosophila* bHLH-PAS proteins Similar (Sima) and Tango (Tgo) are HIF-1α and -β orthologues, respectively (*108, 117, 118*). *Drosophila* orthologues of VHL, or dVHL, and of PHD, encoded by the *hph* gene, have also been identified (*118–121*). Like in mammals, *Drosophila* HIF-1β homologue *tango* is constitutively expressed regardless of oxygen conditions, while *HIF-1α* homologue *sima* is rapidly degraded in normoxia and stabilized in hypoxia (*108, 117, 118*). Following its mammalian homologue pattern, normoxic Sima degradation in flies depends on the activity of a conserved prolyl-4-hydroxylase, *Drosophila* PHD (*118*).

The HIF-dependent transcription is peaking when oxygen is lowered to 5% (vs normal 20.8%). Flies can generally adapt to mild hypoxic conditions and even continue to grow and reproduce (*122*). However, under severe hypoxic conditions (<1% oxygen), they struggle to survive. At 1% oxygen or below, *Drosophila* embryos enter a state of suspended activity, whereas larvae will attempt to escape from the oxygen-poor environment (*123*). The responses to mild hypoxia are under the control of HIF (*124*), whereas the response to severe hypoxia involves distinct controls that are largely independent of gene expression (*125*). Abrupt termination of oxygen supply immediately arrests nearly every metabolic process in flies: cell cycle, cell motility, gene expression, turnover of nucleic acids, and proteins. Embryos typically retain about 75% of their ATP, indicating that they are conserving limited reserves (*126*). Once oxygen is restored, even after several days, embryos with suspended activity resume their development and normal flies can be produced.

Drosophila larvae, deprived of oxygen, exhibit behavioral changes that are related to a larval phenotype governed by a protein kinase G (PKG) allele (*127*). PKG is involved in one pathway with nitric oxide (NO), which is a well-known regulator of hemodynamics in humans, suggesting that cellular, developmental, and behavioral responses to NO mimicked those induced by hypoxia. Genetic and pharmacological tests showed that NO mediates at least some of these responses (*123*).

Powerful unbiased genetic screening methods available in *Drosophila* have been applied to identify hypoxia-sensitive mutants. Genes *hypnos-1*, *hypnos-2*, and *hypnos-3* were isolated in a screen for mutants with slow recovering mobility after a 5-min period of hypoxia (*102*). *hypnos-2*, a pre-mRNA *adenosine deaminase* and *dMRP4*, a homologue of a human *multidrug resistance protein* (*128*, *129*) were also identified in this screen. The *hypnos-2* mutants were implicated in RNA editing, in neuronal function, and in the response to hypoxia (*129*). Knocking down HIF-1α partially restores sensitivity to chemotherapy including levels of adriamycin, etoposide, and others and may be due to the regulation of multidrug resistance proteins by HIF-1α.

Drosophila models may be of particular interest with regards to nutrition-dependent mechanisms of hypoxia tolerance. In flies, the response to chronic hypoxia appears to be not only strongly dependent on diet but is also largely age independent (*130*, *131*). Remarkably, a rich diet (more sugars and proteins) promotes sensitivity to chronic hypoxia in *Drosophila*, whereas starvation increases the life span in hypoxia conditions (*130*, *131*). Although difficult to directly compare to humans, some studies suggest that calorie restriction decreases cancer risk in mammals (*132*, *133*).

V. Blood, Immune Response, and Infectious Disease

Innate immunity is a phylogenetically ancient protection mechanism. It serves as the first line of defense against infection by foreign pathogens. Evolutionary conservation of biochemical pathways involved in innate immunity make *Drosophila* a powerful model to study the prototypical immune response.

Flies, just like humans, also suffer from infectious disease. A fly could not survive without mechanisms to constantly defend itself against pathogens in its native environment. Despite the fact that immune response in flies, like in all invertebrates, does not involve a T-cell response or the production of specific antibodies against foreign proteins, there are many similarities in innate immunity between flies and humans. Many striking parallels can be drawn between functions of circulating cells, like leukocytes, as well as transcription factors and signaling pathways, such as GATA factors, JAK/STAT, or Notch pathways that regulate hematopoesis and immune response [reviewed in (*134*, *135*)].

Mammalian blood contains three distinct groups of cells: red cells, white cells, and platelet. Red blood cells deliver oxygen, the platelet group promotes clotting, while the white blood cells provide immunity and scavenge dying cells. Mammalian hematopoiesis occurs in two waves called primitive and definitive hematopoiesis, equivalent to *Drosophila* embryonic and larval hematopoiesis, respectively. *Drosophila* hemolymph performs all blood-like circulation duties except delivering oxygen.

There are three lineages of *Drosophila* blood cells, or hemocytes: plasmatocytes, crystal cells, and lamellocytes. The plasmatocytes are the predominant *Drosophila* blood cell line and equivalent of the cells from the mammalian monocyte/macrophage lineage. They play a critical role in the phagocytosis of invading microorganisms, engulfment of apoptotic cells, and tissue remodeling (*134*). Other immune functions including melanization and encapsulation are performed by the less frequently distributed crystal cells and lamellocytes in *Drosophila*.

Although fly hematopoiesis is simpler compared with mammalian hematopoiesis, many processes responsible for making blood cells are well preserved throughout evolution. Blood cell development in humans and flies is regulated by several remarkably homologous transcription factors and signaling cascades. *serpent*, a transcription factor and *Drosophila GATA* orthologue is required for the hemocyte development in flies (*136, 137*). The *Friend-of-GATA* homologue *U-shaped* has been found to block crystal-cell development (*138*). The family of Friend-of-GATA multiple zinc-finger proteins is known to regulate GATA activity in mammals [reviewed in (*139*)]. In mice, *Friend-of-Gata-1* is involved in regulation of erythropoiesis and megakaryopoiesis (*140*). In flies and vertebrates, both GATA factors and Friend-of-GATA are under the control of BMP signaling. Dpp, similar to its vertebrate counterpart BMP2/4, regulates *srp* and *ush* transcription (*141*).

lozenge, another transcription factor involved in crystal cell formation (*142, 143*), contains a "RUNT" domain homologous to a human transcription factor AML1/RUNX1. AML1 was originally isolated as a fusion partner in a chromosomal translocation associated with acute myelogenous leukemia and found to be necessary for definitive hematopoiesis [reviewed by (*144*)]. The expression of *lozenge* in larvae appears to be under the control of the Notch pathway (*145*). Notch signaling has been widely implicated in the regulation of hematopoiesis in mammals [reviewed in (*146, 147*)].

The *Drosophila* orthologue of the vertebrate gene encoding early B-cell factor-1, the transcription factor named *collier*, is required for lamellocyte specification (*148*). *Drosophila* larvae mutant for *collier* fail to produce lamellocytes on parasitization (*148*).

The Janus kinase signal transducers and activators of transcription (JAK-STAT) have been implicated in conserved regulation of blood cell development. Hyperactivation of Hopscotch, the *Drosophila* JAK homologue, causes hemocyte overproliferation and melanized tumor formation (*149–151*). In humans, hyperactivation of STAT homologues is associated with various leukemias and lymphomas (*152, 153*).

Normal plasmatocytes development and their migration to the posterior end of the embryo are regulated by receptor tyrosine kinase pathway that requires the activity of the PDGF/VEGF receptor, or PVR (*154*). PVR is a

TABLE I
Drosophila Genes Related to Human Diseases

FLY ID	Fly name	Corresponding disease, OMIM
CG11734	HERC2	Skin/hair/eye pigmentation 1, blond/brown hair, 227220 (3); Skin/hair/eye pigmentation 1, blue/nonblue eyes, 227220 (3)
CG10367	HMG coenzyme A reductase	Statins, attenuated cholesterol lowering by (3)
CG4067	Pugilist	Abruptio placentae, susceptibility to (3); Spina bifida, folate-sensitive, susceptibility to, 601634 (3)
CG34123	CG34123	Amyotrophic lateral sclerosis-parkinsonism/dementia complex 1, susceptibility to, 105500 (3)
CG1954	Protein C kinase 98E	Cerebral infarction, susceptibility to, 601367 (3)
CG6550	CG6550	Diabetes mellitus, noninsulin-dependent, susceptibility to, 125853 (3)
CG8256	lethal (2) k05713	Diabetes mellitus, noninsulin-dependent, 125853 (3)
CG6713	Nitric oxide synthase	Pyloric stenosis, infantile hypertrophic, 1, susceptibility to, 179010 (3)
CG2118	CG2118	3-Methylcrotonyl-CoA carboxylase 1 deficiency, 210200 (3)
CG6871	Catalase	Acatalasemia (3)
CG3725	Calcium ATPase at 60A	Acrokeratosis verruciformis, 101900 (3); Darier disease, 124200 (3)
CG31183	CG31183	Acromesomelic dysplasia, Maroteaux type, 602875 (3)
CG11567	Cytochrome P450 reductase	Adrenal hyperplasia, congenital, due to combined P450C17 and P450C21 deficiency, 201750 (3); Antley-Bixler syndrome-like with disordered steroidogenesis, 201750 (3); Disordered steroidogenesis, isolated, 201750 (3); POR deficiency, 201750 (3)
CG2316	CG2316	Adrenoleukodystrophy, 300100 (3); Adrenomyeloneuropathy, 300100 (3)
CG5594	CG5594	Agenesis of the corpus callosum with peripheral neuropathy, 218000 (3)
CG11089	CG11089	AICA-ribosiduria due to ATIC deficiency, 608688 (3)
CG3752	Aldehyde dehydrogenase	Alcohol intolerance, acute (3); Fetal alcohol syndrome (1)
CG11423	CG11423	Alexander disease, 203450 (3); Leigh syndrome, 256000 (3); Mitochondrial complex I deficiency, 252010 (3)

(*Continues*)

TABLE I (Continued)

FLY ID	Fly name	Corresponding disease, OMIM
CG9140	CG9140	Alexander disease, 203450 (3); Leigh syndrome, 256000 (3); Mitochondrial complex I deficiency, 252010 (3)
CG8987	Tamas	Alpers syndrome, 203700 (3); Progressive external ophthalmoplegia with mitochondrial DNA deletions, 157640 (3); Sensory ataxic neuropathy, dysarthria, and ophthalmoparesis, 607459 (3); Spinocerebellar ataxia with epilepsy, 607459 (3)
CG5670	Na pump α-subunit	Alternating hemiplegia of childhood, 104290 (3); Migraine, familial basilar, 602481 (3); Migraine, familial hemiplegic, 2, 602481 (3)
CG5870	β-Spectrin	Anemia, neonatal hemolytic, fatal, and near-fatal (3); Elliptocytosis-3 (3); Spherocytosis-1 (3)
CG7955	CG7955	Anemia, sideroblastic, with ataxia, 301310 (3)
CG4545	Serotonin transporter	Anxiety-related personality traits, 607834 (3); Obsessive-compulsive disorder 1, 164230 (3)
CG18214	Trio	Arrhythmogenic right ventricular dysplasia 2, 600996 (3); Ventricular tachycardia, stress-induced polymorphic, 604772 (3)
CG7235	Hsp60C	Arrhythmogenic right ventricular dysplasia, familial, 12, 611528 (3); Naxos disease, 601214 (3)
CG1945	Fat facets	Azoospermia, 415000 (3)
CG31547	CG31547	Bartter syndrome, type 1, 601678 (3)
CG9209	Vacuolar peduncle	Basal cell carcinoma, somatic (3); Capillary malformation-arteriovenous malformation, 608354 (3); Parkes Weber syndrome, 608355 (3)
CG2411	Patched	Basal cell carcinoma, somatic, 605462 (3); Basal cell nevus syndrome, 109400 (3); Holoprosencephaly-7, 610828 (3)
CG34157	Dystrophin	Becker muscular dystrophy, 300376 (3); Cardiomyopathy, dilated, 3B, 302045 (3); Duchenne muscular dystrophy, 310200 (3)
CG3736	Okra	Breast cancer, invasive intraductal (3); Colon adenocarcinoma (3); Lymphoma, non-Hodgkin (3)
CG18572	Rudimentary	Carbamoylphosphate synthetase I deficiency, 237300 (3)
CG32019	Bent	Cardiomyopathy, dilated, 1G, 604145 (3); Cardiomyopathy, familial hypertrophic, 9 (3); Muscular dystrophy, limb-girdle, type 2J, 608807 (3); Myopathy, early-onset, with fatal cardiomyopathy, 611705 (3); Myopathy, proximal, with early respiratory muscle involvement, 603689 (3); Tibial muscular dystrophy, tardive, 600334 (3)

(Continues)

TABLE I (*Continued*)

FLY ID	Fly name	Corresponding disease, OMIM
CG15792	Zipper	Cardiomyopathy, dilated, 1S (3); Cardiomyopathy, familial hypertrophic, 1, 192600 (3); Myopathy, Laing distal, 160500 (3); Myopathy, myosin storage, 608358 (3); Scapuloperoneal syndrome, myopathic type, 181430 (3)
CG1522	Cacophony	Cerebellar ataxia, pure (3); Episodic ataxia, type 2, 108500 (3); Hemiplegic migraine, familial, 141500 (3); Spinocerebellar ataxia-6, 183086 (3)
CG3936	Notch	Cerebral arteriopathy with subcortical infarcts and leukoencephalopathy, 125310 (3)
CG9433	Xeroderma pigmentosum D	Cerebrooculofacioskeletal syndrome 2, 610756 (3); Trichothiodystrophy, 601675 (3); Xeroderma pigmentosum, group D, 278730 (3)
CG18102	Shibire	Charcot-Marie-Tooth disease, dominant intermediate B, 606482 (3); Myopathy, centronuclear, 160150 (3)
CG8566	unc-104	Charcot-Marie-Tooth disease, type 2A1, 118210 (3)
CG3869	Mitochondrial assembly regulatory factor	Charcot-Marie-Tooth disease, type 2A2, 609260 (3); Hereditary motor and sensory neuropathy VI, 601152 (3)
CG6778	Glycyl-tRNA synthetase	Charcot-Marie-Tooth disease, type 2D, 601472 (3); Neuropathy, distal hereditary motor, type V, 600794 (3)
		Charcot-Marie-Tooth disease, type 4B1, 601382 (3)
CG6939	SET domain-binding factor	Charcot-Marie-Tooth disease, type 4B2, 604563 (3)
CG17034	CG17034	Cholestasis, benign recurrent intrahepatic, 243300 (3); Cholestasis, progressive familial intrahepatic 1, 211600 (3)
CG10181	Multidrug resistance 65	Cholestasis, familial intrahepatic, of pregnancy, 147480 (3); Cholestasis, progressive familial intrahepatic 3, 602347 (3); Gallbladder disease 1, 600803 (3)
CG10226	CG10226	Cholestasis, familial intrahepatic, of pregnancy, 147480 (3); Cholestasis, progressive familial intrahepatic 3, 602347 (3); Gallbladder disease 1, 600803 (3)
CG3879	Multidrug resistance 49	Cholestasis, familial intrahepatic, of pregnancy, 147480 (3); Cholestasis, progressive familial intrahepatic 3, 602347 (3); Gallbladder disease 1, 600803 (3)

(*Continues*)

TABLE I (*Continued*)

FLY ID	Fly name	Corresponding disease, OMIM
CG8523	Multidrug resistance 50	Cholestasis, familial intrahepatic, of pregnancy, 147480 (3); Cholestasis, progressive familial intrahepatic 3, 602347 (3); Gallbladder disease 1, 600803 (3)
CG10117	Tout-velu	Chondrosarcoma, 215300 (3); Exostoses, multiple, type 1, 133700 (3)
CG2093	CG2093	Choreoacanthocytosis, 200150 (3)
CG15804	Dynein heavy chain at 62B	Ciliary dyskinesia, primary, 3, 608644 (3); Kartagener syndrome, 244400 (3)
CG1842	Dynein heavy chain at 89D	Ciliary dyskinesia, primary, 3, 608644 (3); Kartagener syndrome, 244400 (3)
CG3339	CG3339	Ciliary dyskinesia, primary, 3, 608644 (3); Kartagener syndrome, 244400 (3)
CG3723	Dynein heavy chain at 93AB	Ciliary dyskinesia, primary, 3, 608644 (3); Kartagener syndrome, 244400 (3)
CG5526	Dynein heavy chain at 36C	Ciliary dyskinesia, primary, 3, 608644 (3); Kartagener syndrome, 244400 (3)
CG7092	Dynein heavy chain at 16F	Ciliary dyskinesia, primary, 3, 608644 (3); Kartagener syndrome, 244400 (3)
CG7507	Dynein heavy chain 64C	Ciliary dyskinesia, primary, 3, 608644 (3); Kartagener syndrome, 244400 (3)
CG2139	Aralar1	Citrullinemia, adult-onset type II, 603471 (3); Citrullinemia, type II, neonatal-onset, 605814 (3)
CG17596	RPS6-protein kinase-II	Coffin-Lowry syndrome, 303600 (3); Mental retardation, X-linked nonspecific, type 19 (3)
CG11579	Armadillo	Colorectal cancer (3); Hepatoblastoma (3); Hepatocellular carcinoma, 114550 (3); Ovarian carcinoma, endometrioid type (3); Pilomatricoma, 132600 (3)
CG15319	Nejire	Colorectal cancer, 114500 (3); Rubinstein-Taybi syndrome, 180849 (3)
CG7003	Msh6	Colorectal cancer, hereditary nonpolyposis, type 5 (3); Endometrial cancer, familial, 608089 (3); Mismatch repair cancer syndrome, 276300 (3); Ovarian cancer, endometrial type, 608089 (3)
CG4567	CG4567	Combined oxidative phosphorylation deficiency 1, 609060 (3)
CG17704	Nipped-B	Cornelia de Lange syndrome 1, 122470 (3)
CG6057	SMC1	Cornelia de Lange syndrome 2, 300590 (3)

(*Continues*)

TABLE I (*Continued*)

FLY ID	Fly name	Corresponding disease, OMIM
CG9802	Chromosome-associated protein	Cornelia de Lange syndrome 3, 610759 (3)
CG12891	Mitochondrial carnitine palmitoyltransferase I	CPT deficiency, hepatic, type IA, 255120 (3)
CG9907	Paralytic	Cramps, familial, potassium-aggravated (3); Hyperkalemic periodic paralysis, 170500 (3); Hypokalemic periodic paralysis, 170400 (3); Myasthenic syndrome (3); Myotonia congenita, atypical, acetazolamide-responsive, 608390 (3); Paramyotonia congenita, 168300 (3)
CG1250	sec23	Craniolenticulosutural dysplasia, 607812 (3)
CG18617	Vha100-2	Cutis laxa, autosomal recessive, type II, 219200 (3); Wrinkly skin syndrome, 278250 (3)
CG7678	CG7678	Cutis laxa, autosomal recessive, type II, 219200 (3); Wrinkly skin syndrome, 278250 (3)
CG1886	ATP7	Cutis laxa, neonatal (3); Menkes disease, 309400 (3); Occipital horn syndrome, 304150 (3)
CG17927	Myosin heavy chain	Deafness, autosomal dominant 11, neurosensory, 601317 (3); Deafness, autosomal recessive 2, neurosensory, 600060 (3); Usher syndrome, type 1B, 276900 (3)
CG6976	Myo28B1	Deafness, autosomal dominant 11, neurosensory, 601317 (3); Deafness, autosomal recessive 2, neurosensory, 600060 (3); Usher syndrome, type 1B, 276900 (3)
CG7595	Crinkled	Deafness, autosomal dominant 11, neurosensory, 601317 (3); Deafness, autosomal recessive 2, neurosensory, 600060 (3); Usher syndrome, type 1B, 276900 (3)
CG5695	Jaguar	Deafness, autosomal dominant 22, 606346 (3); Deafness, autosomal recessive 37, 607821 (3); Deafness, sensorineural, with hypertrophic cardiomyopathy, 606346 (3)
CG7438	Myosin 31DF	Deafness, autosomal dominant nonsyndromic sensorineural, 607841 (3)
CG9155	Myosin 61F	Deafness, autosomal dominant nonsyndromic sensorineural, 607841 (3)
CG17941	Dachsous	Deafness, autosomal recessive 12, 601386 (3); Usher syndrome, type 1D, 601067 (3); Usher syndrome, type 1H, 611581 (3)

(*Continues*)

TABLE I (*Continued*)

FLY ID	Fly name	Corresponding disease, OMIM
CG2165	CG2165	Deafness, autosomal recessive 12, modifier of, 601386 (3)
CG2174	Unconventional myosin class XV	Deafness, autosomal recessive 3, 600316 (3)
CG5284	CG5284	Dent disease, 300009 (3); Hypophosphatemic rickets, 300554 (3); Nephrolithiasis, type I, 310468 (3); Proteinuria, low molecular weight, with hypercalciuric nephrocalcinosis, 308990 (3)
CG18402	Insulin-like receptor	Diabetes mellitus, insulin-resistant, with acanthosis nigricans, 610549 (3); Hyperinsulinemic hypoglycemia, familial, 5, 609968 (3); Leprechaunism, 246200 (3); Rabson-Mendenhall syndrome, 262190 (3)
CG4006	Akt1	Diabetes mellitus, type II, 125853 (3)
CG5602	CG5602	DNA ligase I deficiency (3)
CG6146	Topoisomerase 1	DNA topoisomerase I, camptothecin-resistant (3)
CG10223	Topoisomerase 2	DNA topoisomerase II, resistance to inhibition of, by amsacrine (3)
CG33087	CG33087	Donnai-Barrow syndrome, 222448 (3)
CG10505	CG10505	Dubin-Johnson syndrome, 237500 (3)
CG11897	CG11897	Dubin-Johnson syndrome, 237500 (3)
CG14709	CG14709	Dubin-Johnson syndrome, 237500 (3)
CG31792	CG31792	Dubin-Johnson syndrome, 237500 (3)
CG31793	CG31793	Dubin-Johnson syndrome, 237500 (3)
CG4562	CG4562	Dubin-Johnson syndrome, 237500 (3)
CG5789	CG5789	Dubin-Johnson syndrome, 237500 (3)
CG6214	Multidrug-resistance like protein 1	Dubin-Johnson syndrome, 237500 (3)
CG7627	CG7627	Dubin-Johnson syndrome, 237500 (3)
CG8799	Lethal (2) 03659	Dubin-Johnson syndrome, 237500 (3)
CG9270	CG9270	Dubin-Johnson syndrome, 237500 (3)
CG33950	Terribly reduced optic lobes	Dyssegmental dysplasia, Silverman-Handmaker type, 224410 (3); Schwartz-Jampel syndrome, type 1, 255800 (3)
CG17603	TBP-associated factor 1	Dystonia-parkinsonism, X-linked, 314250 (3)
CG6199	CG6199	Ehlers-Danlos syndrome, type VI, 225400 (3); Nevo syndrome, 601451 (3)
CG1977	α-Spectrin	Elliptocytosis-2, 130600 (3); Pyropoikilocytosis, 266140 (3); Spherocytosis, recessive, 270970 (3)

(*Continues*)

TABLE I (*Continued*)

FLY ID	Fly name	Corresponding disease, OMIM
CG3210	Dynamin-related protein 1	Encephalopahty, lethal, due to defective mitochondrial peroxisomal fission (3)
CG10236	Laminin A	Epidermolysis bullosa, generalized atrophic benign, 226650 (3); Epidermolysis bullosa, junctional, Herlitz type, 226700 (3); Laryngoonychocutaneous syndrome, 245660 (3)
CG8433	Ext2	Exostoses, multiple, type 2, 133701 (3)
CG4389	CG4389	Fatty liver, acute, of pregnancy (3); HELLP syndrome, maternal, of pregnancy (3); LCHAD deficiency (3); Trifunctional protein deficiency, type 1 (3)
CG7793	Son of sevenless	Fibromatosis, gingival, 135300 (3); Noonan syndrome 4, 610733 (3)
CG3937	Cheerio	Frontometaphyseal dysplasia, 304120 (3); Heterotopia, periventricular nodular, with frontometaphyseal dysplasia, 300049 (3); Heterotopia, periventricular, 300049 (3); Heterotopia, periventricular, ED variant, 300537 (3); Melnick-needles syndrome, 309350 (3); Otopalatodigital syndrome, type I, 311300 (3); Otopalatodigital syndrome, type II, 304120 (3)
CG4094	Lethal (1) G0255	Fumarase deficiency, 606812 (3); Leiomyomatosis and renal cell cancer, 605839 (3); Multiple cutaneous and uterine leiomyomata, 150800 (3)
CG4095	CG4095	Fumarase deficiency, 606812 (3); Leiomyomatosis and renal cell cancer, 605839 (3); Multiple cutaneous and uterine leiomyomata, 150800 (3)
CG10693	Slowpoke	Generalized epilepsy and paroxysmal dyskinesia, 609446 (3)
CG4357	Sodium chloride cotransporter 69	Gitelman syndrome, 263800 (3)
CG9799	CG9799	Glaucoma 1, open angle, G, 609887 (3)
CG12140	CG12140	Glutaricaciduria, type IIC, 231680 (3)
CG3999	CG3999	Glycine encephalopathy, 605899 (3)
CG6904	CG6904	Glycogen storage disease 0, muscle, 611556 (3)
CG9485	CG9485	Glycogen storage disease IIIa, 232400 (3); Glycogen storage disease IIIb, 232400 (3)
CG33138	CG33138	Glycogen storage disease IV, 232500 (3)
CG7254	Glycogen phosphorylase	Glycogen storage disease VI (3)
CG4001	Phosphofructokinase	Glycogen storage disease VII, 232800 (3)

(*Continues*)

TABLE I (Continued)

FLY ID	Fly name	Corresponding disease, OMIM
CG7766	CG7766	Glycogenosis, X-linked hepatic, type I (3); Glycogenosis, X-linked hepatic, type II (3)
CG6938	CG6938	Gnthodiaphyseal dysplasia, 166260 (3)
CG2146	Dilute class unconventional myosin	Griscelli syndrome, type 1, 214450 (3)
CG32451	Secretory pathway calcium atpase	Hailey-Hailey disease, 169600 (3)
CG2259	Glutamate-cysteine ligase catalytic subunit	Hemolytic anemia due to γ-glutamylcysteine synthetase deficiency, 230450 (3)
CG8251	Phosphoglucose isomerase	Hemolytic anemia due to glucosephosphate isomerase deficiency (3); Hydrops fetalis, one form (1)
CG11427	Ruby	Hermansky-Pudlak syndrome 2, 608233 (3)
CG9565	Neprilysin 3	Hirschsprung disease, cardiac defects, and autonomic dysfunction (3); Hypertension, essential, susceptibility to, 145500 (3)
CG7470	CG7470	Hyperammonemia with hypoornithinemia, hypocitrullinemia, hypoargininemia, and hypoprolinemia (3)
CG5320	Glutamate dehydrogenase	Hyperinsulinism-hyperammonemia syndrome, 606762 (3)
CG7144	CG7144	Hyperlysinemia, 238700 (3); Saccharopinuria, 268700 (1)
CG11654	Adenosylhomocysteinase at 13	Hypermethioninemia with deficiency of S-adenosylhomocysteine hydrolase (3)
CG7145	CG7145	Hyperprolinemia, type II, 239510 (3)
CG4894	Ca[2+]-channel protein α[[1]] subunit D	Hypokalemic periodic paralysis, 170400 (3); Malignant hyperthermia susceptibility 5, 601887 (3); Thyrotoxic periodic paralysis, susceptibility to, 188580 (3)
CG4039	Minichromosome maintenance 6	Hypolactasia, adult type, 223100 (3); Tall stature, susceptibility to (3)
CG2331	TER94	Inclusion body myopathy with early-onset Paget disease and frontotemporal dementia, 167320 (3)
CG1799	Raspberry	Leber congenital amaurosis XI (3); Retinitis pigmentosa-10, 180105 (3)
CG7430	CG7430	Leigh syndrome, 256000 (3); Maple syrup urine disease, type III, 248600 (3)
CG17246	Succinyl coenzyme A synthetase flavoprotein subunit	Leigh syndrome, 256000 (3); Mitochondrial respiratory chain complex II deficiency, 252011 (3)

(Continues)

TABLE I (*Continued*)

FLY ID	Fly name	Corresponding disease, OMIM
CG5718	CG5718	Leigh syndrome, 256000 (3); Mitochondrial respiratory chain complex II deficiency, 252011 (3)
CG4215	Spellchecker1	Leukemia, juvenile myelomonocytic, 607785 (3); Melanoma, desmoplastic neurotropic (2); Neurofibromatosis, familial spinal, 162210 (3); Neurofibromatosis, type 1 (3); Neurofibromatosis-Noonan syndrome, 601321 (3); Pseudarthrosis, tibial, in NF1 (3); Watson syndrome, 193520 (3)
CG4032	Abl tyrosine kinase	Leukemia, Philadelphia chromosome-positive, resistant to imatinib (3)
CG17320	Sterol carrier protein X-related thiolase	Leukoencephalopathy with dystonia and motor neuropathy (3)
CG1913	α-Tubulin at 84B	Lissencephaly 3, 611603 (3)
CG2512	α-Tubulin at 84D	Lissencephaly 3, 611603 (3)
CG8308	α-Tubulin at 67C	Lissencephaly 3, 611603 (3)
CG9476	α-Tubulin at 85E	Lissencephaly 3, 611603 (3)
CG3182	Seizure	Long QT syndrome-2 (3); Short QT syndrome-1, 609620 (3); Long QT syndrome, acquired, susceptibility to (3)
CG17291	Protein phosphatase 2A at 29B	Lung cancer, 211980 (3)
CG5322	CG5322	Mannosidosis, alpha-, types I and II, 248500 (3)
CG6206	CG6206	Mannosidosis, alpha-, types I and II, 248500 (3)
CG9463	CG9463	Mannosidosis, alpha-, types I and II, 248500 (3)
CG9465	CG9465	Mannosidosis, alpha-, types I and II, 248500 (3)
CG9466	CG9466	Mannosidosis, alpha-, types I and II, 248500 (3)
CG9468	CG9468	Mannosidosis, alpha-, types I and II, 248500 (3)
CG32702	CG32702	Megaloblastic anemia-1, Finnish type, 261100 (3)
CG8711	cul-4	Mental retardation syndrome, X-linked, Cabezas type, 300354 (3); Mental retardation-hypotonic facies syndrome, X-linked, 2, 300639 (3)
CG11155	CG11155	Mental retardation, autosomal recessive, 6, 611092 (3)
CG3822	CG3822	Mental retardation, autosomal recessive, 6, 611092 (3)
CG9935	CG9935	Mental retardation, autosomal recessive, 6, 611092 (3)
CG7020	DISCO interacting protein 2	Mental retardation, FRA12A type, 136630 (3)

(*Continues*)

TABLE I (Continued)

FLY ID	Fly name	Corresponding disease, OMIM
CG8732	Lethal (2) 44DEa	Mental retardation, X-linked nonspecific, 63, 300387 (3)
CG17896	CG17896	Methylmalonate semialdehyde dehydrogenase deficiency (3)
CG1970	CG1970	Mitochondrial complex I deficiency, 252010 (3)
CG2286	NADH:ubiquinone reductase 75 kDa subunit precursor	Mitochondrial complex I deficiency, 252010 (3)
CG15288	Wing blister	Muscular dystrophy, congenital merosin-deficient, 607855 (3); Muscular dystrophy, congenital, due to partial LAMA2 deficiency, 607855 (3)
CG8107	Calpain-B	Muscular dystrophy, limb-girdle, type 2A, 253600 (3)
CG32626	CG32626	Myoadenylate deaminase deficiency (3)
CG10067	Actin 57B	Myopathy, actin, congenital, with cores (3); Myopathy, actin, congenital, with excess of thin myofilaments, 161800 (3); Myopathy, congenital, with fiber-type disporportion 1, 255310 (3); Myopathy, nemaline, 3, 161800 (3)
CG12051	Actin 42A	Myopathy, actin, congenital, with cores (3); Myopathy, actin, congenital, with excess of thin myofilaments, 161800 (3); Myopathy, congenital, with fiber-type disporportion 1, 255310 (3); Myopathy, nemaline, 3, 161800 (3)
CG18290	Actin 87E	Myopathy, actin, congenital, with cores (3); Myopathy, actin, congenital, with excess of thin myofilaments, 161800 (3); Myopathy, congenital, with fiber-type disporportion 1, 255310 (3); Myopathy, nemaline, 3, 161800 (3)
CG4027	Actin 5C	Myopathy, actin, congenital, with cores (3); Myopathy, actin, congenital, with excess of thin myofilaments, 161800 (3); Myopathy, congenital, with fiber-type disporportion 1, 255310 (3); Myopathy, nemaline, 3, 161800 (3)
CG5178	Actin 88F	Myopathy, actin, congenital, with cores (3); Myopathy, actin, congenital, with excess of thin myofilaments, 161800 (3); Myopathy, congenital, with fiber-type disporportion 1, 255310 (3); Myopathy, nemaline, 3, 161800 (3)

(Continues)

TABLE I (*Continued*)

FLY ID	Fly name	Corresponding disease, OMIM
CG7478	Actin 79B	Myopathy, actin, congenital, with cores (3); Myopathy, actin, congenital, with excess of thin myofilaments, 161800 (3); Myopathy, congenital, with fiber-type disporportion 1, 255310 (3); Myopathy, nemaline, 3, 161800 (3)
CG4376	α-Actinin	Myopathy, actin, congenital, with cores (3); Myopathy, actin, congenital, with excess of thin myofilaments, 161800 (3); Myopathy, congenital, with fiber-type disporportion 1, 255310 (3); Myopathy, nemaline, 3, 161800 (3)
CG9115	Myotubularin	Myotubular myopathy, X-linked, 310400 (3)
CG8439	T-complex chaperonin 5	Neuropathy, hereditary sensory, with spastic paraplegia, 256840 (3)
CG12092	NPC1b	Niemann-Pick disease, type C1, 257220 (3); Niemann-Pick disease, type D, 257220 (3)
CG5722	Niemann-Pick type C-1	Niemann-Pick disease, type C1, 257220 (3); Niemann-Pick disease, type D, 257220 (3)
CG11144	Metabotropic glutamate receptor	Night blindness, congenital stationary, type 1B, 257270 (3)
CG8479	Optic atrophy 1-like	Optic atrophy 1, 165500 (3); Optic atrophy and deafness, 125250 (3); Glaucoma, normal tension, susceptibility to, 606657 (3)
CG8380	Dopamine transporter	Orthostatic intolerance, 604715 (3)
CG7578	sec71	Periventricular heterotopia with microcephaly, 608097 (3)
CG8475	CG8475	Phosphorylase kinase deficiency of liver and muscle, autosomal recessive, 261750 (3)
CG6518	Inactivation no afterpotential C	Pituitary tumor, invasive (3)
CG6622	Protein C kinase 53E	Pituitary tumor, invasive (3)
CG1511	Eph receptor tyrosine kinase	Prostate cancer, progression and metastasis of, 603688 (3)
CG1516	CG1516	Pyruvate carboxylase deficiency, 266150 (3)
CG17369	Vacuolar H[+]-ATPase 55 kDa B subunit	Renal tubular acidosis with deafness, 267300 (3)
CG12602	CG12602	Renal tubular acidosis, distal, autosomal recessive, 602722 (3)
CG1709	Vha100-1	Renal tubular acidosis, distal, autosomal recessive, 602722 (3)
CG4675	Na[+]-driven anion exchanger 1	Renal tubular acidosis, proximal, with ocular abnormalities, 604278 (3)

(*Continues*)

TABLE I (Continued)

FLY ID	Fly name	Corresponding disease, OMIM
CG8877	prp8	Retinitis pigmentosa-13, 600059 (3)
CG10253	CG10253	Rhizomelic chondrodysplasia punctata, type 3, 600121 (3)
CG4252	Meiotic 41	Seckel syndrome 1, 210600 (3)
CG8363	PAPS synthetase	SEMD, Pakistani type (3)
CG8585	I[[h]] channel	Sinus bradycardia syndrome, 163800 (3)
CG7765	Kinesin heavy chain	Spastic paraplegia 10, 604187 (3)
CG12101	Heat-shock protein 60	Spastic paraplegia-13, 605280 (3)
CG2658	CG2658	Spastic paraplegia-7, 607259 (3)
CG12272	CG12272	Spastic paraplegia-8, 603563 (3)
CG1651	Ankyrin	Spherocytosis-2 (3)
CG1782	Ubiquitin-activating enzyme 1	Spinal muscular atrophy, infantile X-linked
CG6235	Twins	Spinocerebellar ataxia 12, 604326 (3)
CG33715	Muscle-specific protein 300	Spinocerebellar ataxia, autosomal recessive 8, 610743 (3)
CG1718	CG1718	Surfactant metabolism dysfunction, pulmonary, 3, 610921 (3)
CG6052	CG6052	Surfactant metabolism dysfunction, pulmonary, 3, 610921 (3)
CG2194	Suppressor of rudimentary	Thymine-uraciluria (3); Fluorouracil toxicity, sensitivity to (3)
CG8019	Haywire	Trichothiodystrophy, 601675 (3); Xeroderma pigmentosum, group B, 610651 (3)
CG7461	CG7461	VLCAD deficiency, 201475 (3)
CG12311	Protein O-mannosyl-transferase 2	Walker–Warburg syndrome, 236670 (3)
CG7642	Rosy	Xanthinuria, type I, 278300 (3)
CG12703	CG12703	Zellweger syndrome-2 (3)

receptor tyrosine kinase that is specifically localized in embryonic hemocytes (155–157). In *pvr*-mutant embryos, the migration of plasmatocytes is halted and many of them become apoptotic (155). The *Drosophila* genome contains three ligands of PVR, PVF1, PVF2, and PVF3, that are closely related to mammalian VEGF (154).

Many questions regarding *Drosophila* and human hematopoieses remain unanswered. However, based on phenotypic and genetic analyses, the parallels between innate immune response and blood cell development in insects and mammals are remarkable. The availability of fly mutations in members of evolutionarily conserved signaling pathways makes *Drosophila* a convenient system to characterize these pathways as well as cross talks between them. Once our understanding of *Drosophila* blood system gets more sophisticated, many questions on hematopoiesis and immune response will find their answers.

VI. Future Candidates

There are a large number of *Drosophila* genes that would potentially contribute to unraveling the basic biology of human disorders. Sequence comparison databases are regularly updated and integrated into a more synergistic view of the model system. A current list of *Drosophila* genes that are homologues to human disease genes to an e-value of $<10^{-100}$ is presented in Table I.

In conclusion, ongoing research on *Drosophila* models is very valuable and promising. A number of genes have been identified in genetic screens, studied, and characterized, and many of these conserved pathways are directly relevant to human diseases. We are just beginning to realize that the types of genes amenable to study in fly models is limited mostly by the creativity of the investigators. Still a tremendous amount of work is yet to be done. A combination of various multidisciplinary approaches and mammalian model systems will potentially increase our chances for developing therapeutic strategies for these disorders in our lifetime.

References

1. Reiter, L. T., Potocki, L., Chien, S., Gribskov, M., and Bier, E. (2001). A systematic analysis of human disease-associated gene sequences in *Drosophila* melanogaster. *Genome Res.* **11**, 1114–1125.
2. Bilen, J., and Bonini, N. M. (2005). *Drosophila* as a model for human neurodegenerative disease. *Annu. Rev.. Genet.* **39**, 153–171.
3. Marsh, J. L., and Thompson, L. M. (2006). *Drosophila* in the study of neurodegenerative disease. *Neuron* **52**, 169–178.
4. Chien, S., Reiter, L. T., Bier, E., and Gribskov, M. (2002). Homophila: Human disease gene cognates in *Drosophila. Nucleic Acids Res.* **30**, 149–151.
5. Birkmayer, W., and Hornykiewicz, O. (1961). [The L-3,4-dioxyphenylalanine (DOPA)-effect in Parkinson-akinesia.]. *Wien. Klin. Wochenschr.* **73**, 787–788.
6. Braak, H., Del Tredici, K., Rub, U., de Vos, R. A., Jansen Steur, E. N., and Braak, E. (2003). Staging of brain pathology related to sporadic Parkinson's disease. *Neurobiol. Aging.* **24**, 197–211.
7. Forno, L. S., DeLanney, L. E., Irwin, I., and Langston, J. W. (1996). Electron microscopy of Lewy bodies in the amygdala-parahippocampal region. Comparison with inclusion bodies in the MPTP-treated squirrel monkey. *Adv. Neurol.* **69**, 217–228.

8. Feany, M. B., and Bender, W. W. (2000). A *Drosophila* model of Parkinson's disease. *Nature* **404**, 394–398.
9. Park, S. S., and Lee, D. (2006). Selective loss of dopaminergic neurons and formation of Lewy body-like aggregations in alpha-synuclein transgenic fly neuronal cultures. *Eur. J. Neurosci.* **23**, 2908–2914.
10. Pendleton, R. G., Parvez, F., Sayed, M., and Hillman, R. (2002). Effects of pharmacological agents upon a transgenic model of Parkinson's disease in *Drosophila* melanogaster. *J. Pharmacol. Exp. Ther.* **300**, 91–96.
11. Pendleton, R. G., Rasheed, A., Sardina, T., Tully, T., and Hillman, R. (2002). Effects of tyrosine hydroxylase mutants on locomotor activity in *Drosophila*: A study in functional genomics. *Behav. Genet.* **32**, 89–94.
12. Clark, I. E., Dodson, M. W., Jiang, C., Cao, J. H., Huh, J. R., Seol, J. H., Yoo, S. J., Hay, B. A., and Guo, M. (2006). *Drosophila* pink1 is required for mitochondrial function and interacts genetically with parkin. *Nature* **441**, 1162–1166.
13. Park, J., Lee, S. B., Lee, S., Kim, Y., Song, S., Kim, S., Bae, E., Kim, J., Shong, M., Kim, J. M., and Chung, J. (2006). Mitochondrial dysfunction in *Drosophila* PINK1 mutants is complemented by parkin. *Nature* **441**, 1157–1161.
14. Greene, J. C., Whitworth, A. J., Kuo, I., Andrews, L. A., Feany, M. B., and Pallanck, L. J. (2003). Mitochondrial pathology and apoptotic muscle degeneration in *Drosophila* parkin mutants. *Proc. Natl. Acad. Sci. USA* **100**, 4078–4083.
15. Pesah, Y., Pham, T., Burgess, H., Middlebrooks, B., Verstreken, P., Zhou, Y., Harding, M., Bellen, H., and Mardon, G. (2004). *Drosophila* parkin mutants have decreased mass and cell size and increased sensitivity to oxygen radical stress. *Development* **131**, 2183–2194.
16. Yang, Y., Nishimura, I., Imai, Y., Takahashi, R., and Lu, B. (2003). Parkin suppresses dopaminergic neuron-selective neurotoxicity induced by Pael-R in *Drosophila*. *Neuron* **37**, 911–924.
17. Auluck, P. K., Chan, H. Y., Trojanowski, J. Q., Lee, V. M., and Bonini, N. M. (2002). Chaperone suppression of alpha-synuclein toxicity in a *Drosophila* model for Parkinson's disease. *Science* **295**, 865–868.
18. LaFerla, F. M., Green, K. N., and Oddo, S. (2007). Intracellular amyloid-beta in Alzheimer's disease. *Nat. Rev.. Neurosci.* **8**, 499–509.
19. Turner, R. S. (2006). Alzheimer's disease. *Semin. Neurol.* **26**, 499–506.
20. Struhl, G., and Greenwald, I. (2001). Presenilin-mediated transmembrane cleavage is required for Notch signal transduction in *Drosophila*. *Proc. Natl. Acad. Sci. USA* **98**, 229–234.
21. Gunawardena, S., and Goldstein, L. S. (2001). Disruption of axonal transport and neuronal viability by amyloid precursor protein mutations in *Drosophila*. *Neuron* **32**, 389–401.
22. Rusu, P., Jansen, A., Soba, P., Kirsch, J., Lower, A., Merdes, G., Kuan, Y. H., Jung, A., Beyreuther, O., Kjaerulff, O., and Kins, S. (2007). Axonal accumulation of synaptic markers in APP transgenic *Drosophila* depends on the NPTY motif and is paralleled by defects in synaptic plasticity. *Eur. J. Neurosci.* **25**, 1079–1086.
23. Torroja, L., Chu, H., Kotovsky, I., and White, K. (1999). Neuronal overexpression of APPL, the *Drosophila* homologue of the amyloid precursor protein (APP), disrupts axonal transport. *Curr. Biol.* **9**, 489–492.
24. Crowther, D. C., Page, R., Chandraratna, D., and Lomas, D. A. (2006). A *Drosophila* model of Alzheimer's disease. *Methods Enzymol.* **412**, 234–255.
25. Finelli, A., Kelkar, A., Song, H. J., Yang, H., and Konsolaki, M. (2004). A model for studying Alzheimer's Abeta42-induced toxicity in *Drosophila* melanogaster. *Mol. Cell. Neurosci.* **26**, 365–375.
26. Greeve, I., Kretzschmar, D., Tschape, J. A., Beyn, A., Brellinger, C., Schweizer, M., Nitsch, R. M., and Reifegerste, R. (2004). Age-dependent neurodegeneration and Alzheimer-amyloid plaque formation in transgenic *Drosophila*. *J. Neurosci.* **24**, 3899–3906.

27. Iijima, K., Liu, H. P., Chiang, A. S., Hearn, S. A., Konsolaki, M., and Zhong, Y. (2004). Dissecting the pathological effects of human Abeta40 and Abeta42 in *Drosophila*: A potential model for Alzheimer's disease. *Proc. Natl. Acad. Sci. USA* **101**, 6623–6628.
28. Seidner, G. A., Ye, Y., Faraday, M. M., Alvord, W. G., and Fortini, M. E. (2006). Modeling clinically heterogeneous presenilin mutations with transgenic *Drosophila*. *Curr. Biol.* **16**, 1026–1033.
29. Ganguly, A., Feldman, R. M., and Guo, M. (2008). ubiquilin antagonizes presenilin and promotes neurodegeneration in *Drosophila*. *Hum. Mol. Genet.* **17**, 293–302.
30. Blurton-Jones, M., and Laferla, F. M. (2006). Pathways by which Abeta facilitates tau pathology. *Curr. Alzheimer Res.* **3**, 437–448.
31. Rhein, V., and Eckert, A. (2007). Effects of Alzheimer's amyloid-beta and tau protein on mitochondrial function—role of glucose metabolism and insulin signalling. *Arch. Physiol. Biochem.* **113**, 131–141.
32. Jackson, G. R., Wiedau-Pazos, M., Sang, T. K., Wagle, N., Brown, C. A., Massachi, S., and Geschwind, D. H. (2002). Human wild-type tau interacts with wingless pathway components and produces neurofibrillary pathology in *Drosophila*. *Neuron* **34**, 509–519.
33. Mudher, A., Shepherd, D., Newman, T. A., Mildren, P., Jukes, J. P., Squire, A., Mears, A., Drummond, J. A., Berg, S., MacKay, D. *et al.* (2004). GSK-3beta inhibition reverses axonal transport defects and behavioural phenotypes in *Drosophila*. *Mol. Psychiatry* **9**, 522–530.
34. Khurana, V., Lu, Y., Steinhilb, M. L., Oldham, S., Shulman, J. M., and Feany, M. B. (2006). TOR-mediated cell-cycle activation causes neurodegeneration in a *Drosophila* tauopathy model. *Curr. Biol.* **16**, 230–241.
35. Mutsuddi, M., and Rebay, I. (2005). Molecular genetics of spinocerebellar ataxia type 8 (SCA8). *RNA Biol.* **2**, 49–52.
36. Jackson, G. R., Salecker, I., Dong, X., Yao, X., Arnheim, N., Faber, P. W., MacDonald, M. E., and Zipursky, S. L. (1998). Polyglutamine-expanded human huntingtin transgenes induce degeneration of *Drosophila* photoreceptor neurons. *Neuron* **21**, 633–642.
37. Warrick, J. M., Paulson, H. L., Gray-Board, G. L., Bui, Q. T., Fischbeck, K. H., Pittman, R. N., and Bonini, N. M. (1998). Expanded polyglutamine protein forms nuclear inclusions and causes neural degeneration in *Drosophila*. *Cell* **93**, 939–949.
38. Fayazi, Z., Ghosh, S., Marion, S., Bao, X., Shero, M., and Kazemi-Esfarjani, P. (2006). A *Drosophila* ortholog of the human MRJ modulates polyglutamine toxicity and aggregation. *Neurobiol. Dis* **24**, 226–244.
39. Iijima-Ando, K., Wu, P., Drier, E. A., Iijima, K., and Yin, J. C. (2005). cAMP-response element-binding protein and heat-shock protein 70 additively suppress polyglutamine-mediated toxicity in *Drosophila*. *Proc. Natl. Acad. Sci. USA* **102**, 10261–10266.
40. Warrick, J. M., Chan, H. Y., Gray-Board, G. L., Chai, Y., Paulson, H. L., and Bonini, N. M. (1999). Suppression of polyglutamine-mediated neurodegeneration in *Drosophila* by the molecular chaperone HSP70. *Nat. Genet.* **23**, 425–428.
41. Branco, J., Al-Ramahi, I., Ukani, L., Perez, A. M., Fernandez-Funez, P., Rincon-Limas, D., and Botas, J. (2008). Comparative analysis of genetic modifiers in *Drosophila* points to common and distinct mechanisms of pathogenesis among polyglutamine diseases. *Hum. Mol. Genet.* **17**, 376–390.
42. Brumby, A. M., and Richardson, H. E. (2005). Using *Drosophila* melanogaster to map human cancer pathways. *Nat. Rev.. Cancer* **5**, 626–639.
43. Chia, W., Somers, W. G., and Wang, H. (2008). *Drosophila* neuroblast asymmetric divisions: Cell cycle regulators, asymmetric protein localization, and tumorigenesis. *J. Cell Biol.* **180**, 267–272.
44. Jang, A. C., Starz-Gaiano, M., and Montell, D. J. (2007). Modeling migration and metastasis in *Drosophila*. *J. Mammary Gland Biol. Neoplasia* **12**, 103–114.

45. Lilly, M. A., and Duronio, R. J. (2005). New insights into cell cycle control from the *Drosophila* endocycle. *Oncogene* **24**, 2765–2775.
46. Saucedo, L. J., and Edgar, B. A. (2007). Filling out the Hippo pathway. *Nat. Rev.. Mol. Cell Biol.* **8**, 613–621.
47. Vidal, M., and Cagan, R. L. (2006). *Drosophila* models for cancer research. *Curr. Opin. Genet. Dev.* **16**, 10–16.
48. Dominguez, M. (2006). Interplay between Notch signaling and epigenetic silencers in cancer. *Cancer Res.* **66**, 8931–8934.
49. Giebel, B., and Wodarz, A. (2006). Tumor suppressors: Control of signaling by endocytosis. *Curr. Biol.* **16**, R91–R92.
50. Huangfu, D., and Anderson, K. V. (2006). Signaling from Smo to Ci/Gli: Conservation and divergence of Hedgehog pathways from *Drosophila* to vertebrates. *Development* **133**, 3–14.
51. Kanwar, R., and Fortini, M. E. (2004). Notch signaling: A different sort makes the cut. *Curr. Biol.* **14**, R1043–R1045.
52. Moon, R. T., Bowerman, B., Boutros, M., and Perrimon, N. (2002). The promise and perils of Wnt signaling through beta-catenin. *Science* **296**, 1644–1646.
53. Perrimon, N. (1998). Creating mosaics in *Drosophila*. *Int. J. Dev. Biol.* **42**, 243–247.
54. Moberg, K. H., Bell, D. W., Wahrer, D. C., Haber, D. A., and Hariharan, I. K. (2001). Archipelago regulates cyclin E levels in *Drosophila* and is mutated in human cancer cell lines. *Nature* **413**, 311–316.
55. Kwak, E. L., Moberg, K. H., Wahrer, D. C., Quinn, J. E., Gilmore, P. M., Graham, C. A., Hariharan, I. K., Harkin, D. P., Haber, D. A., and Bell, D. W. (2005). Infrequent mutations of Archipelago (hAGO, hCDC4, Fbw7) in primary ovarian cancer. *Gynecol. Oncol.* **98**, 124–128.
56. Strohmaier, H., Spruck, C. H., Kaiser, P., Won, K. A., Sangfelt, O., and Reed, S. I. (2001). Human F-box protein hCdc4 targets cyclin E for proteolysis and is mutated in a breast cancer cell line. *Nature* **413**, 316–322.
57. Spruck, C. H., Strohmaier, H., Sangfelt, O., Muller, H. M., Hubalek, M., Muller-Holzner, E., Marth, C., Widschwendter, M., and Reed, S. I. (2002). hCDC4 gene mutations in endometrial cancer. *Cancer Res.* **62**, 4535–4539.
58. Moberg, K. H., Schelble, S., Burdick, S. K., and Hariharan, I. K. (2005). Mutations in erupted, the *Drosophila* ortholog of mammalian tumor susceptibility gene 101, elicit non-cell-autonomous overgrowth. *Dev. Cell* **9**, 699–710.
59. Tapon, N., Harvey, K. F., Bell, D. W., Wahrer, D. C., Schiripo, T. A., Haber, D. A., and Hariharan, I. K. (2002). Salvador Promotes both cell cycle exit and apoptosis in *Drosophila* and is mutated in human cancer cell lines. *Cell* **110**, 467–478.
60. Gaggioli, C., and Sahai, E. (2007). Melanoma invasion—current knowledge and future directions. *Pigment Cell Res.* **20**, 161–172.
61. Kedrin, D., van Rheenen, J., Hernandez, L., Condeelis, J., and Segall, J. E. (2007). Cell motility and cytoskeletal regulation in invasion and metastasis. *J. Mammary Gland Biol. Neoplasia* **12**, 143–152.
62. Dow, L. E., and Humbert, P. O. (2007). Polarity regulators and the control of epithelial architecture, cell migration, and tumorigenesis. *Int. Rev. Cytol.* **262**, 253–302.
63. Kunwar, P. S., Siekhaus, D. E., and Lehmann, R. (2006). In vivo migration: A germ cell perspective. *Annu. Rev. Cell Dev. Biol.* **22**, 237–265.
64. Naora, H., and Montell, D. J. (2005). Ovarian cancer metastasis: Integrating insights from disparate model organisms. *Nat. Rev. Cancer* **5**, 355–366.
65. Wood, W., and Jacinto, A. (2007). *Drosophila melanogaster* embryonic haemocytes: Masters of multitasking. *Nat. Rev. Mol. Cell Biol.* **8**, 542–551.

66. Bai, J., Uehara, Y., and Montell, D. J. (2000). Regulation of invasive cell behavior by taiman, a *Drosophila* protein related to AIB1, a steroid receptor coactivator amplified in breast cancer. *Cell* **103**, 1047–1058.
67. Anzick, S. L., Kononen, J., Walker, R. L., Azorsa, D. O., Tanner, M. M., Guan, X. Y., Sauter, G., Kallioniemi, O. P., Trent, J. M., and Meltzer, P. S. (1997). AIB1, a steroid receptor coactivator amplified in breast and ovarian cancer. *Science* **277**, 965–968.
68. Fisher, B. J., Perera, F. E., Cooke, A. L., Opeitum, A., and Stitt, L. (1998). Long-term follow-up of axillary node-positive breast cancer patients receiving adjuvant tamoxifen alone: Patterns of recurrence. *Int. J. Radiat. Oncol. Biol. Phys.* **42**, 117–123.
69. Yoshida, H., Liu, J., Samuel, S., Cheng, W., Rosen, D., and Naora, H. (2005). Steroid receptor coactivator-3, a homolog of Taiman that controls cell migration in the *Drosophila* ovary, regulates migration of human ovarian cancer cells. *Mol. Cell. Endocrinol.* **245**, 77–85.
70. Beccari, S., Teixeira, L., and Rorth, P. (2002). The JAK/STAT pathway is required for border cell migration during *Drosophila* oogenesis. *Mech. Dev.* **111**, 115–123.
71. Ghiglione, C., Devergne, O., Georgenthum, E., Carballes, F., Medioni, C., Cerezo, D., and Noselli, S. (2002). The *Drosophila* cytokine receptor Domeless controls border cell migration and epithelial polarization during oogenesis. *Development* **129**, 5437–5447.
72. Silver, D. L., and Montell, D. J. (2001). Paracrine signaling through the JAK/STAT pathway activates invasive behavior of ovarian epithelial cells in *Drosophila*. *Cell* **107**, 831–841.
73. Brown, S., Hu, N., and Hombria, J. C. (2001). Identification of the first invertebrate interleukin JAK/STAT receptor, the *Drosophila* gene domeless. *Curr. Biol.* **11**, 1700–1705.
74. Chen, H. W., Chen, X., Oh, S. W., Marinissen, M. J., Gutkind, J. S., and Hou, S. X. (2002). mom identifies a receptor for the *Drosophila* JAK/STAT signal transduction pathway and encodes a protein distantly related to the mammalian cytokine receptor family. *Genes Dev.* **16**, 388–398.
75. Shen, Y., Devgan, G., Darnell, J. E., Jr., and Bromberg, J. F. (2001). Constitutively activated Stat3 protects fibroblasts from serum withdrawal and UV-induced apoptosis and antagonizes the proapoptotic effects of activated Stat1. *Proc. Natl. Acad. Sci. USA* **98**, 1543–1548.
76. Geisbrecht, E. R., and Montell, D. J. (2004). A role for *Drosophila* IAP1-mediated caspase inhibition in Rac-dependent cell migration. *Cell* **118**, 111–125.
77. Fulda, S. (2007). Inhibitor of apoptosis proteins as targets for anticancer therapy. *Expert Rev. Anticancer Ther.* **7**, 1255–1264.
78. Hunter, A. M., LaCasse, E. C., and Korneluk, R. G. (2007). The inhibitors of apoptosis (IAPs) as cancer targets. *Apoptosis* **12**, 1543–1568.
79. Fraser, M., Leung, B. M., Yan, X., Dan, H. C., Cheng, J. Q., and Tsang, B. K. (2003). p53 is a determinant of X-linked inhibitor of apoptosis protein/Akt-mediated chemoresistance in human ovarian cancer cells. *Cancer Res.* **63**, 7081–7088.
80. Sasaki, H., Sheng, Y., Kotsuji, F., and Tsang, B. K. (2000). Down-regulation of X-linked inhibitor of apoptosis protein induces apoptosis in chemoresistant human ovarian cancer cells. *Cancer Res.* **60**, 5659–5666.
81. Chen, J. K., Taipale, J., Cooper, M. K., and Beachy, P. A. (2002). Inhibition of Hedgehog signaling by direct binding of cyclopamine to smoothened. *Genes Dev.* **16**, 2743–2748.
82. Taipale, J., Chen, J. K., Cooper, M. K., Wang, B., Mann, R. K., Milenkovic, L., Scott, M. P., and Beachy, P. A. (2000). Effects of oncogenic mutations in smoothened and patched can be reversed by cyclopamine. *Nature* **406**, 1005–1009.
83. Borzillo, G. V., and Lippa, B. (2005). The Hedgehog signaling pathway as a target for anticancer drug discovery. *Curr. Top. Med. Chem.* **5**, 147–157.
84. Kiselyov, A. S. (2006). Targeting the hedgehog signaling pathway with small molecules. *Anticancer Agents Med. Chem.* **6**, 445–449.
85. Kaelin, W. G. (2007). Von hippel-lindau disease. *Annu. Rev. Pathol.* **2**, 145–173.
86. Maxwell, P. H. (2005). The HIF pathway in cancer. *Semin. Cell Dev. Biol.* **16**, 523–530.

87. Maynard, M. A., and Ohh, M. (2007). The role of hypoxia-inducible factors in cancer. *Cell. Mol. Life Sci.* **64**, 2170–2180.
88. Airley, R. E., and Mobasheri, A. (2007). Hypoxic regulation of glucose transport, anaerobic metabolism and angiogenesis in cancer: Novel pathways and targets for anticancer therapeutics. *Chemotherapy* **53**, 233–256.
89. Greco, O., and Scott, S. (2007). Tumor hypoxia and targeted gene therapy. *Int. Rev. Cytol.* **257**, 181–212.
90. Bickler, P. E., and Donohoe, P. H. (2002). Adaptive responses of vertebrate neurons to hypoxia. *J. Exp. Biol.* **205**, 3579–3586.
91. Driscoll, M., and Gerstbrein, B. (2003). Dying for a cause: Invertebrate genetics takes on human neurodegeneration. *Nat. Rev. Genet.* **4**, 181–194.
92. Donohoe, P. H., and Boutilier, R. G. (1998). The protective effects of metabolic rate depression in hypoxic cold submerged frogs. *Respir. Physiol.* **111**, 325–336.
93. Hochachka, P. W. (1986). Defense strategies against hypoxia and hypothermia. *Science* **231**, 234–241.
94. Padilla, P. A., Nystul, T. G., Zager, R. A., Johnson, A. C., and Roth, M. B. (2002). Dephosphorylation of cell cycle-regulated proteins correlates with anoxia-induced suspended animation in Caenorhabditis elegans. *Mol. Biol. Cell* **13**, 1473–1483.
95. Padilla, P. A., and Roth, M. B. (2001). Oxygen deprivation causes suspended animation in the zebrafish embryo. *Proc. Natl. Acad. Sci. USA* **98**, 7331–7335.
96. Storey, K. B. (2007). Anoxia tolerance in turtles: Metabolic regulation and gene expression. *Comp. Biochem. Physiol. A Mol. Integr. Physiol.* **147**, 263–276.
97. Blackstone, E., Morrison, M., and Roth, M. B. (2005). H2S induces a suspended animation-like state in mice. *Science* **308**, 518.
98. Blackstone, E., and Roth, M. B. (2007). Suspended animation-like state protects mice from lethal hypoxia. *Shock* **27**, 370–372.
99. Foe, V. E., and Alberts, B. M. (1985). Reversible chromosome condensation induced in *Drosophila* embryos by anoxia: Visualization of interphase nuclear organization. *J. Cell Biol.* **100**, 1623–1636.
100. Zalokar, M., and Erk, I. (1977). Phase-partition fixation and staining of *Drosophila* eggs. *Stain Technol.* **52**, 89–95.
101. Haddad, G. G. (2000). Enhancing our understanding of the molecular responses to hypoxia in mammals using *Drosophila* melanogaster. *J. Appl. Physiol.* **88**, 1481–1487.
102. Haddad, G. G., Wyman, R. J., Mohsenin, A., Sun, Y., and Krishnan, S. N. (1997). Behavioral and electrophysiologic responses of *Drosophila* melanogaster to prolonged periods of anoxia. *J. Insect. Physiol.* **43**, 203–210.
103. Jarecki, J., Johnson, E., and Krasnow, M. A. (1999). Oxygen regulation of airway branching in *Drosophila* is mediated by branchless FGF. *Cell* **99**, 211–220.
104. Krishnan, K. S., Chakravarty, S., Rao, S., Raghuram, V., and Ramaswami, M. (1996). Alleviation of the temperature-sensitive paralytic phenotype of shibire(ts) mutants in *Drosophila* by sub-anesthetic concentrations of carbon dioxide. *J. Neurogenet.* **10**, 221–238.
105. Semenza, G. L. (2001). HIF-1 and mechanisms of hypoxia sensing. *Curr. Opin. Cell Biol.* **13**, 167–171.
106. Waters, K., and Gozal, D. (2004). Developmental and metabolic implications of the hypoxic ventilatory response. *Paediatr. Respir. Rev.* **5**, 173–181.
107. Ke, Q., and Costa, M. (2006). Hypoxia-inducible factor-1 (HIF-1). *Mol. Pharmacol.* **70**, 1469–1480.
108. Bruick, R. K., and McKnight, S. L. (2001). A conserved family of prolyl-4-hydroxylases that modify HIF. *Science* **294**, 1337–1340.

109. Ivan, M., Kondo, K., Yang, H., Kim, W., Valiando, J., Ohh, M., Salic, A., Asara, J. M., Lane, W. S., and Kaelin, W. G., Jr. (2001). HIFalpha targeted for VHL-mediated destruction by proline hydroxylation: Implications for O2 sensing. *Science* **292**, 464–468.
110. Epstein, A. C., Gleadle, J. M., McNeill, L. A., Hewitson, K. S., O'Rourke, J., Mole, D. R., Mukherji, M., Metzen, E., Wilson, M. I., Dhanda, A., Tian, Y. M., Masson, N. *et al.* (2001). C. elegans EGL-9 and mammalian homologs define a family of dioxygenases that regulate HIF by prolyl hydroxylation. *Cell* **107**, 43–54.
111. Jaakkola, P., Mole, D. R., Tian, Y. M., Wilson, M. I., Gielbert, J., Gaskell, S. J., Kriegsheim, A., Hebestreit, H. F., Mukherji, M., Schofield, C. J., Maxwell, P. H., Pugh, C. W. *et al.* (2001). Targeting of HIF-alpha to the von Hippel-Lindau ubiquitylation complex by O2-regulated prolyl hydroxylation. *Science* **292**, 468–472.
112. Cockman, M. E., Masson, N., Mole, D. R., Jaakkola, P., Chang, G. W., Clifford, S. C., Maher, E. R., Pugh, C. W., Ratcliffe, P. J., and Maxwell, P. H. (2000). Hypoxia inducible factor-alpha binding and ubiquitylation by the von Hippel-Lindau tumor suppressor protein. *J. Biol. Chem.* **275**, 25733–25741.
113. Kamura, T., Sato, S., Iwai, K., Czyzyk-Krzeska, M., Conaway, R. C., and Conaway, J. W. (2000). Activation of HIF1alpha ubiquitination by a reconstituted von Hippel-Lindau (VHL) tumor suppressor complex. *Proc. Natl. Acad. Sci. USA* **97**, 10430–10435.
114. Tanimoto, K., Makino, Y., Pereira, T., and Poellinger, L. (2000). Mechanism of regulation of the hypoxia-inducible factor-1 alpha by the von Hippel-Lindau tumor suppressor protein. *EMBO J.* **19**, 4298–4309.
115. Wang, G. L., and Semenza, G. L. (1993). Characterization of hypoxia-inducible factor 1 and regulation of DNA binding activity by hypoxia. *J. Biol. Chem.* **268**, 21513–21518.
116. Wang, G. L., and Semenza, G. L. (1993). General involvement of hypoxia-inducible factor 1 in transcriptional response to hypoxia. *Proc. Natl. Acad. Sci. USA* **90**, 4304–4308.
117. Gorr, T. A., Tomita, T., Wappner, P., and Bunn, H. F. (2004). Regulation of *Drosophila* hypoxia-inducible factor (HIF) activity in SL2 cells: Identification of a hypoxia-induced variant isoform of the HIFalpha homolog gene similar. *J. Biol. Chem.* **279**, 36048–36058.
118. Lavista-Llanos, S., Centanin, L., Irisarri, M., Russo, D. M., Gleadle, J. M., Bocca, S. N., Muzzopappa, M., Ratcliffe, P. J., and Wappner, P. (2002). Control of the hypoxic response in *Drosophila* melanogaster by the basic helix-loop-helix PAS protein similar. *Mol. Cell. Biol.* **22**, 6842–6853.
119. Adryan, B., Decker, H. J., Papas, T. S., and Hsu, T. (2000). Tracheal development and the von Hippel-Lindau tumor suppressor homolog in *Drosophila*. *Oncogene* **19**, 2803–2811.
120. Aso, T., Yamazaki, K., Aigaki, T., and Kitajima, S. (2000). *Drosophila* von Hippel-Lindau tumor suppressor complex possesses E3 ubiquitin ligase activity. *Biochem. Biophys. Res. Commun.* **276**, 355–361.
121. Frei, C., and Edgar, B. A. (2004). *Drosophila* cyclin D/Cdk4 requires Hif-1 prolyl hydroxylase to drive cell growth. *Dev. Cell* **6**, 241–251.
122. Zhou, D., Xue, J., Chen, J., Morcillo, P., Lambert, J. D., White, K. P., and Haddad, G. G. (2007). Experimental selection for *Drosophila* survival in extremely low O2 environment. *PLoS ONE* **2**, e490.
123. Wingrove, J. A., and O'Farrell, P. H. (1999). Nitric oxide contributes to behavioral, cellular, and developmental responses to low oxygen in *Drosophila*. *Cell* **98**, 105–114.
124. Centanin, L., Ratcliffe, P. J., and Wappner, P. (2005). Reversion of lethality and growth defects in Fatiga oxygen-sensor mutant flies by loss of hypoxia-inducible factor-alpha/Sima. *EMBO Rep.* **6**, 1070–1075.
125. Teodoro, R. O., and O'Farrell, P. H. (2003). Nitric oxide-induced suspended animation promotes survival during hypoxia. *EMBO J.* **22**, 580–587.

126. DiGregorio, P. J., Ubersax, J. A., and O'Farrell, P. H. (2001). Hypoxia and nitric oxide induce a rapid, reversible cell cycle arrest of the *Drosophila* syncytial divisions. *J. Biol. Chem.* **276**, 1930–1937.
127. Osborne, K. A., Robichon, A., Burgess, E., Butland, S., Shaw, R. A., Coulthard, A., Pereira, H. S., Greenspan, R. J., and Sokolowski, M. B. (1997). Natural behavior polymorphism due to a cGMP-dependent protein kinase of *Drosophila*. *Science* **277**, 834–836.
128. Huang, H., and Haddad, G. G. (2007). *Drosophila* dMRP4 regulates responsiveness to O2 deprivation and development under hypoxia. *Physiol. Genomics* **29**, 260–266.
129. Ma, E., Gu, X. Q., Wu, X., Xu, T., and Haddad, G. G. (2001). Mutation in pre-mRNA adenosine deaminase markedly attenuates neuronal tolerance to O2 deprivation in *Drosophila melanogaster*. *J. Clin. Invest.* **107**, 685–693.
130. Vigne, P., and Frelin, C. (2006). A low protein diet increases the hypoxic tolerance in *Drosophila*. *PLoS ONE* **1**, e56.
131. Vigne, P., and Frelin, C. (2007). Plasticity of the responses to chronic hypoxia and dietary restriction in an aged organism: Evidence from the *Drosophila* model. *Exp. Gerontol.* **42**, 1162–1166.
132. Hanf, V., and Gonder, U. (2005). Nutrition and primary prevention of breast cancer: Foods, nutrients and breast cancer risk. *Eur. J. Obstet. Gynecol. Reprod. Biol.* **123**, 139–149.
133. Jolly, C. A. (2005). Diet manipulation and prevention of aging, cancer and autoimmune disease. *Curr. Opin. Clin. Nutr. Metab. Care* **8**, 382–387.
134. Crozatier, M., and Meister, M. (2007). *Drosophila* haematopoiesis. *Cell. Microbiol.* **9**, 1117–1126.
135. Evans, C. J., Hartenstein, V., and Banerjee, U. (2003). Thicker than blood: Conserved mechanisms in *Drosophila* and vertebrate hematopoiesis. *Dev. Cell* **5**, 673–690.
136. Lebestky, T., Chang, T., Hartenstein, V., and Banerjee, U. (2000). Specification of *Drosophila* hematopoietic lineage by conserved transcription factors. *Science* **288**, 146–149.
137. Rehorn, K. P., Thelen, H., Michelson, A. M., and Reuter, R. (1996). A molecular aspect of hematopoiesis and endoderm development common to vertebrates and *Drosophila*. *Development* **122**, 4023–4031.
138. Fossett, N., Tevosian, S. G., Gajewski, K., Zhang, Q., Orkin, S. H., and Schulz, R. A. (2001). The Friend of GATA proteins U-shaped, FOG-1, and FOG-2 function as negative regulators of blood, heart, and eye development in *Drosophila*. *Proc. Natl. Acad. Sci. USA* **98**, 7342–7347.
139. Cantor, A. B., and Orkin, S. H. (2005). Coregulation of GATA factors by the Friend of GATA (FOG) family of multitype zinc finger proteins. *Semin. Cell Dev. Biol.* **16**, 117–128.
140. Tsang, A. P., Fujiwara, Y., Hom, D. B., and Orkin, S. H. (1998). Failure of megakaryopoiesis and arrested erythropoiesis in mice lacking the GATA-1 transcriptional cofactor FOG. *Genes Dev.* **12**, 1176–1188.
141. Mandal, L., Banerjee, U., and Hartenstein, V. (2004). Evidence for a fruit fly hemangioblast and similarities between lymph-gland hematopoiesis in fruit fly and mammal aorta-gonadal-mesonephros mesoderm. *Nat. Genet.* **36**, 1019–1023.
142. Daga, A., Karlovich, C. A., Dumstrei, K., and Banerjee, U. (1996). Patterning of cells in the *Drosophila* eye by Lozenge, which shares homologous domains with AML1. *Genes Dev.* **10**, 1194–1205.
143. Rizki, T. M., and Rizki, R. M. (1981). Genetics of tumor-W in *Drosophila melanogaster*: Mapping a gene with incomplete penetrance. *J. Hered.* **72**, 78–80.
144. Blyth, K., Cameron, E. R., and Neil, J. C. (2005). The RUNX genes: Gain or loss of function in cancer. *Nat. Rev. Cancer* **5**, 376–387.
145. Lebestky, T., Jung, S. H., and Banerjee, U. (2003). A Serrate-expressing signaling center controls *Drosophila* hematopoiesis. *Genes Dev.* **17**, 348–353.

146. Allman, D., Aster, J. C., and Pear, W. S. (2002). Notch signaling in hematopoiesis and early lymphocyte development. *Immunol. Rev.* **187,** 75–86.
147. Ohishi, K., Katayama, N., Shiku, H., Varnum-Finney, B., and Bernstein, I. D. (2003). Notch signalling in hematopoiesis. *Semin. Cell Dev. Biol.* **14,** 143–150.
148. Crozatier, M., Ubeda, J. M., Vincent, A., and Meister, M. (2004). Cellular immune response to parasitization in *Drosophila* requires the EBF orthologue collier. *PLoS Biol.* **2,** E196.
149. Harrison, D. A., Binari, R., Nahreini, T. S., Gilman, M., and Perrimon, N. (1995). Activation of a *Drosophila* Janus kinase (JAK) causes hematopoietic neoplasia and developmental defects. *EMBO J.* **14,** 2857–2865.
150. Hou, S. X., Zheng, Z., Chen, X., and Perrimon, N. (2002). The Jak/STAT pathway in model organisms: Emerging roles in cell movement. *Dev. Cell* **3,** 765–778.
151. Luo, H., Hanratty, W. P., and Dearolf, C. R. (1995). An amino acid substitution in the *Drosophila* hopTum-l Jak kinase causes leukemia-like hematopoietic defects. *EMBO J.* **14,** 1412–1420.
152. Bromberg, J. (2002). Stat proteins and oncogenesis. *J. Clin. Invest.* **109,** 1139–1142.
153. Rayet, B., and Gelinas, C. (1999). Aberrant rel/nfkb genes and activity in human cancer. *Oncogene* **18,** 6938–6947.
154. Duchek, P., Somogyi, K., Jekely, G., Beccari, S., and Rorth, P. (2001). Guidance of cell migration by the *Drosophila* PDGF/VEGF receptor. *Cell* **107,** 17–26.
155. Bruckner, K., Kockel, L., Duchek, P., Luque, C. M., Rorth, P., and Perrimon, N. (2004). The PDGF/VEGF receptor controls blood cell survival in *Drosophila*. *Dev. Cell* **7,** 73–84.
156. Cho, N. K., Keyes, L., Johnson, E., Heller, J., Ryner, L., Karim, F., and Krasnow, M. A. (2002). Developmental control of blood cell migration by the *Drosophila* VEGF pathway. *Cell* **108,** 865–876.
157. Heino, T. I., Karpanen, T., Wahlstrom, G., Pulkkinen, M., Eriksson, U., Alitalo, K., and Roos, C. (2001). The *Drosophila* VEGF receptor homolog is expressed in hemocytes. *Mech. Dev.* **109,** 69–77.

Applications of Fluorescence Correlation Spectroscopy to the Study of Nucleic Acid Conformational Dynamics

KAUSHIK GURUNATHAN[*,†] AND
MARCIA LEVITUS[*,†,‡]

[*]*Department of Chemistry and Biochemistry, Arizona State University, Tempe, Arizona 85287*

[†]*The Biodesign Institute, Arizona State University, Tempe, Arizona 85287*

[‡]*Department of Physics, Arizona State University, Tempe, Arizona 85287*

I. Introduction: Fluorescence and Energy Transfer	34
A. Measuring Conformational Dynamics	36
II. Fluorescence Correlation Spectroscopy	37
A. Introduction	37
B. Measuring Fluorescence Fluctuations	38
C. Quantifying Fluorescence Fluctuations	39
D. Contributions to the FCS Signal	42
III. Conformational Dynamics of Nucleic Acids	46
A. Dealing with the Overlap of Diffusion and Conformational Dynamics	49
B. Selected Applications	56
IV. Experimental Techniques	60
A. Fluorophores	60
B. Nucleic Acid Labeling	62
C. FCS Instrumentation	62
D. Calibration	64
E. Signal-To-Noise Ratio	64
F. Two-Photon Excitation FCS	65
V. Concluding Remarks	65
References	66

Fluorescence correlation spectroscopy (FCS) is a quantitative technique where temporal fluctuations in fluorescence intensity are analyzed to yield information about physical processes that contribute to the fluctuations. These include diffusion through the observation volume, conformational dynamics, and chemical or photophysical reactions. To date, FCS applications in biochemistry and biophysics have been mostly focused on the investigation of molecular diffusion. In this chapter, we present a look into the recent applications of FCS to the study of conformational dynamics of nucleic acids. We first survey the basic theoretical and experimental aspects of the technique, and then focus on the results obtained in studies with DNA hairpins, nucleosomes, and single-stranded and duplex DNA.

I. Introduction: Fluorescence and Energy Transfer

Fluorescence is the spontaneous emission of light due to transitions from excited singlet states to various vibrational levels of the electronic ground state of a molecule. Figure 1 shows a schematic representation of the spontaneous relaxation processes that follow excitation to a higher electronic state. Fluorescence is one of the several processes that can follow the absorption of a photon. The efficiency of fluorescence emission depends on how efficient other processes that originate from the same excited state are. Usually the most efficient competing processes are internal conversion (IC) and intersystem crossing (ISC) to the triplet state. IC is a nonradiative transition between two electronic states of the same spin multiplicity and is more efficient in large, flexible molecules with many vibrational modes. In solution, the excess energy is transferred to the solvent during collisions of the excited molecule with the surrounding solvent molecules. ISC is a nonradiative spin-forbidden transition between electronic states of different multiplicities (see Fig. 1). The probability of ISC depends on the singlet and triplet states involved. For instance, it can be very efficient in molecules containing heavy atoms (e.g., bromonaphtalene) or

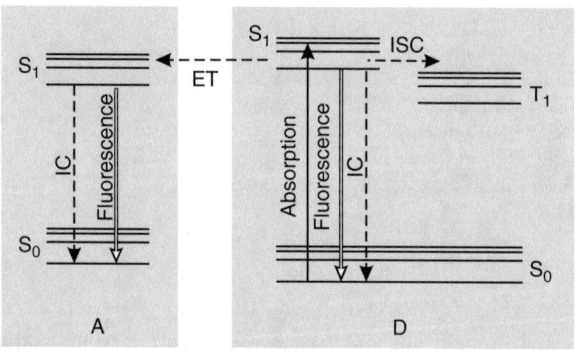

FIG. 1. Simplified Jablonski diagram showing the electronic energy levels of a fluorophore (S_0, S_1 and T_1), illustrating absorption, fluorescence, internal conversion (IC), intersystem crossing (ISC), and energy transfer (ET). The levels on the right represent the donor (D) and the levels on the left the acceptor (A). The triplet states of the acceptor have been omitted for clarity. Dashed lines symbolize nonradiative transitions. Solid lines symbolize transitions involving the absorption or emission of a photon.

possessing (n, π*) states (e.g., some compounds containing carbonyl groups such as benzophenone). Both IC and ISC should be minimized in order to maximize the efficiency of fluorescence emission. From the structural point of view, this is achieved with rigid compounds with aromatic functional groups with low (π,π*) states (see Fig. 11 for some examples).

In some cases, the properties of fluorescent emission are sensitive to the environment and can be used to some extent to study conformational changes in some systems. For example, the environmental sensitivity of the dye Cy3 was used by Luo et al. to investigate conformational changes in the bacteriophage T7 DNA polymerase at the single-molecule level (1). However, more information can be obtained when the distance-dependent interactions between the excited states of two different fluorophores are observed and analyzed. This is exploited in the technique now commonly known as FRET (Förster resonance energy transfer). Here, a fluorescent donor is excited at its specific excitation wavelength, and the energy of the excited state is partially transferred to a second molecule called the acceptor, which then emits at its characteristic fluorescence frequency. The term "fluorescence resonance energy transfer" is often used, but this long-rage dipole–dipole coupling mechanism is nonradiative, so energy is not actually transferred by fluorescence.

For this interaction to be efficient, the absorption spectrum of the acceptor should overlap with the emission spectrum of the donor. For a given donor–acceptor pair, the rate of energy transfer is highly dependent on the distance between the molecules. The efficiency of energy transfer, E, is expressed as

$$E = \frac{R_0^6}{R_0^6 + r^6} \quad (1)$$

where r is the distance between the donor and acceptor molecules, and R_0 is the Förster distance, which depends on the spectral characteristics of the donor–acceptor pair, the efficiency of fluorescence of the donor in the absence of acceptor, and the relative orientation of the fluorophores. R_0 is usually expressed in angstroms (Å) and, as can be seen from Eq. (1), represents the distance at which the efficiency of energy transfer is 50%. The efficiency of FRET can be thus used to measure molecular distances in macromolecules using fluorescent probes bound to specific sites. This technique was pioneered by Stryer, who coined the term "spectroscopic ruler" (2). Typical effective distances between the donor and acceptor molecules are in the 10–100 Å range, representing roughly the distance between 3 and 30 nucleotides in the DNA double helix. The main drawback of this technique is the fact that the relative orientation of the transition dipole moments of the donor–acceptor pair is rarely known. This introduces an uncertainty in the value of R_0 that translates into an uncertainty in the measured donor–acceptor distance.

However, we will show that this is not a critical problem in fluorescence correlation spectroscopy applications, because FRET is not used to determine distances but rather to quantify fluctuations in donor and acceptor intensities due to conformational dynamics.

FRET measurements have played an important role in the study of the structure and function of nucleic acids for several decades. Applications include the determination of the global structure and dynamics of DNA and RNA four-way junctions (3) and other branched and bulged structures (4), the study of the handedness of hybrid DNA (5), the investigation of DNA flexibility and bending (6), and the detection of hybridization and denaturation of DNA both *in vivo* (7) and *in vitro* (8).

A. Measuring Conformational Dynamics

The distance dependence of the resonance energy transfer process can be exploited to investigate conformational dynamics in biopolymers. Figure 2 represents a generic example, where a donor- and acceptor-labeled biopolymer is in equilibrium between two different conformations. In a single-molecule experiment, where one molecule is observed at the time, the fluorescence emission of both the donor and the acceptor is expected to fluctuate with time as the molecule changes conformation. However, because of the intrinsically stochastic nature of the process, individual fluctuations are averaged out when a large number of molecules are measured simultaneously. In other words, the donor and acceptor intensities measured in thermodynamic equilibrium are not expected to show temporal variations when millions of unsynchronized molecules are measured simultaneously. Flow and relaxation methods were developed in the 1960s to overcome this problem. Relaxation methods are based on the analysis of the return to equilibrium of a system after a small perturbation is applied. Temperature and pressure jumps are the most frequently used biochemical perturbation techniques, and provided a wealth of the information we have today on nucleic acid conformational dynamics (9–11).

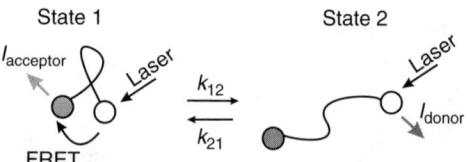

FIG. 2. Schematic representation of a fluorescently labeled biopolymer in equilibrium between two conformations. The kinetic rates of interconversion between states 1 and 2 are represented by k_{12} and k_{21}, respectively. The biomolecule contains a donor (white) fluorescent probe that is excited directly by a laser. A fraction of the excitation energy is transferred to an acceptor (gray) molecule by FRET, with an efficiency that depends strongly on the distance between the dyes.

For example, Wetmur and Davidson reported the renaturation rates of double-stranded DNA by ultraviolet absorption after a rapid change in temperature in a pioneering manuscript published in 1968 (*12*). Relaxation techniques have also been instrumental in elucidating the mechanisms of several processes including the interaction of nucleic acids with proteins and other ligands, protein conformation changes, and helix-coil transitions (*13–16*).

However, not all biological systems can be perturbed in a suitable manner, particularly in studies in biological environments such as membranes and living cells. Moreover, it is obviously advantageous to perform measurements in thermodynamic equilibrium without the need of disturbing the system. Single-molecule techniques emerged in the 1990s as a promising approach to observe directly time-dependent reactions without the need of synchronizing a population of molecules (*17*). The first demonstration of energy transfer between a single donor and single acceptor (*18*) practically revolutionized the way we think about and investigate conformational dynamics in biopolymers (*19–21*). However, the extremely low signals obtained in single-molecule experiments (~5 photons/ms) limit their application to the study of slow (approximately >50 ms) kinetic process. Moreover, single-molecule experiments are extremely hard to perform in living cells, where endogenous fluorescence and scattered light overwhelm the small number of photons emitted by a single fluorophore. Fluorescence correlation spectroscopy (FCS), a technique traditionally used to study diffusion properties of biomolecules, has recently emerged as an alternative to study conformational fluctuations in a wide range of timescales (*22, 23*). The accessible timescales span almost seven orders of magnitude, all the way from submicroseconds to seconds. In this technique, the spontaneous fluorescence intensity fluctuations of a small number of molecules are analyzed statistically in order to extract dynamic information of the system. An advantage of FCS relative to relaxation methods is that measurements can be carried out on systems in thermodynamic equilibrium without the need of disturbing the system by changes in temperature or pressure.

In this chapter, we will provide an overview of the FCS technique, and we will focus on its applications to the investigation of conformational dynamics of nucleic acids.

II. Fluorescence Correlation Spectroscopy

A. Introduction

Fluorescence correlation spectroscopy (FCS) is a technique based on the measurement of the spontaneous fluctuations of the fluorescence signal of a small number of molecules. Fluorescence fluctuations are typically measured in an optically restricted submicron observation volume, and then analyzed statistically to reveal kinetic information about the processes that lead to these fluctuations. Such processes include concentration fluctuations via molecular diffusion, chemical reactions, photophysical processes, and so on. The concept of FCS was introduced by Magde *et al.* in 1972 (24), and first demonstrated experimentally by the same authors in 1974 by studying the interaction of DNA with ethidium bromide (25). However, the measurement of fluorescence fluctuations was an extremely challenging endeavor until stable laser sources, photodetectors with high sensitivity, and new microscope techniques were introduced in the 1990s.

Note that in contrast to all other fluorescence techniques, the relevant variable in FCS is the fluctuation in intensity, and not the fluorescence intensity itself. Intensity fluctuations are defined as deviations from the mean: $\delta I(t) = I(t) - \langle I(t) \rangle$, where $\langle I(t) \rangle$ represents the average intensity over the whole measurement, typically lasting for minutes to hours. Intensity fluctuations can be negative or positive depending on whether the intensity at a particular time is higher or lower than the mean. Because fluctuations are due to transient changes in the number of fluorescent molecules in the observation volume, the mean number of molecules has to be kept small. In a typical FCS setup, a small sample volume of about 1 fl (1 fl = 10^{-15} liter = 1 μm^3) is achieved by a combination of a tightly focused laser and a pinhole in the emission path (Fig. 3). Typically, FCS experiments are performed at concentrations in the nanomolar range, which results in a mean of about 10 molecules in the observation volume.

B. Measuring Fluorescence Fluctuations

A typical setup for FCS measurements is shown in Fig. 3. The light of a commercial laser source is first expanded so as to almost fill the back aperture of the microscope objective, and then reflected by a dichroic mirror into the objective. High numerical aperture objectives are preferred for efficient fluorescence collection and to minimize the size of the observation volume and thus contributions from the background. Fluorophores excited by the laser emit in all directions, and a fraction is collected by the same objective used for excitation. The emitted signal is then passed through a pinhole assembly, which confines the observation volume in the axial direction by rejecting fluorescence light coming from planes distant from the focal region (usually a few microns). This region, illustrated as a dark ellipse in the right side of Fig. 3, will be referred to as the confocal or observation volume in the remainder of this chapter. Note that a much larger region is illuminated by the laser, but only emission coming from this restricted region reaches the detection system. Finally, if the sample contains a FRET pair, the donor and acceptor

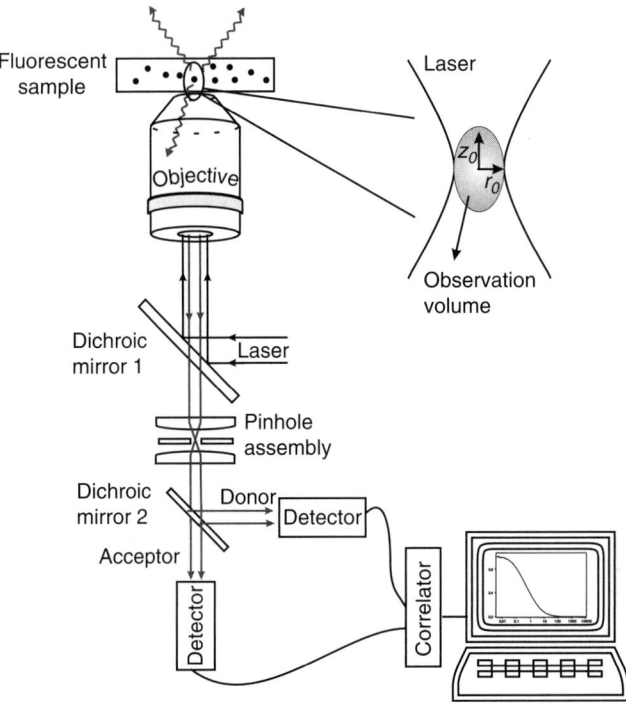

FIG. 3. Schematic diagram of an FCS instrument. A laser is expanded (not shown), reflected on a dichroic mirror (mirror 1), and focused into the sample. A fraction of the emitted fluorescence is collected by the same objective used for excitation and passed through the dichroic mirror. The fluorescence light is focused on a pinhole and recollimated by a second lens. A second dichroic mirror (mirror 2) separates the donor and acceptor contributions, which are measured by independent detectors. The electronic output of the detectors is processed by a data acquisition board. This optical arrangement defines an observation volume (dark ellipse) that is approximately Gaussian with radial and axial semiaxes r_0 and z_0, respectively (see text).

contributions should be separated by a dichroic mirror so they can be detected by two independent photodetectors. A more detailed description of the instrument and experimental conditions is presented in Section III.

C. Quantifying Fluorescence Fluctuations

In FCS, fluorescence fluctuations are analyzed in order to extract dynamic information about the processes that lead to the measured fluctuations. Intensity fluctuations are quantified by their normalized auto- ($x = y$) or cross- ($x \neq y$) correlation function, which compares the fluorescence intensity at a given time with the intensity after a certain lag time τ. As we will discuss shortly, the variables x and y refer to the detectors in the setup (most commonly donor and acceptor).

Mathematically, the normalized correlation functions are defined as:

$$G_{xy}(\tau) = \frac{\langle I_x(t)I_y(t+\tau)\rangle}{\langle I_x(t)\rangle\langle I_y(t)\rangle} \qquad (2)$$

Here, the fluorescence intensity measured at detector x and time t is represented by $I_x(t)$, and the angular brackets represent the time average over the data accumulation time. Three correlation functions can be calculated in measurements where the donor and acceptor are measured simultaneously (Fig. 3): G_{DD}, G_{AA}, and G_{DA}. These represent the autocorrelation function of the signal measured in the donor (G_{DD}) or acceptor (G_{AA}) detector, and the cross-correlation function between the signals measured in the donor and acceptor detectors (G_{DA}).

Some confusion arises in the biophysical literature because some authors define $G(\tau)$ not in terms of the intensities [Eq. (2)] but in terms of the fluorescence fluctuations: $\delta I_{x,y}(t) = I_{x,y}(t) - \langle I_{x,y}(t)\rangle$.

$$G'_{xy} = \frac{\langle \delta I_x(t)\delta I_y(t+\tau)\rangle}{\langle I_x(t)\rangle\langle I_y(t)\rangle} \qquad (3)$$

It is straightforward to prove that $G_{xy}(\tau) = G'_{xy}(\tau) + 1$ (26), so both definitions provide exactly the same physical information. However, it is important to note that while in most experiments $G'_{xy}(\tau)$ decays to zero at long lag times, $G_{xy}(\tau)$ decays to one. We will use the definition in terms of the fluctuations [Eq. (3)], and refer to this function as the "autocorrelation function" or "cross-correlation function" in the remainder of the chapter. The procedure followed to calculate the correlation decays according to Eq. (3) is illustrated in Fig. 4. Although this procedure is useful to understand the physical meaning of the correlation functions, it is important to note that calculating correlation that decays in this way is not computationally efficient, and other algorithms are typically used instead.

Consider the two-detector setup described in Fig. 3. The intensity of fluorescence in each detector is measured as a function of time, typically over minutes to hours (Fig. 4, left). Fluctuations in fluorescence intensity in each detector can be then calculated point-by-point as the difference between the actual intensity and the mean. Three correlation decays can then be calculated from these measurements. It is important to stress that the correlation functions are calculated as a function of the lag time (τ), which is different from the actual experiment time (t). To calculate the autocorrelation decay of the donor channel (G_{DD}) at a particular lag time, one would multiply all pairs of values of

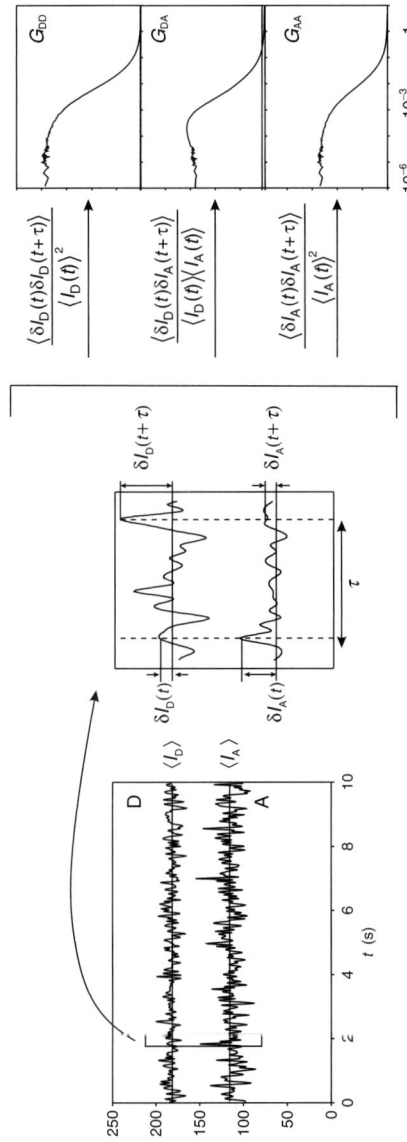

FIG. 4. Schematic representation of how the correlation functions (G_{DD}, G_{AA}, and G_{DA}) are calculated from the measured donor and acceptor intensities. The intensity of fluorescence in each detector is measured as a function of time, typically over minutes to hours (left). Fluctuations in fluorescence intensity in each detector can be then calculated point-by-point as the difference between the actual intensity and the mean (middle). Three correlation decays can then be calculated from these measurements according to Eq. (3) (right).

δI separated by a time τ (Fig. 4, middle), take the average of all these pairs, and finally normalize the result by dividing by $\langle I_D(t)\rangle^2$. This procedure would render a point in the autocorrelation decay (Fig. 4, right). To calculate the full autocorrelation decay, the same procedure would be repeated at different values of τ. Since the timescales in FCS usually span several orders of magnitude, it is customary to plot the correlation functions using a logarithmic abscissa (Fig. 4, right). To calculate the donor–acceptor cross-correlation decay (G_{DA}), the same procedure is followed except that the values of intensity fluctuation in the donor detector are multiplied by values in the acceptor channel separated by τ. In this case, the average is normalized by dividing by the product $\langle I_D(t)\rangle \langle I_A(t)\rangle$.

To understand the physical meaning of the autocorrelation functions, consider a fluorescently labeled particle diffusing freely in solution. At short lag times, the displacement of the molecule is small, and $I(t)$ and $I(t+\tau)$ are likely to be similar. This results in a large value of $\langle \delta I_D(t)\delta I_D(t+\tau)\rangle$ and a corresponding large value of G. At long lag times, molecules diffuse into and out of the observation volume and the signals $I(t)$ and $I(t+\tau)$ are not likely to be related. In this case, the measured fluorescence intensity is equally likely to show a positive or negative deviation from the mean, rendering $\langle \delta I_D(t)\delta I_D(t+\tau)\rangle_{\tau\to\infty} = 0$. From this argument, it is evident that the autocorrelation decay is a measure of the similarity of the signal with itself after a time τ has elapsed, and it is thus useful to study processes that cause fluctuations in fluorescence intensity. These fluctuations reduce the correlation of the fluorescence signal in the timescales of the processes that give rise to the measured fluctuations.

D. Contributions to the FCS Signal

1. Translational Diffusion

In order to obtain dynamic or kinetic information from FCS experiments, the experimentally acquired auto- and cross-correlation functions have to be compared with the decays predicted by the appropriate physical models. The simplest FCS experiment involves measuring the intensity fluctuations of a fluorescent particle diffusing freely in three dimensions. If the observation volume is approximated by a 3D Gaussian with semiaxes r_0 and z_0 (Fig. 3), the autocorrelation function takes the form:

$$G(\tau) = \frac{1}{\langle N\rangle}\frac{1}{[1+(4D\tau)/r_0^2]}\frac{1}{\sqrt{1+(4D\tau)/z_0^2}} \quad (4)$$

Here, D represents the diffusion coefficient of the diffusing particle, and $\langle N\rangle$ is the mean number of molecules in the observation volume. Note that the amplitude of the decay is inversely proportional to the sample concentration as expected from the fact that the relative magnitude of the fluctuations decreases as the number of particles increases. The average residence time of the diffusing molecules in the observation volume is characterized by the term $\tau_D = r_0^2/4D$, usually known as the diffusion time.

The assumption that the observation volume is Gaussian is important because it simplifies the mathematical description of the problem to the point that it is possible to obtain the analytical expression of $G(\tau)$ described in Eq. (4). However, the validity of this assumption depends largely on optical considerations of the experimental setup, and should be carefully evaluated. Hess et al. performed an extensive analysis of the artifacts introduced by non-Gaussian observation volumes, finding that the FCS decay can contain significant systematic artifacts (27). These artifacts can be erroneously interpreted as additional diffusing species or conformational dynamics, so a proper characterization of the optical setup is critical to verify the validity of the assumptions behind Eq. (4).

To date, FCS has been primarily used to determine mobility-related parameters of biologically relevant molecules in solution, membranes, and living cells (28–31). For instance, Politz et al. studied the intranuclear diffusion of fluorescently labeled oligodeoxynucleotides, obtaining the surprising result that oligos can move at rates comparable to those in aqueous solution (31). It is worth mentioning that Eq. (4) assumes free three-dimensional diffusion, and has to be modified to include other transport processes such as anomalous diffusion, convection, or active transport (29, 32, 33). The potential of FCS to study diffusion phenomena is illustrated in Fig. 5, which shows the autocorrelation decay of the fluorescent dye 5-carboxytetramethylrhodamine (TMR) in water and water–glycerol mixtures of increasing viscosity. The diffusion coefficient is inversely proportional to the viscosity of the solution, so the characteristic diffusion time, τ_D increases linearly with viscosity as the fraction of glycerol in the mixture increases. Assuming a diffraction-limited spot, the radial dimension (r_0) of the observation volume created with a green laser is approximately 250 nm. For TMR in water, the diffusion coefficient is approximately $D = 2.8 \times 10^{-10}$ m^2/s (23), yielding a residence time $\tau_D = r_0^2/4D \sim 50$ μs. This is in fact the timescale over which correlation is lost when the autocorrelation decay of TMR is measured in water (Fig. 5, leftmost decay). An aqueous solution of glycerol 80% (v/v) has a viscosity coefficient approximately 60 times larger than pure water, so the corresponding FCS curve decays in the millisecond timescale ($\tau_D \sim 3$ ms, rightmost decay in Fig. 5). It is interesting to note that in order to decrease the diffusion coefficient by the same amount by changing the size of the diffusing particle instead of solvent viscosity, the volume of the molecule would have to increase by a factor of about 2×10^5 (i.e., 60^3).

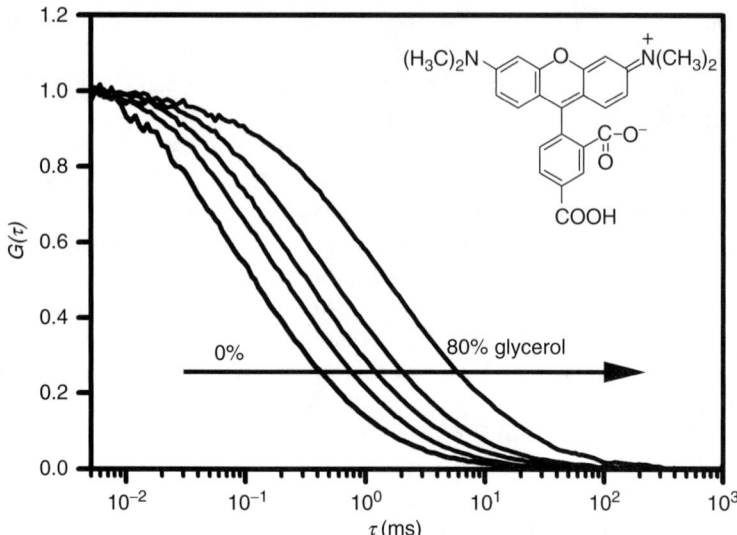

Fig. 5. Autocorrelation decays of the fluorophore TMR in water–glycerol mixtures (0%, 20%, 40%, 60%, and 80%). The autocorrelation function was normalized for clarity.

The dependence of the residence time with the diffusion coefficient has been exploited to study biomolecular binding. For instance, Kinjo and Rigler first used this concept to study the binding kinetics of short fluorescently labeled DNA probes to longer DNA targets (34). Since then, FCS has been extensively used to study a variety of ligand-receptor systems (34–36). Following the same idea, one could imagine using the same principle to study conformational dynamics of systems undergoing conformational changes that lead to changes in diffusion coefficient. However, the differences in the hydrodynamic radius between two folding states are generally quite small, and hard to measure by FCS (37). Yet, this idea was successfully applied to the investigation of DNA condensation (38) and DNA flexibility (39), as discussed in more detail in Sections II.B.4 and II.B.5.

2. Photophysical Processes

Fluorescence fluctuations due to photophysical processes such as triplet dynamics or photoisomerizations produce features in the FCS decay that should be thoroughly understood to avoid misinterpreting these features as fluctuations caused by conformational dynamics. Widengren et al. investigated the singlet–triplet dynamics of rhodamine 6G and fluorescein isothiocyanate by FCS (40). Nonradiative singlet–triplet transitions are spin-forbidden, and thus slower than the singlet–singlet radiative transitions. Thus, the system can be regarded as a dye fluctuating between a light-emitting state (excited singlet state) and a dark state (excited triplet state). Triplet dynamics produce a series of dark intervals that interrupt fluorescence emission as the molecule traverses the confocal volume. In this case, the correlation function shows an additional shoulder in the measured curves at short timescales. The overall decay is described by Eq. (5):

$$G(\tau) = G_{\text{diff}}(\tau) \cdot G_{\text{triplet}}(\tau) = G_{\text{diff}}(\tau)\left(1 + \frac{f}{1-f}e^{-\tau/\tau_R}\right) \quad (5)$$

Here, G_{diff} represents the diffusion term [Eq. (4)], f is the fraction of molecules in the triplet (dark) state, and τ_R is the relaxation time of the singlet↔triplet process. Both τ_R and f depend on the excitation power, providing a useful way of distinguishing these contributions to the autocorrelation decay from those arising from conformational dynamics. In particular, triplet population increases and relaxation times are shortened at higher excitation intensities (40).

Other photophysical processes that produce fluorescence fluctuations in the millisecond timescale include *cis-trans* photoisomerizations. Schwille *et al.* performed a thorough study of the popular cyanine dye Cy5 (usually used as an acceptor in combination with the donor Cy3), and showed that under typical FCS conditions, the photoinduced isomerization and back-isomerization processes produce millisecond timescale contributions to the FCS curve (41). The relaxation time in this case was shown to depend on excitation power in a complex way. This dependence can be very small if the cross section of the two isomers at the excitation wavelength is not too different, so great caution should be used when using cyanine dyes to study conformational dynamics. In principle, experiments with control samples should help distinguishing between fluctuations arising from photophysics from those due to conformational dynamics. However, recent reports have shown that Cy5 blinking can be influenced by the proximity of Cy3, producing anticorrelated Cy3–Cy5 fluctuations in the donor–acceptor double-labeled sample (42). These fluctuations in FRET efficiency have a photophysical origin, and can be difficult to discern form variations in donor–acceptor distance. The most obvious way to avoid artifacts due to fluorophore photophysics is to select fluorescent probes with low triplet and isomerization yields, particularly when investigating conformational changes in the micro- to millisecond timescales. However, these processes cannot be completely eliminated, so the proper control experiments should always be carried out.

3. Quenching and FRET

For FCS to be useful in the study of conformational dynamics, the intensity of fluorescence of the probe has to be influenced by the conformation of the biopolymer being studied. The most common approach is to label the biopolymer with a fluorophore–quencher or donor–acceptor FRET pair, so that the efficiency of emission of the fluorophore is a direct measure of the distance between the two tags.

For instance, the fluorescence of oxazine and rhodamine fluorophores is known to be quenched selectively upon complex formation with deoxyguanosines (dG) or tryptophan residues due to photoinduced electron transfer (PET) (43, 44). In this case, the reporter probe is in a fluorescent or nonfluorescent state depending on its proximity to dG or tryptophan residues. It is important to note that in contrast with FRET, PET occurs only when fluorophore and quencher are in close proximity (within van der Waals contact). Thus, small conformational changes that affect the distance between the fluorophore and quencher will be evidenced by fluctuations in the fluorescent signal of the dye. In contrast, donor–acceptor interactions in FRET occur *via* dipole–dipole interactions over distances in the 50 Å scale, therefore this technique can be used to study longer-range interactions. FRET pairs also have the benefit of increasing the amount of information that can be extracted from the FCS measurement. In this case, it is possible to calculate not only the autocorrelation decay of the donor but also the autocorrelation of the acceptor and the cross-correlation between the donor and acceptor detectors. As we will discuss in detail later on, this allows the diffusion contributions to be deconvoluted from the contributions due to conformational dynamics. On the down side, these measurements require chemical labeling with two extrinsic fluorophores instead of one.

Specific examples of PET and FRET will be discussed in the next section in the context of their applications to the study of conformational dynamics of nucleic acids by FCS.

III. Conformational Dynamics of Nucleic Acids

The analysis and interpretation of the auto- and cross-correlation curves depend on the relative timescales of the dynamics one wishes to study and the residence time of the biomolecule in the confocal volume, which is dictated by translational diffusion. The residence time is characterized by the parameter $\tau_D = r_0^2/4D$, and it is typically of the order of 100 μs–1 ms in aqueous solutions. The timescale for dynamics is characterized by the relaxation time of the reaction, τ_R, and depends on the nature of the process that causes the fluctuation in fluorescence (i.e., triplet dynamics, photoisomerization, conformational dynamics, etc). If fluctuations arise from conformational dynamics of a molecule that exists in two states in equilibrium (Fig. 2), then $\tau_R = (k_{21}+k_{12})^{-1}$.

The simplest case arises when $\tau_R \ll \tau_D$ because the contributions to the correlation decay due to diffusion and conformational dynamics are temporally separated. In this case, diffusion is much slower than conformational dynamics, so the different features in the correlation decay can be easily separated and analyzed independently. For small biological molecules ($D \sim 10^{-9}$–10^{-10} m^2/s), the residence time in the confocal volume is of the order of 100 μs. Larger macromolecules such as nucleosomes have diffusion coefficients of the order of 10^{-11} m^2/s and spend on average about 1 ms in the observation volume (see Fig. 8). Thus, relaxation times faster than approximately 1 μs give rise to fluorescence fluctuations that can be easily separated from those arising from translational diffusion. This is illustrated schematically in Fig. 6. When the relaxation time is much faster than diffusion, the two contributions to the autocorrelation decay can be easily distinguished and analyzed independently. However, as the relevant timescales for dynamics decrease and approach the characteristic timescales for diffusion, the two parts of the curve merge together and it is not possible to separate the different contributions to the decay.

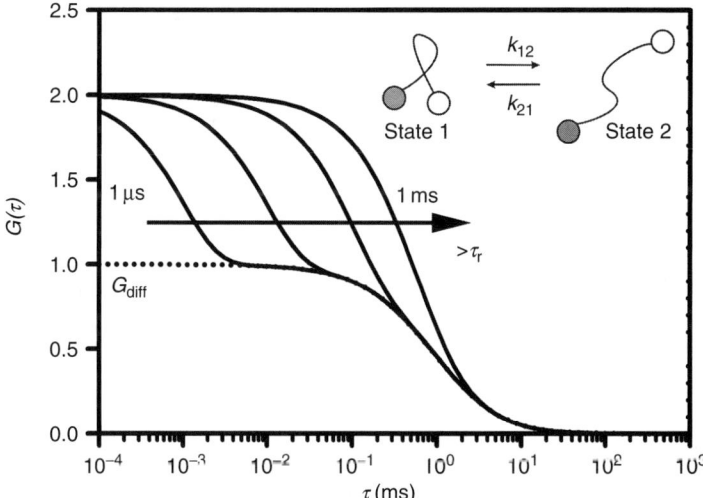

FIG. 6. Predicted donor autocorrelation function for a two-state system diffusing freely in solution. The dotted line represents the diffusion contribution for a molecule with $\tau_D = 1$ ms. Conformational dynamics also produce fluctuations in intensity, which result in additional features in the autocorrelation decay. The solid lines represent the expected autocorrelation decay for a system with $\eta = 0$ and $\tau_R = 1$ μs, 10 μs, 100 μs, and 1 ms (see Eq. 6). The diffusion and kinetic contributions are readily distinguishable when $\tau_R \ll \tau_D$.

Examples of studies in the regime $\tau_R \ll \tau_D$ include the work of Sauer *et al.*, who studied the dynamics of DNA hairpins containing a poly(dT)$_5$ loop with a short stem containing one or two dC–dG pairs (45). In this work, DNA hairpin dynamics was monitored by quenching of the oxazine derivative MR121 by intrinsic guanosine residues. Relaxation times were in the microsecond timescale, allowing a straightforward distinction between hairpin dynamics and translational diffusion in the autocorrelation decay of the fluorescent dye (Fig. 7). A similar approach was used by the same group to study protein folding in the nana- to microsecond timescale. In this case, folding dynamics was probed by taking advantage of selective quenching of the probe MR121 by the amino acid tryptophan (46).

Hairpins containing longer stems show dynamics at longer timescales, so the contributions of dynamics and diffusion are not longer readily distinguishable. In fact, most of the applications of FCS to the study of conformational dynamics in nucleic acids to date fall in this regime, posing the question of how to separate the contributions of diffusion and dynamics to the correlation decays. This important problem is analyzed in the Section III.A in detail.

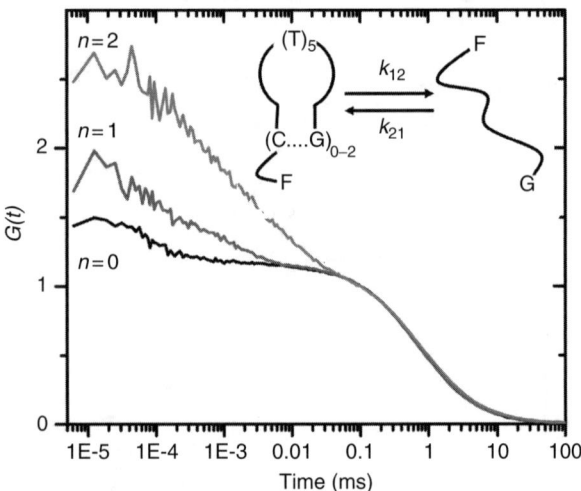

FIG. 7. Autocorrelation functions for single-stranded DNA ($n = 0$) and DNA hairpins with one ($n = 1$) or two ($n = 2$) dC–dG pairs. A fluorescent MR121 molecule (F) was attached to one of the ends of the strand. Quenching with deoxyguanosines (dG) occurs via electron transfer in the closed conformation. The submillisecond contributions to the autocorrelation decay are due to conformational dynamics. [Adapted with permission from reference (45), copyright by Oxford University Press.]

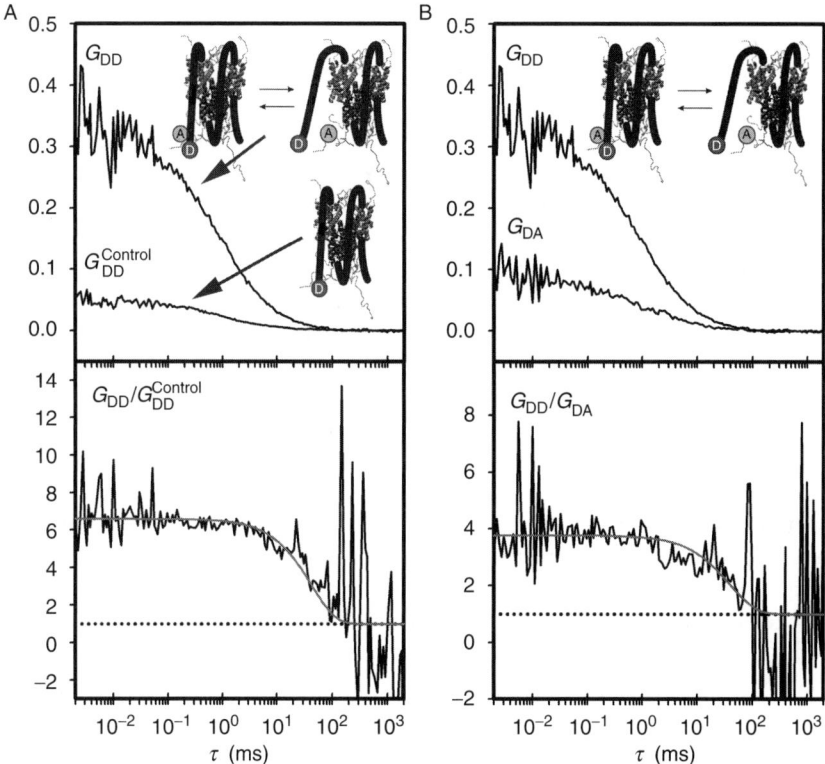

FIG. 8. FCS analysis of conformational dynamics in nucleosomes. (A) Analysis of the donor autocorrelation function in a donor–acceptor labeled sample (G_{DD}) and donor-only sample ($G_{DD}^{control}$). The ratio $G_{DD}/G_{DD}^{control}$ (bottom) isolates the contributions due to conformational dynamics. Solid gray line: least-squares fit to the kinetic term in Eq. (6), from which k_{12} and k_{21} were obtained. [Reprinted from reference (50) with permission from Nature Structural and Molecular Biology.] (B) Experimental donor autocorrelation (G_{DD}) and donor–acceptor cross-correlation (G_{DA}) decays obtained with a donor–acceptor double-labeled nucleosome in the same experiment. Bottom: ratio G_{DD}/G_{DA} isolates the contributions due to conformational dynamics (see text). The gray solid line represents the fit to a kinetic from which we obtained $k_{12} = 0.8\ s^{-1}$ and $k_{21} = 20\ s^{-1}$. See reference (48) for a complete quantitative analysis. [Adapted from reference (48), copyright by American Chemical Society.]

A. Dealing with the Overlap of Diffusion and Conformational Dynamics

1. A Donor-Only Sample as a Control

Bonnet et al. reported the first study where FCS was used to characterize conformational fluctuations in biopolymers (47). In this work, the authors examined a DNA hairpin containing a variable number of poly-T or poly-A loops and a 5-bp stem using a rhodamine 6G-dabcyl quencher modification. When the hairpin is in the closed state, the quencher and fluorophore are in proximity and the molecule is not fluorescent. In the open state, the quencher and fluorophore are far apart and fluorescence is restored. In this landmark paper, the authors assume that the hairpin fluctuates between two states: a "bright" open state and a completely "dark" closed state. Subsequent reports have shown that more states are probably involved (see Section II.B.1). Measured reaction times were in the 5 μs–1 ms range, where diffusion dominates the correlation decay. Under these conditions, the autocorrelation decay of the rhodamine probe contains a term describing diffusion and another one describing the fluctuations due to conformational dynamics (47, 48):

$$G(\tau) = G_{\text{diff}}(\tau) \cdot G_{\text{kinetics}}(\tau) = G_{\text{diff}}(\tau)\left[1 + \frac{f_1 f_2 (\eta - 1)^2}{(f_1 \eta + f_2)^2} e^{-\tau/\tau_R}\right] \quad (6)$$

Here, f_1 and f_2 represent the fraction of molecules in states 1 and 2, respectively, and η is the relative fluorescence efficiency of the closed and open states (i.e., the ratio of their fluorescence quantum yields). The relaxation time, τ_R, is given by $(k_{12} + k_{21})^{-1}$, and G_{diff} represents the diffusion contributions [Eq. (4)]. The terms f_1 and f_2 can be expressed in terms of the rate constants k_{12} and k_{21} as $k_{21}/(k_{21} + k_{12})$ and $k_{12}/(k_{21} + k_{12})$, respectively. If the closed state is not fluorescent, $\eta = 0$, and the preexponential factor in Eq. (6) simplifies to $f_1 f_2 = k_{21}/k_{12} = 1/K_{\text{eq}}$, where K_{eq} represents the equilibrium constant of the process as described in Fig. 2. In principle, one could envision fitting the autocorrelation decay with three parameters: τ_D, k_{12}, and k_{21}. However, this is practically impossible when the timescales of diffusion and kinetics overlap. To resolve this problem, Bonnet et al. compared the autocorrelation decay of the fluorophore–quencher double-labeled hairpin with the corresponding decay for a fluorophore-only sample (control). For the control, the autocorrelation function consists of the diffusion contribution only [Eq. (4)]. The kinetic term, G_{kinetics}, can be then isolated from the ratio of the autocorrelation decay of the two samples: $G(\tau)/G_{\text{control}}(\tau) = G_{\text{kinetics}}(\tau)$. The kinetic term can be then fitted with the two parameters of interest: k_{12} and k_{21}.

The derivation of Eq. (6) involves several important assumptions. First, the diffusion coefficients of states 1 and 2 are assumed to be the same (i.e., $D_1 = D_2 = D$). This simplifies the set of differential equations that are solved to obtain this equation. Otherwise, the system of equations cannot be solved in closed form (i.e., there is no analytical solution), and it is not possible to extract physical information from FCS measurements in a direct way. The validity of this assumption should be tested independently whenever possible. For example, it is often possible to shift the equilibrium toward one or the other conformation by changing chemical conditions such as ionic strength and pH.

In addition, Eq. (6) assumes that only two states exist in equilibrium, or in other words, the lifetime of any transient state is assumed to be much shorter than the relaxation time. Whether this is a good approximation or not depends on the particular system being studied. For instance, DNA hairpin folding was initially reported to be a two-state system, but subsequent experiments showed that more states were involved (49). In this regard, it is important to note that the analysis of the FCS decays is model-dependent. Kinetic parameters are extracted from the fit of the correlation decays to equations that were derived under certain assumptions.

In collaboration with Widom et al., we applied the same methodology to study the dynamics of DNA wrapping–unwrapping in nucleosomes (50). In this case, the fluorophore–quencher combination used by Bonnet et al. was replaced by a donor–acceptor FRET pair. Nucleosomes were prepared centered on a 147-bp nucleosome positioning DNA sequence. The donor dye was attached the DNA 5′ end, whereas the acceptor dye was attached to a unique cystein residue on one of the histones nearby the 5′ end of the wrapped DNA. The autocorrelation function of the donor in the double-labeled sample was compared to the corresponding decay of a donor-only sample, so fluctuations due to the wrapping–unwrapping process could be separated from fluctuations due to diffusion. Results are shown in Fig. 8A. The plots in the top panel represent the autocorrelation decay of the donor in the donor-only sample ($G_{DD}^{control}$, lower amplitude decay) and the autocorrelation decay of the donor in the donor–acceptor double-labeled nucleosome (G_{DD}, higher amplitude decay). The ratio of the two decays is shown in the bottom panel that represents the term $G_{kinetics}$. Since the acceptor acts as a quencher for the donor, Eq. (6) is still valid. In this case, the term η can be expressed as a function of the FRET efficiencies of states 1 and 2 as: $\eta = (1 - E_1)/(1 - E_2)$. Based on this analysis, the rate of spontaneous unwrapping (k_{12}) and rewrapping (k_{21}) were reported as 4 and 20 s^{-1}, respectively. This was the first experimental observation of spontaneous DNA fluctuations in nucleosomes. Later on, Kelbauskas et al. used the same approach to study nucleosome dynamics with other DNA sequences (51, 52).

The use of a FRET pair as opposed to the fluorophore–quencher combination used by Bonnet et al. had the benefit of providing a direct way of corroborating that the nucleosome particles were intact. Dissociation of the DNA–protein complex would lead to a loss of FRET, so the stability of the sample could be assessed from the acceptor intensity. It is worth noting that the relaxation time of the dynamics measured in this work ($\tau \sim 40$ ms) is two orders of magnitude larger than the residence time (~ 0.4 ms). As a consequence, the contributions to the autocorrelation decay due to conformational dynamics lie at the tail of the decay, where the value of G_{DD} is low. This gives rise to the large noise observed in the ratio $G_{DD}/G_{DD}^{control}$ at long correlation times. It is obvious that the poor signal-to-noise ratio in this region limits the precision of the measurement, and efforts to increase the residence time of the particle in the observation volume should be made to improve it. Yet, the signal-to-noise ratio in this experiment is enough to determine the rates of the process with about 50% uncertainty.

2. Simultaneous Analysis of the Auto- and Cross-Correlation Decays

The procedure outlined above, although seemingly simple, suffers from two major drawbacks that limit its applicability. First, it relies on the preparation of an identical sample containing only the donor. Second, because the kinetic information is obtained from the ratio of two autocorrelation decays (donor only and donor–acceptor, or fluorophore only and fluorophore–quencher samples), it is extremely critical that the two samples are measured under identical fluorophore concentration and optical conditions. The amplitude of the autocorrelation decay depends on the shape and dimensions of the confocal volume, which are very difficult to control precisely between measurements, particularly when working in biological media. Preparing two samples with identical concentration in the nanometer range is also difficult, especially when measurements are performed inside artificial vesicles, membranes, or living cells. To overcome these issues, we developed a new approach based on the simultaneous measurement and analysis of the three correlation decays: $G_{DD}(\tau)$, $G_{AA}(\tau)$, and $G_{DA}(\tau)$ (48). We showed that under the same set of assumptions used by Bonnet et al. in their original paper, the diffusion term cancels out when the ratio between any two correlation functions is taken. In other words, the ratio between the donor autocorrelation and the donor–acceptor cross-correlation functions measured with the double-labeled sample, $G_{DD}(\tau)/G_{DA}(\tau)$, can be fitted to an equation that depends on the rates of interest (k_{12} and k_{21}) and the FRET efficiencies of the involved states only. The ratios $G_{DA}(\tau)/G_{AA}(\tau)$ and $G_{AA}(\tau)/G_{DD}(\tau)$ can be used in the same way. The advantage of this approach is that kinetic information is obtained independently of the diffusion terms in a single experiment using the donor–acceptor labeled sample. This approach eliminates the need of conducting two experiments (sample and control) in identical conditions, and it is thus a promising methodology to study conformational dynamics in biological media.

To test this approach, we reanalyzed the results of the experiments with nucleosomes described above. In contrast to our original work, where a donor-only sample was used as a control to measure the diffusion contributions, we now show that the same kinetic information can be obtained from the analysis of the donor autocorrelation (G_{DD}) and the donor–acceptor cross-correlation (G_{DA}) of the donor–acceptor double-labeled sample (Fig. 8B, top panel).

The ratio of these decays (bottom panel), which were measured simultaneously in a single experiment, is independent of the diffusion term and can be fitted with an equation that depends on k_{12}, k_{21}, and η only [see reference (48) for a detailed mathematical description of this procedure].

3. Spatially Offset Observation Volumes

Van Orden and Jung addressed the problem of separating the diffusion and kinetic terms by devising a technique in which the molecules flow sequentially through two spatially offset observation volumes positioned a few microns apart (53). In this technique, two spatially offset laser beams are focused into the center of a microcapillary tube (Fig. 9). The sample flows through the tube under pressure, and the fluorescence signal from each observation region is observed and analyzed using auto- and cross-correlation functions. It is important to stress that in this case, the cross-correlation function refers to the cross-correlation of the signals originated in the two offset confocal volumes. This should not be confused with the cross-correlation of a donor–acceptor pair measured within the same observation volume. The flow velocity and the distance between detection volumes are selected so as to attain conditions where the conformational fluctuations make a negligible contribution to the cross-correlation function of the fluorescence from the two detection volumes. This allows the characterization of the translational diffusion properties of the biomolecule, which are subsequently used to isolate the kinetic terms in the autocorrelation analysis. An advantage of this approach is that it can be applied to systems labeled with a single dye–quencher pair because it does not rely on the measurement of the acceptor intensity. It presents, however, some technical challenges to the average user because it requires that two identical observation volumes offset by a few microns are formed near the center of the capillary tube. Slow to moderate flow velocities must be used, and that the focal volumes should centered inside the capillary tube, near the maximum of the parabolic flow velocity profile of the flowing solution.

Figure 9B shows the auto- and cross-correlation functions obtained with the hairpin R6G-AACCC-(T)$_{30}$-GGGTT-dabcyl (53). R6G is a fluorescent rhodamine that is efficiently quenched by dabcyl when the two molecules are in proximity. The inset shows the cross-correlation function together with the fit from which the authors measure the diffusion coefficient of the molecule. This value is used in the analysis of the autocorrelation function, from which the kinetic parameters are extracted. Curve A in Fig. 9 shows the autocorrelation function that one would expect if diffusion and flow were the only process contributing to the fluorescence fluctuations. It is obvious that other processes contribute to the actual autocorrelation functions, including triplet dynamics and fluorescence fluctuations due to conformational changes. To investigate the nature of these contributions, the authors first fit the experimental results to

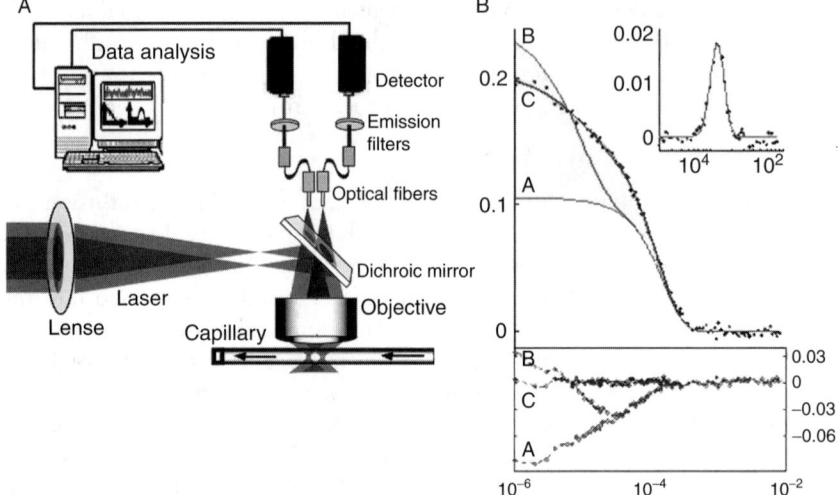

FIG. 9. (A) Schematics of a two-beam FCS setup used to separate contributions from diffusion and conformational dynamics. Molecules flow sequentially through two spatially offset observation volumes positioned a few microns apart. Fluorescence form each focus is analyzed using auto- and cross-correlation and PCH. (B) Experimental autocorrelation data (black circles) and fitting curves for the R6G-AACCC-$(T)_{30}$-GGGTT-dabcyl DNA sample. The inset shows the corresponding cross-correlation data and fitting curves, from which the diffusion parameters are determined. Curve A shows the autocorrelation function that one would expect if diffusion and flow were the only processes contributing to the fluorescence fluctuations. Curve B represents a fit to the data based on a model that incorporates triplet dynamics. Curve C shows the result of the fit when conformational dynamics is also included in the model. The bottom panel represents the residuals for these three types of analysis. [Adapted with permission from reference (53), copyright by American Chemical Society.]

a model that includes not only diffusion and flow, but also triplet dynamics (curve B). From these results, they conclude that there are significant contributions due to conformational dynamics of the DNA hairpin structure. Curve C shows the result of the fit when all these processes are included in the model, from which the relevant rate constants can be obtained.

4. Autocorrelation Analysis of Proximity Factors

Klenerman et al. tackled the problem by analyzing the fluctuations in the so-called proximity factor (54). The idea behind this approach is that although the intensities of the donor and acceptor fluctuate due to diffusion, the efficiency of transfer fluctuates only due to conformational dynamics. The calculation of the FRET efficiency from the intensities of fluorescence of the donor and acceptor requires the measurement of the relative sensitivity with which the two dyes are measured. This factor includes the relative fluorescence efficiency of the two dyes, but also instrumental factors such as the wavelength-dependent detector sensitivity, and transmission efficiency of filters and other optical components. While performing this calibration is possible, the same information can be obtained by using an operational definition of FRET efficiency, known as the "proximity factor." This quantity is defined as $P = I_A/(I_D + I_A)$, where I_D and I_A are the donor and acceptor fluorescence intensities, respectively. Fluctuations in the proximity factor occur as a consequence of distance-dependent changes in FRET efficiency, but not because of diffusion. In this way, the autocorrelation function of the proximity factor $G_P(\tau)$ can be used to study conformational dynamics independently of diffusion.

Wallace et al. used this approach to reinvestigate the dynamics of DNA hairpins (54, 55). For example, the autocorrelation decays of the calculated proximity factors for the 40-base oligonucleotide 5'-GGGTT-(A)$_{30}$-AACCC-3' at three different temperatures are shown in Fig. 10. The time resolution of this approach is limited by the smallest binning time that allows the calculation of $P(t)$ with an acceptable signal-to-noise ratio. The proximity ratio is calculated from a ratio between two fluorescence intensities, so that it can only be computed with a reasonable signal-to-noise ratio when enough photons are measured in both channels. For example, Wallace et al. used a 20 μs binning time to measure both I_A and I_D, so G_P cannot be calculated at shorter lag times. If the binning time is reduced, it will eventually reach the point where zero or one photon per bin is registered in each detector, making the computation of $P(t)$ impossible. In contrast, the calculation of G_{DD}, G_{DA}, and G_{AA} is still possible under these conditions, making it possible to calculate these functions at shorter times. The computation of the correlation function of a vector composed of zeros and ones is possible, and yields a meaningful result, provided that the calculation is performed over a large number of photons. In this way, in contrast to G_P, the correlation decays G_{DD}, G_{DA}, and G_{AA} can be measured in the submicrosecond timescale even if the average signal in this timescale is less than one photon per bin. Another advantage of analyzing G_{DD}, G_{DA}, and G_{AA} relies on the fact that correlator boards measure these functions directly, but cannot measure or generate G_P, making the approach presented by Wallace et al. unfeasible with conventional FCS instrumentation.

4. Sample Immobilization

Maybe the most obvious way of getting rid of the diffusion term is by eliminating diffusion altogether. Various immobilization strategies have been used to observe single-molecule dynamics, the most popular being the use of biotin–streptavidin interactions. Most commonly, a biotinilated biopolymer interacts with streptavidin molecules that are bound to biotin–BSA molecules nonspecifically attached to a slide or a coverslip. Other types of immobilization techniques have been used and reviewed elsewhere (56).

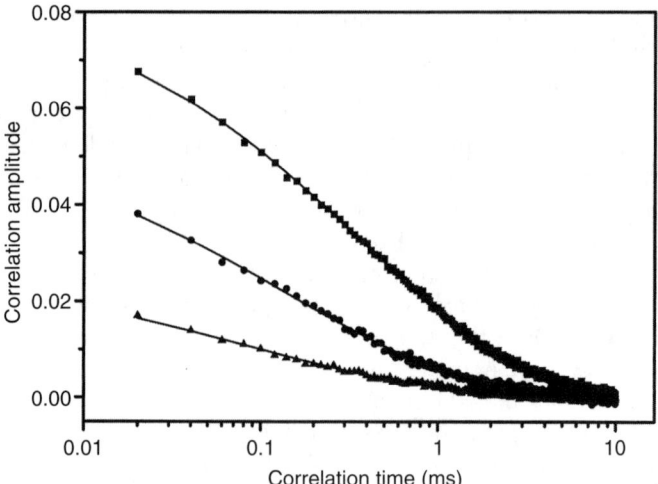

Fig. 10. Autocorrelation functions of the proximity ratio (G_P) calculated for a donor–acceptor double-labeled DNA hairpin. Solid squares, −0.6 °C; solid circles, 19.9 °C; solid triangles, 40.4 °C. The solid lines are fits to Eq. (7). [Adapted with permission from reference (55), copyright by the National Academy of Sciences of the United State of America.]

Although there are many careful experimental studies that show that the dynamic behavior of several DNA and RNA constructs is not influenced by surface interactions (57), finding an appropriate surface for each particular biochemical system can be a daunting experience. An interesting example of a successful application of FCS on immobilized molecules is the study of the Mg^{2+}-facilitated conformational change of an RNA three-helix junction (58). Correlation functions were used to determine transition rates between a folded and open conformation as a function of Mg^{2+} or Na^+ concentration. Cosa et al. used a similar approach to investigate the secondary structure dynamics of DNA hairpins complexed with the HIV-1 protein (59).

Even when immobilization can be successfully achieved in many cases, it is not practical in many situations, including the rapidly growing field of in vivo single-molecule fluorescence studies. For this reason, the problem of dealing with the overlap of diffusion and conformational dynamics in FCS measurements with diffusing molecules continues to be a topic of active research.

B. Selected Applications

So far we have concentrated on the discussion on the experimental approaches used to study conformational dynamics by means of FCS. In this section, we present the results of selected applications where FCS was used to investigate different aspects of nucleic acid conformational dynamics.

1. DNA Hairpins

Since the original study by Bonnet et al. (47), several authors have reinvestigated the dynamics of DNA hairpins by means of FCS. For instance, Wallace et al. analyzed fluctuations in the calculated proximity factor (Section II.A.4) of the same sequences investigated by Bonnet et al., but found that the simple single-exponential decay [Eq. (6)] used in the original paper could not fit their results. Instead, the autocorrelation function of the proximity factor had to be fitted to a stretched exponential with $\beta \sim 0.5$ (55):

$$G_P(\tau) = G_P(0)\exp\left[-\left(\frac{\tau}{\tau_R}\right)^\beta\right] \tag{7}$$

In this equation, β represents a stretched parameter describing the heterogeneity of the sample. This parameter takes the value $\beta = 1$ in systems displaying two-state Arrhenius kinetics with one discrete energy barrier. It is important to stress that the stretched exponential in Eq. (7) is only a phenomenological description of the kinetics, and it is not sufficient to determine a particular mechanism for the conformational fluctuation (60). Interestingly, Sauer et al. found that a single base pair in the stem is sufficient to induce this behavior (Fig. 7) (45). Whereas results for oligonucleotides without complementary base pairs could be well described by a monoexponential decay (Fig. 7, $n = 0$), the introduction of as few as one dC–dG pair was enough to require a stretched parameter in the fit (Fig. 7, $n = 1, 2$). This reveals a distribution of rate constants even for $n = 1$, suggesting that the distribution of transition states cannot be caused solely by mismatches in the stem, but probably by stacking interactions between nucleotides in the stem and the loop.

Finally, Van Orden et al. used their dual-beam approach to investigate the folding trajectories of DNA hairpins by combining FCS with the photon-counting histogram (PCH) technique (49, 53, 61). The latter is a fluorescence fluctuation method that allows the characterization of the brightness and concentration of different fluorescent species present in a sample (62, 63). In this case, PCH was used to analyze the equilibrium distributions and fluorescence intensities of the different populations in the sample. On the basis of this analysis, the authors showed that the equilibrium constants measured by fluctuation spectroscopy did not correspond to the equilibrium distributions of the fully open and closed conformations determined by melting curve analysis (49). To explain these results, a three-state mechanism was proposed to include a dark intermediate state connecting the fully open and closed conformations. The relaxation times and equilibrium constants measured by FCS and PCH would then characterize the dynamic equilibrium between the open and intermediate state, whereas the dynamics of the fully closed state would occur at longer timescales and thus appear static on the FCS timescale.

To address this issue, the authors repeated the studies with a hairpin containing a shorter stem, with the hope of shifting the slow fluctuations to faster timescales so they can be captured by FCS (61). In fact, a DNA hairpin containing a 21-nucleotide loop and 4-bp stem exhibited double relaxation kinetics with time constants 84 and 393 μs. The slow time constant was assigned to dynamics of the fully closed hairpin, and was probably missed in the previous study with a 5-bp stem sample because it was too slow for FCS timescales.

2. Nucleosome Dynamics

The first experimental demonstration of spontaneous conformational fluctuations in nucleosomes was described by Li et al., as described above (see Fig. 8) (50). Previous studies had quantified the equilibrium aspects of this process (64), but these FCS measurements provided the first experimental determination of the rates of site exposure and rewrapping. These rates are of interest because they provide insight into the mechanisms by which remodeling factors and other proteins gain access to DNA buried inside nucleosomes. In this work, individual nucleosomes were prepared centered on a 147-bp nucleosome positioning sequence. It is important to stress that the use of a sequence for which a single dominant position is known exactly is instrumental in the analysis of the FRET signal. A Cy3 dye donor molecule was placed a the DNA 5′ end, whereas a Cy5 dye acceptor molecule was covalently attached to a unique cystein on histone H3 V35C C110A, located on the histone core nearby the 5′ end of the wrapped DNA. Results showed that the correlation decays are consistent with a model where two states are in equilibrium with $k_{21} = 20$ s^{-1} and $k_{12} = 3.6$ s^{-1}. Similar values were obtained from stopped flow experiments reported in the same work.

Kelbauskas et al. used a similar methodology to investigate sequence-dependent nucleosome dynamics (51, 52). Here, the authors compared the correlation decays of nucleosomes reconstituted from three different DNA sequences: a TATA-containing sequence from yeast GAL10, a sequence containing four of the six glucocorticoid-receptor response elements from the MMTV promoter, and the sea urchin 5S rDNA fragment. In contrast to the previous studies by Li et al., the donor and acceptor dyes were placed on the DNA 80 bp apart, so that FRET does not occur unless the two dyes are brought together by the nucleosomal wrap. Also, an interesting aspect of this work is that measurements were performed in 3% agarose gels so as to decrease the diffusion time and improve the signal-to-noise ratio of the quotient $G_{DD}/G_{DD}^{control}$. The reported lifetimes for the closed states ($1/k_{12}$) were about one order of magnitude shorter than the one measured in reference (50). This is consistent with the fact that the three sequences used by Kelbauskas et al. are known to bind histones less strongly than the synthetic sequence used by Li et al.

3. Breathing Fluctuations in dsDNA

Altan-Bonnet et al. used the same experimental approach used in their previous investigation of hairpin dynamics, discussed above, to investigate breathing fluctuations in dsDNA (65). Breathing fluctuations are thermal excitations that lead to local denaturation and reclosing of the double-stranded structure. In order to verify that the dyes do not induce significant disruption of the constructs' secondary structure, the authors measured the melting curves of the labeled and nonlabeled constructs by UV absorption with identical results. The correlation decay in this case could not be fitted with the single-exponential decay of Eq. (6), indicating the existence of a distribution of excited modes. The results of this work are consistent with a picture where bubbles of 2–10 bp open spontaneously with a lifetime in the 50-μs timescale.

4. DNA Mobility and Flexibility

So far we have made a clear distinction between diffusion and conformational dynamics. However, the flexibility of semiflexible polymers such as ss-DNA can be investigated by determining the radius of gyration of the polymer from diffusion measurements. For instance, Doose et al. investigated the length-dependent diffusion coefficient of polythymine (66). Depending on the ionic strength of the solution, inverse diffusion constants scale with the number of bases (N) as N^v, with v between 0.5 and 0.7. These results yield an experimental confirmation that polythymine with a few tens of residues resembles an unstructured polyelectrolyte that can be described as a semiflexible polymer.

Similarly, Tatarkova and Berk (67) and Petrov et al. (39) investigated the mobility of dsDNA. While Tatarkova and Berk's data follow a $D \sim L^{-1/2}$ power law, Petrov et al.'s data suggest that this is true only for dsDNA lengths exceeding 10^5 bp. In fact, the diffusion coefficients measured in reference (67) for DNA lengths in the 500–2000 bp range seem to be smaller than the corresponding values measured in reference (39, 68), which are in better agreement with values measured by other techniques. It is a possibility that the diffusion coefficients measured by Tatarkova and Berk are somewhat affected by the large amounts of intercalating dye used in the experiment. In contrast, Petrov et al. used a single fluorescent label on the 5' end of the DNA, which prevents DNA distortion.

5. DNA Condensation and Packaging

Kral et al. investigated DNA condensation by the multivalent cationic compounds spermine and hexadecyltrimethyl ammonium bromide by measuring the change in diffusion coefficient that accompanies the changes in conformation due to condensation (37). The intercalating dyes ethidium bromide and propidium iodide were used in this work to label the plasmids. Changes in diffusion coefficient of up to an order of magnitude were observed upon condensation. However, subsequent studies revealed that intercalating dyes affect the structure of DNA and the condensation process. In fact, the same authors investigated this issue and found that the dye PicoGreen labels nucleic acids more uniformly without concentration-dependent artifacts (69).

Along the same lines, Sabanayagam et al. studied the DNA packaging machinery of the bacteriophage T4 (70). In this case, the ATP-dependent translocation kinetics of labeled DNA was measured by monitoring the decrease in DNA diffusibility. DNAs with a single 5'-R6G fluorophore were used as substrates to avoid the problems encountered with intercalating dyes described above.

More recently, Spring and Clegg used FCS to investigate the formation of DNA aggregates under conditions favorable for M-DNA formation (71) (millimolar zinc and pH = 8.6). M-DNA is a metal complex of DNA that forms cooperatively in the presence of high concentrations of Zn^{2+}, Co^{2+}, or Ni^{2+} cations and pH in the range of 8–9. The results of this work show that conversion to M-DNA is concomitant with aggregation of the dsDNA molecules.

IV. Experimental Techniques

A. Fluorophores

Highly fluorescent nucleotide base analogues such as 2-aminopurine and 6-methylisoxanthopterin have been successfully used for studying the structure, dynamics, and interactions of nucleic acids using fluorescence methods (72). These analogues are structurally similar to the naturally occurring nucleotides, so they form thermodynamically equivalent base pairs and do not perturb the nucleic acid structure. Unfortunately these compounds absorb and emit in the UV region of the spectrum, where scattering and background fluorescence from organelles, membranes, proteins, and buffer components overwhelm the small fluorescence signal from the molecules of interest. Extrinsic fluorescent probes that absorb in the green-red region of the spectrum are usually preferred for this reason. Chromophores that absorb at longer wavelengths have also higher dipole strengths, which results in an increased efficiency of absorption and emission.

Dyes that perform well in confocal laser-scanning microscopy are usually among the best choices for FCS applications. Dyes with large extinction coefficients and fluorescence quantum yields are best because they yield high emission intensities at low laser excitations. Photochemical stability is another important parameter to consider. All fluorescent molecules have some finite probability of undergoing an irreversible photochemical reaction from the excited state. These photochemical processes are usually referred to as "photobleaching" because they result in a nonfluorescent product. Photobleaching limits the total number of photons that are emitted by each individual molecule, and can be somewhat minimized by the addition of oxygen-scavenging systems. These systems most typically use a mixture of glucose oxidase and catalase that converts glucose and O_2 into gluconic acid and water, resulting in the net loss of O_2 in solution (73). However, oxygen-scavenging systems can introduce intensity fluctuations in the millisecond timescale presumably due the increased triplet-state lifetime upon O_2 removal. Photobleaching can also be avoided by decreasing the excitation intensity at the cost of a longer measurement time.

Fluorescent probes for FCS should also have a low efficiency of triplet formation. This is true for all fluorescence applications because triplet formation decreases the efficiency of fluorescence, but it is particularly important in FCS applications when measuring dynamics in the micro- to millisecond timescale. As described above, triplet dynamics gives rise to features in the autocorrelation decay that might overlap with the features produced by the kinetic processes under study (40). Other undesired photophysical processes that produce features in the autocorrelation decay include *cis–trans* photoisomerization transitions in cyanine dyes (e.g., Cy3 and Cy5) (41). In addition, fluorescent probes should be small to prevent distortions in the nucleic acid structure and to avoid increasing the viscous drag of the molecule.

Figure 11 shows the structures of the most widely used fluorophores in the field. Fluorescent proteins are also widely used, but mainly for intracellular diffusion studies. It is important to stress that there is a much larger selection of

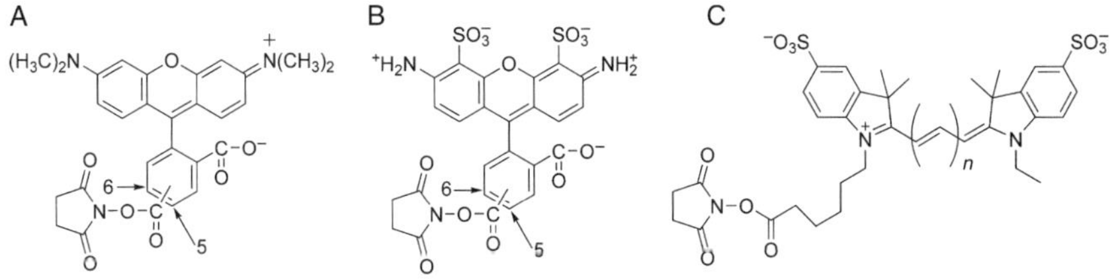

FIG. 11. Chemical structures of some of the most widely used fluorophores used in FCS. All structures are shown as succinimidyl ester derivatives. (A) Carboxytetramethylrhodamine, succinimidyl ester ($\lambda_{ex} = 547$ nm, $\lambda_{em} = 573$ nm, Molecular probes, Eugene OR); (B): Alexa Fluor® 488 carboxylic acid, succinimidyl ester ($\lambda_{ex} = 494$ nm, $\lambda_{em} = 517$ nm, Molecular probes, Eugene OR); (C): Cy3 (n=1) and Cy5 (n=2) mono reactive NHS esters (Cy3: $\lambda_{ex} = 550$ nm, $\lambda_{em} = 570$ nm, Cy5: $\lambda_{ex} = 645$ nm, $\lambda_{em} = 665$ nm, GE Healthcare, Piscataway, NJ). λ_{ex} and λ_{em} represent the absorption and emission maxima, respectively.

fluorescent probes available from a variety of companies, and that fluorophores with improved photophysical and photochemical properties are constantly introduced in the market.

B. Nucleic Acid Labeling

Today, almost all synthetic oligonucleotides are prepared by automated solid phase phosphoramidite techniques. Automated fluorescent labeling can be performed using the same chemistry (74). Several commercial companies offer a variety of fluorescent labels that can be incorporated directly into the synthetic sequence, most commonly at the 5'- or 3'-terminus. Although extremely convenient, these offerings are limited to a small fraction of the large variety of fluorescent probes that are otherwise available as other type of chemical derivatives. Also, these commercial modifications are mostly limited to 3'- or 5'–terminal labeling, while many applications require that the fluorescent probe is covalently attached internally in the nucleic acid sequence. To work around these limitations, manual labeling can be performed on nucleotides bearing functional groups such as primary amines or thiols. These functional groups can be incorporated during automated synthesis and be conjugated with a fluorescently labeled postsynthesis. This allows more versatility in terms of the location of the probe in the oligonucleotide and increases dramatically the selection of fluorescent probes that can be used. Companies such as Molecular Probes, Inc. (Eugene, Oregon, an Invitrogen company) have an extensive selection of fluorescent probes suitable for conjugation to amines, thiols, carboxylic acids, aldehydes, etc. The most commonly used modification to produce internally labeled oligonucleotides is the so-called amino-modifier C6 dT, which is commercialized as a phosphoramidite by Glen Research (Sterling, Virginia). This modification consists of a 2'-deoxyuridine with a 6-carbon linker ending with a primary amino group, and can be incorporated during automated synthesis at either terminus or internally. Modified oligonucleotides are then used in a reaction with fluorescent isothiocyanates or N-hydroxysuccinimidyl esters (NHS) (Fig. 11), which produce covalent bonds with amino groups. NHS esters are favored as they give higher yields of labeled oligonucleotides and produce more stable linkages for long-term storage. The literature provided by Molecular Probes, Inc., is an excellent source of information and protocols on manual fluorescent labeling (http://molecularprobes.com/handbook/).

C. FCS Instrumentation

FCS instruments are commercially available from a number of companies. However, building a home-made instrument is relatively simple, providing more flexibility and significant cost savings. A comprehensive review by Sengupta *et al*. provides a detailed discussion of the optical components required to build an FCS instrument in the laboratory (75). Most commonly, a laser light source and appropriate detectors are incorporated into a research-grade biological microscope, which is used as the body of the instrument. A well-characterized laser beam is important to allow a reliable description of the focal volume. Low power (~1 mW) continuous wave lasers are sufficient for one-photon excitation FCS. Commercial argon ion lasers (488, 514 nm), helium-neon (633 nm), and solid-state lasers (e.g., Nd:Yag, 532 nm) are common choices. A pulsed laser such as a Ti:sapphire is required for two-photon FCS. These lasers provide short pulses over a wide range of wavelengths (700–1000 nm), which can be doubled in frequency to obtain pulses in the visible range for two-photon excitation of visible dyes.

The laser beam should be expanded but should not overfill the back aperture of the microscope objective to define a Gaussian confocal volume (27). Oil-immersion objectives are characterized by the highest numerical aperture (~1.4), but because they are designed to focus and collect light in a high refractive index environment, they generate optical aberrations when used to focus the laser deeper inside aqueous samples. Water-immersion objectives have smaller numerical apertures, but are more efficient in focusing the excitation light and collecting the emission deeper in the solution. The collected fluorescence is separated from the excitation light by a dichroic mirror (Fig. 3, dichroic mirror 1), and focused into a pinhole placed in the image plane. The pinhole is used to reject light form other regions other than the focal plane. Typically, pinhole diameters are about 30–50 μm, although larger pinholes can be used when large detection volumes are needed. The optimization of pinhole size was thoroughly considered by Rigler et al. (76) and Hess and Webb (27). Alternatively, fluorescence can be focused into a multimode fiber that acts as a pinhole. Changing the fiber allows easy adjustment of the pinhole size without the need for substantial realignment of the instrument.

Photomultipliers and Avalanche photodiodes (APDs) are the most commonly used detectors in FCS experiments. Avalanche photodiodes are more sensitive in the visible and near infrared region, but photomultipliers perform better in the blue region of the spectrum. An important consideration when choosing detectors for FCS is the problem of afterpulsing. Afterpulses are spurious dark count pulses that follow the detection of a photon typically in the 100 ns–1 μs timescale. Detector afterpulsing is a very common and often overlooked instrumental artifact that leads to a distortion of the autocorrelation function at short lag times. A fast initial decay of an autocorrelation function can be easily misinterpreted as a triplet contribution. The most elegant way to remove afterpulsing is to mount a 50/50 beam splitter in the detection path, add an additional detector, and cross-correlate the signal of the two detectors. The cross-correlation of the signal with itself is by definition equal to the autocorrelation, but because afterpulses are not correlated between the two detectors, they do not contribute to the cross-correlation decay.

Detected photons can be processed in two different ways. Hardware digital correlator boards take the signal from the detector and calculate the correlation functions electronically. These boards are very convenient, but they provide access only to the auto- or cross-correlation function and not the complete time sequence of photon counts. The most popular correlator boards for FCS are commercialized by ALV-GmbH (Langen, Germany) and Correlator.com (Bridgewater, NJ). Single-photon counting boards on the other hand can measure the arrival time of each detected photon, from which the autocorrelation functions can be calculated subsequently by software. The advantage of this approach is that it allows other kinds of analysis on the raw data such as PCHs, autocorrelation functions of proximity factors, and higher-order correlations. The main developers of this type of boards are PicoQuant GmbH (Berlin, Germany) and Becker & Hickl GmbH (Berlin, Germany).

D. Calibration

The most common way to calibrate the observation volume is to measure the autocorrelation curve of a sample of known concentration and diffusion coefficient. Equation (4) is used to fit the experimental FCS decay using r_0 and z_0 as fitting parameters. The effective volume can be defined as $V_{eff} = \pi^{3/2} \cdot r_0^2 \cdot z_0$, so $\langle N \rangle$ can be calculated as $\langle C \rangle V_{eff}$ with $\langle C \rangle$ being the bulk sample concentration. Note that the shape and size of the observation volume are different in two-photon excitation setups, and the residence time in this case is $\tau_D = r_0^2/8D$ (29). Rhodamine 6G is widely used as a calibration standard ($D = 2.8 \times 10^{-10}$ m^2/s). Fluorescein in a high-pH buffer ($D = 3.0 \times 10^{-10}$ m^2/s) and rhodamine 110 are also commonly used (23). Fluorescent spheres have also been used for this purpose, but they tend to aggregate as a function of time. To verify that the instrument is working properly, it is recommended that measurements of solutions with different concentrations of a suitable dye and different solvent viscosities are performed.

E. Signal-To-Noise Ratio

Shot noise affects primarily the initial part of the correlation curve, whereas short sampling times are used to calculate G at short lag times. The signal-to-noise ratio under these conditions is proportional to the product of the square root of the total measurement time and the count rate per detected molecule (77, 78). It is important to note that the critical parameter is the count rate per detected molecule, and not the total amount of measured fluorescence. The count rate per detected molecule depends on the brightness of the fluorophore (extinction coefficient and fluorescence quantum yield), the detection efficiency of the system (including detectors, filters, optics, etc), and the excitation intensity.

The count rate per detected molecule increases linearly with excitation intensity at low values, but it eventually plateaus due to saturation effects and can actually decrease at high laser powers due to photobleaching.

Equation (4) shows that the amplitude of the FCS decays depends inversely on $\langle N \rangle$. However, background signal dominates the detected signal at very low values, making the measurement difficult. On the other hand, the maximum number of particles is restricted by the maximum photon rate that the detector can count in the linear response regime (~500 kHz). Ideally, the count rate per detected molecule should be maximized to improve the signal-to-noise ratio, and $\langle N \rangle$ adjusted so as to remain within the linear regime of the detector.

F. Two-Photon Excitation FCS

The phenomenon of two-photon excitation (TPE) arises from the simultaneous absorption of two photons in a single quantized event. Because the energy of a photon is inversely proportional to its wavelength, the two absorbed photons must have a wavelength about twice that required for one-photon excitation. The probability of the near-simultaneous absorption of two photons is extremely low, so a high flux of excitation photons is typically required. Ti-sapphire lasers allow the high photon density and flux required for TPE. Two-photon excitation provides inherent spatial confinement of excitation, diminished photobleaching and phototoxicity, less scattering, and better optical penetration in turbid media. For these reasons, TPE has been successfully applied in studies with cells, tissues, and membranes (28, 29). Although TPE applications to date have been mostly limited to the study of molecular diffusion and ligand-receptor binding, it represents a promising approach for investigating the conformational dynamics of nucleic acids in living cells.

V. Concluding Remarks

FCS has become a powerful tool to investigate conformational dynamics of nucleic acids in a wide range of timescales. The technique is based on the analysis of the fluctuations in fluorescence that arise when the efficiency of emission of a fluorescent probe is affected by changes in the conformation of the biopolymer. Most commonly, this is achieved by taking advantage of a specific quencher (such as a proximal guanosine residue) or by introducing a donor–acceptor FRET pair in the system. Much of the effort in the field has been focused on dealing with the problem of separating the contributions due to conformational dynamics from those due to molecular diffusion. The recent developments in this regard, combined with the latest advances in two-photon excitation FCS, represent a promising approach for the investigation of conformational dynamics of nucleic acids in living cells.

REFERENCES

1. Luo, G., Wang, M., Konigsberg, W. H., and Xie, X. S. (2007). Single-molecule and ensemble fluorescence assays for a functionally important conformational change in T7 DNA polymerase. *Proc. Natl. Acad. Sci. USA* **104**, 12610–12615.
2. Stryer, L., and Haugland, R. P. (1967). Energy transfer—a spectroscopic ruler. *Proc. Natl. Acad. Sci. USA* **58**, 719–726.
3. Lilley, D. M. J., and Clegg, R. M. (1993). The structure of the 4-way junction in DNA. *Annu. Rev. Biophys. Biomol. Struct.* **22**, 299–328.
4. Stuhmeier, F., Welch, J. B., Murchie, A. I. H., Lilley, D. M. J., and Clegg, R. M. (1997). Global structure of three-way DNA junctions with and without additional unpaired bases: A fluorescence resonance energy transfer analysis. *Biochemistry* **36**, 13530–13538.
5. JaresErijman, E. A., and Jovin, T. M. (1996). Determination of DNA helical handedness by fluorescence resonance energy transfer. *J. Mol. Biol.* **257**, 597–617.
6. Parkhurst, L. J., Parkhurst, K. M., Powell, R., Wu, J., and Williams, S. (2001). Time-resolved fluorescence resonance energy transfer studies of DNA bending in double-stranded oligonucleotides and in DNA-protein complexes. *Biopolymers* **61**, 180–200.
7. Sixou, S., Szoka, F. C., Green, G. A., Giusti, B., Zon, G., and Chin, D. J. (1994). Intracellular oligonucleotide Hybridization detected by fluorescence resonance energy-transfer (FRET). *Nucleic Acids Res.* **22**, 662–668.
8. Hiyoshi, M., and Hosoi, S. (1994). Assay of DNA denaturation by polymerase chain reaction-driven fluorescent label incorporation and fluorescence resonance energy-transfer. *Anal. Biochem.* **221**, 306–311.
9. Craig, M. E., Crothers, D. M., and Doty, P. (1971). Relaxation kinetics of dimer formation by self complementary oligonucleotides. *J. Mol. Biol.* **62**, 383–401.
10. Lecuyer, K. A., and Crothers, D. M. (1994). Kinetics of an RNA conformational switch. *Proc. Natl. Acad. Sci. USA* **91**, 3373–3377.
11. Shen, Y. Q., Kuznetsov, S. V., and Ansari, A. (2001). Loop dependence of the dynamics of DNA hairpins. *J. Phys. Chem. B* **105**, 12202–12211.
12. Wetmur, J. G., and Davidson, N. (1968). Kinetics of renaturation of DNA. *J. Mol. Biol.* **31**, 349–370.
13. Callender, R., and Dyer, R. B. (2002). Probing protein dynamics using temperature jump relaxation spectroscopy. *Curr. Opin. Struct. Biol.* **12**, 628–633.
14. Eftink, M. R. (1997). Fluorescence methods for studying equilibrium macromolecule-ligand interactions. *Fluoresc. Spectrosc.* **278**, 221–257.
15. Li, H. J., and Crothers, D. M. (1969). Relaxation studies of proflavine-DNA complex—kinetics of an intercalation reaction. *J. Mol. Biol.* **39**, 461–477.
16. Pasternack, R. F., Gibbs, E. J., and Villafranca, J. J. (1983). Interactions of porphyrins with nucleic-acids. *Biochemistry* **22**, 5409–5417.
17. Weiss, S. (2000). Measuring conformational dynamics of biomolecules by single molecule fluorescence spectroscopy. *Nat. Struct. Biol.* **7**, 724–729.
18. Ha, T., Enderle, T., Ogletree, D. F., Chemla, D. S., Selvin, P. R., and Weiss, S. (1996). Probing the interaction between two single molecules: Fluorescence resonance energy transfer between a single donor and a single acceptor. *Proc. Natl. Acad. Sci. USA* **93**, 6264–6268.
19. Cornish, P. V., and Ha, T. (2007). A survey of single-molecule techniques in chemical biology. *ACS Chem. Biol.* **2**, 53–61.
20. Ditzler, M. A., Aleman, E. A., Rueda, D., and Walter, N. G. (2007). Focus on function: Single molecule RNA enzymology. *Biopolymers* **87**, 302–316.
21. Seidel, R., and Dekker, C. (2007). Single-molecule studies of nucleic acid motors. *Curr. Opin. Struct. Biol.* **17**, 80–86.

22. Gosch, M., and Rigler, R. (2005). Fluorescence correlation spectroscopy of molecular motions and kinetics. *Adv. Drug Deliv. Rev.* **57**, 169–190.
23. Muller, J. D., Chen, Y., and Gratton, E. (2003). Fluorescence correlation spectroscopy. *Biophotonics Pt B* **361**, 69–92.
24. Magde, D., Webb, W. W., and Elson, E. (1972). Thermodynamic fluctuations in a reacting system—measurement by fluorescence correlation spectroscopy. *Phys. Rev. Lett.* **29**, 705–708.
25. Magde, D., Elson, E. L., and Webb, W. W. (1974). Fluorescence correlation spectroscopy. 2. Experimental realization. *Biopolymers* **13**, 29–61.
26. Widengren, J., and Mets, Ü. (2003). *In* "Single Molecule Detection in Solution" (C. Zander, J. Enderlein, and R. A. Keller, Eds.), pp. 69–120. Wiley-VCH, Berlin.
27. Hess, S. T., and Webb, W. W. (2002). Focal volume optics and experimental artifacts in confocal fluorescence correlation spectroscopy. *Biophys. J.* **83**, 2300–2317.
28. Kim, S. A., Heinze, K. G., and Schwille, P. (2007). Fluorescence correlation spectroscopy in living cells. *Nat. Methods* **4**, 963–973.
29. Schwille, P., Haupts, U., Maiti, S., and Webb, W. W. (1999). Molecular dynamics in living cells observed by fluorescence correlation spectroscopy with one- and two-photon excitation. *Biophys. J.* **77**, 2251–2265.
30. Weiss, M., Hashimoto, H., and Nilsson, T. (2003). Anomalous protein diffusion in living cells as seen by fluorescence correlation spectroscopy. *Biophys. J.* **84**, 4043–4052.
31. Politz, J. C., Browne, E. S., Wolf, D. E., and Pederson, T. (1998). Intranuclear diffusion and hybridization state of oligonucleotides measured by fluorescence correlation spectroscopy in living cells. *Proc. Natl. Acad. Sci. USA* **95**, 6043–6048.
32. Hess, S. T., Huang, S. H., Heikal, A. A., and Webb, W. W. (2002). Biological and chemical applications of fluorescence correlation spectroscopy: A review. *Biochemistry* **41**, 697–705.
33. Schwille, P. (2001). Fluorescence correlation spectroscopy and its potential for intracellular applications. *Cell Biochem. Biophys.* **34**, 383–408.
34. Kinjo, M., and Rigler, R. (1995). Ultrasensitive hybridization analysis using fluorescence correlation spectroscopy. *Nucleic Acids Res.* **23**, 1795–1799.
35. Allen, N. W., and Thompson, N. L. (2006). Ligand binding by estrogen receptor beta attached to nanospheres measured by fluorescence correlation spectroscopy. *Cytometry Part A* **69A**, 524–532.
36. Grunwald, D., Cardoso, M. C., Leonhardt, H., and Buschmann, V. (2005). Diffusion and binding properties investigated by fluorescence correlation spectroscopy (FCS). *Curr. Pharm. Biotechnol.* **6**, 381–386.
37. Kral, T., Langner, M., Benes, M., Baczynska, D., Ugorski, M., and Hof, M. (2002). The application of fluorescence correlation spectroscopy in detecting DNA condensation. *Biophys. Chem.* **95**, 135–144.
38. Haustein, E., and Schwille, P. (2003). Ultrasensitive investigations of biological systems by fluorescence correlation spectroscopy. *Methods* **29**, 153–166.
39. Petrov, E. P., Ohrt, T., Winkler, R. G., and Schwille, P. (2006). Diffusion and segmental dynamics of double-stranded DNA. *Phys. Rev. Lett.* **97**, 258101.
40. Widengren, J., Mets, U., and Rigler, R. (1995). Fluorescence correlation spectroscopy of triplet-states in solution – a theoretical and experimental-study. *J. Phys. Chem.* **99**, 13368–13379.
41. Widengren, J., and Schwille, P. (2000). Characterization of photoinduced isomerization and back-isomerization of the cyanine dye Cy5 by fluorescence correlation spectroscopy. *J. Phys. Chem. A* **104**, 6416–6428.
42. Sabanayagam, C. R., Eid, J. S., and Meller, A. (2005). Long time scale blinking kinetics of cyanine fluorophores conjugated to DNA and its effect on Forster resonance energy transfer. *J. Chem. Phys.* **123**, 224708.

43. Heinlein, T., Knemeyer, J. P., Piestert, O., and Sauer, M. (2003). Photoinduced electron transfer between fluorescent dyes and guanosine residues in DNA-hairpins. *J. Phys. Chem. B* **107**, 7957–7964.
44. Seidel, C. A. M., Schulz, A., and Sauer, M. H. M. (1996). Nucleobase-specific quenching of fluorescent dyes.1. Nucleobase one-electron redox potentials and their correlation with static and dynamic quenching efficiencies. *J. Phys. Chem.* **100**, 5541–5553.
45. Kim, J., Doose, S., Neuweiler, H., and Sauer, M. (2006). The initial step of DNA hairpin folding: A kinetic analysis using fluorescence correlation spectroscopy. *Nucleic Acids Res.* **34**, 2516–2527.
46. Neuweiler, H., Doose, S., and Sauer, M. (2005). A microscopic view of miniprotein folding: Enhanced folding efficiency through formation of an intermediate. *Proc. Natl. Acad. Sci. USA* **102**, 16650–16655.
47. Bonnet, G., Krichevsky, O., and Libchaber, A. (1998). Kinetics of conformational fluctuations in DNA hairpin-loops. *Proc. Natl. Acad. Sci. USA* **95**, 8602–8606.
48. Torres, T., and Levitus, M. (2007). Measuring conformational dynamics: A new FCS-FRET approach. *J. Phys. Chem. B* **111**, 7392–7400.
49. Jung, J. Y., and Van Orden, A. (2006). A three-state mechanism for DNA hairpin folding characterized by multiparameter fluorescence fluctuation spectroscopy. *J. Am. Chem. Soc.* **128**, 1240–1249.
50. Li, G., Levitus, M., Bustamante, C., and Widom, J. (2005). Rapid spontaneous accessibility of nucleosomal DNA. *Nat. Struct. Mol. Biol.* **12**, 46–53.
51. Kelbauskas, L., Chan, N., Bash, R., DeBartolo, P., Sun, J., Woodbury, N., and Lohr, D. (2008). Sequence-dependent variations associated with H(2)A/H2B depletion of nucleosomes. *Biophys. J.* **94**, 147–158.
52. Kelbauskas, L., Chan, N., Bash, R., Yodh, J., Woodbury, N., and Lohr, D. (2007). Sequence-dependent nucleosome structure and stability variations detected by Forster resonance energy transfer. *Biochemistry* **46**, 2239–2248.
53. Jung, J. M., and Van Orden, A. (2005). Folding and unfolding kinetics of DNA hairpins in flowing solution by multiparameter fluorescence correlation spectroscopy. *J. Phys. Chem. B* **109**, 3648–3657.
54. Wallace, M. I., Ying, L. M., Balasubramanian, S., and Klenerman, D. (2000). FRET fluctuation spectroscopy: Exploring the conformational dynamics of a DNA hairpin loop. *J. Phys. Chem. B* **104**, 11551–11555.
55. Wallace, M. I., Ying, L. M., Balasubramanian, S., and Klenerman, D. (2001). Non-Arrhenius kinetics for the loop closure of a DNA hairpin. *Proc. Natl. Acad. Sci. USA* **98**, 5584–5589.
56. Rasnik, I., Mckinney, S. A., and Ha, T. (2005). Surfaces and orientations: Much to FRET about? *Acc. Chem. Res.* **38**, 542–548.
57. Okumus, B., Wilson, T. J., Lilley, D. M. J., and Ha, T. (2004). Vesicle encapsulation studies reveal that single molecule ribozyme heterogeneities are intrinsic. *Biophys. J.* **87**, 2798–2806.
58. Kim, H. D., Nienhaus, G. U., Ha, T., Orr, J. W., Williamson, J. R., and Chu, S. (2002). Mg2+-dependent conformational change of RNA studied by fluorescence correlation and FRET on immobilized single molecules. *Proc. Natl. Acad. Sci. USA* **99**, 4284–4289.
59. Cosa, G., Harbron, E. J., Zeng, Y. N., Liu, H. W., O'Connor, D. B., Eta-Hosokawa, C., Musier-Forsyth, K., and Barbara, P. F. (2004). Secondary structure and secondary structure dynamics of DNA hairpins complexed with HIV-1NC protein. *Biophys. J.* **87**, 2759–2767.
60. Metzler, R., Klafter, J., Jortner, J., and Volk, M. (1998). Multiple time scales for dispersive kinetics in early events of peptide folding. *Chem. Phys. Lett.* **293**, 477–484.
61. Jung, J., Ihly, R., Scott, E., Yu, M., and Van Orden, A. (2008). Probing the complete folding trajectory of a DNA hairpin using dual beam fluorescence fluctuation spectroscopy. *J. Phys. Chem. B* **112**, 127–133.

62. Chen, Y., Muller, J. D., So, P. T. C., and Gratton, E. (1999). The photon counting histogram in fluorescence fluctuation spectroscopy. *Biophys. J.* **77**, 553–567.
63. Perroud, T. D., Huang, B., Wallace, M. I., and Zare, R. N. (2003). Photon counting histogram for one-photon excitation. *Chemphyschem* **4**, 1121–1123.
64. Li, G., and Widom, J. (2004). Nucleosomes facilitate their own invasion. *Nat. Struct. Mol. Biol.* **11**, 763–769.
65. Altan-Bonnet, G., Libchaber, A., and Krichevsky, O. (2003). Bubble dynamics in double-stranded DNA. *Phys. Rev. Lett.* **90**, 138101.
66. Doose, S., Barsch, H., and Sauer, M. (2007). Polymer properties of polythymine as revealed by translational diffusion. *Biophys. J.* **93**, 1224–1234.
67. Tatarkova, S. A., and Berk, D. A. (2005). Probing single DNA mobility with fluorescence correlation microscopy. *Phys. Rev. E* **71**, 041913.
68. Petrov, E. P., Ohrt, T., Winkler, R. G., and Schwille, P. (2006). Diffusion and segmental dynamics of double-stranded DNA. *Phys. Rev. Lett.* **97**, 258101.
69. Kra, T., Widerak, K., Langner, M., and Hof, M. (2005). Propidium iodide and PicoGreen as dyes for the DNA fluorescence correlation spectroscopy measurements. *J. Fluoresc.* **15**, 179–183.
70. Sabanayagam, C. R., Oram, M., Lakowicz, J. R., and Black, L. W. (2007). Viral DNA packaging studied by fluorescence correlation spectroscopy. *Biophys. J.* **93**, L17–L19.
71. Spring, B. Q., and Clegg, R. M. (2007). Fluorescence measurements of duplex DNA oligomers under conditions conducive for forming M-DNA (a metal-DNA complex). *J. Phys. Chem. B* **111**, 10040–10052.
72. Rist, M. J., and Marino, J. P. (2002). Fluorescent nucleotide base analogs as probes of nucleic acid structure, dynamics and interactions. *Curr. Org. Chem.* **6**, 775–793.
73. Rasnik, I., McKinney, S. A., and Ha, T. (2006). Nonblinking and longlasting single-molecule fluorescence imaging. *Nat. Methods* **3**, 891–893.
74. Davies, M. J., Shah, A., and Bruce, I. J. (2000). Synthesis of fluorescently labelled oligonucleotides and nucleic acids. *Chem. Soc. Rev.* **29**, 97–107.
75. Sengupta, P., Balaji, J., and Maiti, S. (2002). Measuring diffusion in cell membranes by fluorescence correlation spectroscopy. *Methods* **27**, 374–387.
76. Rigler, R., Mets, U., Widengren, J., and Kask, P. (1993). Fluorescence correlation spectroscopy with high count rate and low-background: Analysis of translational diffusion. *Eur. Biophys. J. Biophys. Lett.* **22**, 169–175.
77. Kask, P., Gunther, R., and Axhausen, P. (1997). Statistical accuracy in fluorescence fluctuation experiments. *Eur. Biophys. J. Biophys. Lett.* **25**, 163–169.
78. Qian, H. (1990). On the statistics of fluorescence correlation spectroscopy. *Biophys. Chem.* **38**, 49–57.

RNA Structure and Modeling: Progress and Techniques

> DINGGENG CHAI
>
> *Department of Biological Sciences,*
> *University of Calgary, Calgary, Alberta,*
> *Canada T2N 1N4*

I. Introduction	72
A. Structural Elements of RNA	72
B. Differences Between RNA and Protein Folding	73
C. Representative Classes of RNAs and Solved Structures	75
II. Chemical and Enzymatic Methods	78
A. Enzyme and Chemical Probing	78
B. Hydroxyl Radical Mapping	78
C. Nucleotide Analogues	80
D. Cross-Linking	81
III. Physical Approaches to Study RNA Folding and Structure	82
A. X-Ray and Nuclear Magnetic Resonance	82
B. Single-Molecule Studies	83
C. Development of Microscopies	85
IV. A Molecular Dynamic View of RNA Molecules	87
A. Free Energy	87
B. Ionic Environment	88
V. Computer-Assisted Modeling	89
A. *Ab Initio* Tertiary Modeling	90
B. RNA Secondary Structure Prediction	90
C. Tertiary Structure Modeling	91
VI. Conclusion	93
References	93

RNA modeling has become an increasingly attractive field for researchers as new functions for RNA are identified and characterized. However, our progress in determining three-dimensional structures is still behind our discovery of functional RNA molecules. Continuous development of experimental methods has enabled us to characterize biochemical and physical properties of the RNA molecules. Advancement in computer simulation and modeling is bringing us closer to the all-atomic-detail modeling. But there is still a big gap between our current achievement and our goal of predicting the three-dimensional structure based on their sequence information. In this chapter, we will go over the important progresses and techniques of structure characterization of nucleic acids, as an introduction for readers to wider range of approaches.

I. Introduction

As the most versatile macromolecule in the cell, RNA is attracting increasing interest. Many different types of RNAs have been found to possess enzymatic activity. In addition to acting as an intermediate molecule passing information to the coded protein, RNA molecules in the cell also work as functioning units. Such kinds of RNAs are commonly called non-coding RNAs. The tertiary structures for these RNAs have been shown experimentally to be important for function, as in the case of RNase P (1) and group II introns (2). Detailed knowledge of RNA structures will certainly expand our understanding of their biochemical functions in the cellular environment. In this chapter, we will summarize the most widely used methods in RNA structure modeling and the progress in recent years.

A. Structural Elements of RNA

To understand the structure of RNA, it is necessary to first understand some of the unique biochemical properties associated with RNA and its structural elements. These properties directly lead to the formation of structural elements and motifs in tertiary structures. These include not only basic structure elements like helices, loops, bulges, and junctions but also less common elements like ribose zipper and tetraloop-receptor that contribute to the difficulty of tertiary structure prediction. Some well-known elements are listed below.

> *Helices*: The 2′-hydroxyl (OH) group of the ribose sugar is responsible for all the differences between RNA and DNA molecules. In addition, there is uridine in RNA, which is the counterpart of thymidine of DNA. RNA forms an A-form helix instead of a B-form helix. In A-form helices, the ribose nucleotides adopt the C3′-endo sugar pucker, so that the flanking phosphates are closer and helices are more compact than the B-form DNA. There are also 11 base pairs per turn and the major groove is narrower and deeper, whereas the minor groove is wider and flatter, compared with B-form DNA helix (3). Because of the structural properties mentioned above, RNA helices can accommodate pairings between almost any two nucleotides. In addition to the canonical Watson–Crick base pairs, a variety of base pairing patterns were observed in crystal structures (4). The most common example is the G:U wobble pair.
>
> *Multiple interactions*: It is quite common to find RNA nucleotides in multiple interactions. Highly structured RNAs could have triple interactions, in which a single-stranded nucleotide interacts with a base pair (5), or even larger arrays of interactions. These interactions involve both the base and sugar (6). The A-minor motif is also a ubiquitously found theme; it involves an adenosine in contact with the minor groove of a Watson–Crick base pair (7, 8).

Pseudoknot: A Pseudoknot is a tertiary structure containing two stem-loop structures, where the loop of the first stem-loop forms part of the stem for the second one. There has been no computationally effective way to predict pseudoknots yet (9).

Ribose zipper: Because the 2′-OH group in ribonucleic acids can serve as both a hydrogen donor and a receptor, a ribose zipper can form at the interface between two RNA duplexes, where the interdigitated 2′-OH groups line up by hydrogen bonding (10, 11).

Tetraloop-receptor: A tetraloop-receptor is a long-range interaction. It involves a specific arrangement of base stacking and hydrogen bonds between a GNRA tetraloop (a structurally conserved loop closed by a guanine–adenine base pair, where the guanine is 5′ to the helix and the adenine is 3′ to the helix) and a conserved stem-loop helix (11).

Coaxial stacking: Coaxial stacking is a common stabilizing force in RNA. It can be described as the tendency for two adjacent helices to stack coaxially. It was first observed in the four-way helical junctions of DNA (12), and subsequently in many RNA structures including tRNA, the hammerhead ribozyme, and pseudoknots (13). In Walter's free-energy minimization experiment, the thermodynamic contribution of coaxial stacking was tested by measuring the binding of an oligomer to a 4-nt overhang at the 5′ end of a hairpin stem. The oligomer forms a new helix with the 4-nt overhang. It is shown that the melting temperature for the complex with hairpin is about 20 °C higher than that of the duplex alone, and coaxial stacking makes oligomers bind approximately 1000-folds more tightly than binding to a free tetramer at 37 °C (14). In another molecular dynamic simulation, it is indicated that the propeller twist is slightly increased and coaxial stacking is slightly twisted (39°) because of the absence of the phosphate group. But the hydrogen bond and stacking interactions are strong enough to keep the RNA structure (15).

These properties make RNA molecules structurally versatile. The possible interactions and conformations for a particular large RNA sequence can easily go beyond the capacity of our computers. Thus, the direct prediction of three-dimensional (3D) structures from RNA sequence alone is not currently feasible.

B. Differences Between RNA and Protein Folding

The study of RNA structure started later than protein structure research, people followed the path of protein studies for experience and methods, such as crystallography and molecular dynamic simulation. However, there are substantial differences between RNA and protein folding pathway, which deserve our precaution in RNA structure research.

RNA is transcribed linearly and begins to fold directly after initiation of transcription. Some domains can form even before the transcription of other domains is complete. Energies involved in secondary structure formation are generally greater than those in tertiary structure, so most secondary structures can remain stable without tertiary interactions, and there is a natural hierarchy in the folding process (16). In light of this, we need to be cautious about the results from *in vitro* folding, as most *in vitro* foldings are initiated by adding metal ions into the ready-made RNA instead of allowing RNA to fold into intermediates in the process of transcription. In contrast, protein folding is very cooperative and usually occurs after the entire polypeptide chain is synthesized (Fig. 1). The energies stabilizing protein secondary structure are comparable with energies for their tertiary structure (17).

The unique properties of RNA as well as the many stabilizing interactions described above allow RNA to fold and form stable tertiary structures. There are, however, some obvious differences between RNA and protein structure in addition to those described above. First, there are only 4 nucleotides to build RNA molecules, rather than the 20 amino acids for proteins. So there are less possible combinations of interactions. Second, the structures of individual nucleotides are very similar to each other, and there are only two conformations for the ribose. In contrast, amino acids are not only structurally different from each other but also more versatile in their conformational choice. The third

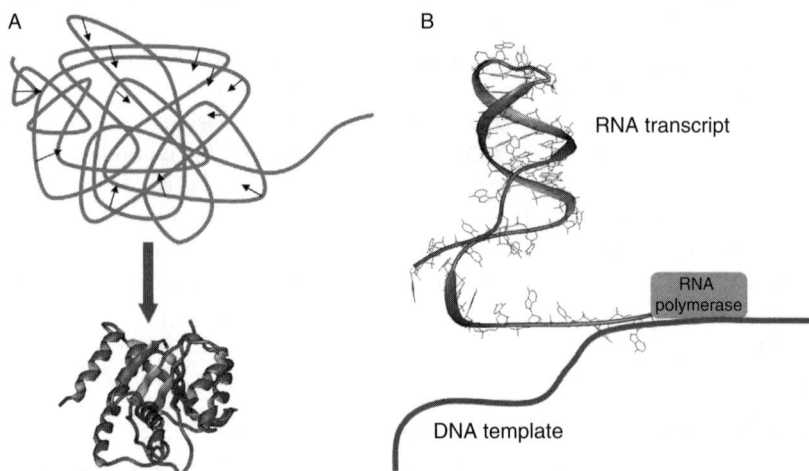

Fig. 1. Comparison between protein and RNA folding. (A) Most proteins do not have a definite structure in the translation process. The hydrophobic interaction drives their folding afterward. (B) RNA transcripts start folding into secondary structures right after they get out of the RNA polymerases.

difference is their driving force of structure formation. For protein, the driving force of folding is the burial of hydrophobic groups (sometimes called hydrophobic interaction), whereas for RNA, the most significant driving forces are hydrogen bonding, base stacking, and ionic interactions.

Comparatively, folding of RNA seems easier than that of protein because of the reasons mentioned above. But there are fewer RNA structures solved than protein, probably reflecting the fact that RNA research is relatively new and has not attracted enough attention from structural researchers. Despite that, there has been some progress in recent years, which will be discussed below.

C. Representative Classes of RNAs and Solved Structures

A group of RNA molecules that have been used for structural studies are the ribozymes. A ribozyme is an RNA molecule that has the ability to catalyze a reaction. The first found ribozyme RNase P entitled Sidney Altman to the Nobel Prize in 1989 (18). RNase P was found to have the catalytic activity of cleavaging the tRNA precursors. This discovery had great influence on the way we view the origin of life. For the first time, a molecule was shown to have both catalytic activity and the ability to store information. This eventually leaded to the RNA world hypothesis (19). RNase P and the ribosome are called universal ribozymes, as they are found in all living organisms (20). RNase P has been one of the most-studied and best-characterized ribozymes, and the 3D structure of it has been resolved by crystallography (1, 21–23).

There are different kinds of ribozymes. After RNase P, many ribozymes have been identified such as the hammerhead ribozyme, the hairpin ribozyme, and the hepatitis delta virus ribozyme. The first detailed 3D structure of a hammerhead ribozyme appeared in 1994. It was solved by X-ray crystallography, and it is an RNA–DNA ribozyme–inhibitor complex (24). Soon after, a minimal all-RNA structure of the hammerhead ribozyme was published by Scott et al. in Cell in early 1995 (25). It was not until 2006 that a 2.2 Å resolution crystal structure of a full-length hammerhead was obtained (26). The crystal structure of hepatitis delta virus ribozyme was solved by engineering the RNA to bind a small protein that does not affect its activity. The cocrystal structure diffracts X-rays to 2.3 Å resolution, and shows that the core comprises five helical segments connected as a double pseudoknot (27). The crystal structure of hairpin ribozyme, one of the four known natural catalytic RNAs that carry out sequence-specific cleavage of RNA, is solved as a hairpin ribozyme–inhibitor complex at a resolution of 2.4 Å (28). The structure of the GlmS ribozyme (glucosamine-6-phosphate activated ribozyme that functions as a catalytic riboswitch in regulating amino sugar metabolism) was also determined by X-ray crystallography (29, 30).

Group I, II, and III introns are also ribozymes. They are capable of catalyzing their own splicing out of a primary RNA transcript. Group I, II, and III introns are relatively rare compared to spliceosomal introns. Group I introns are found in organelles and nuclear rRNA of plants, fungi, protests, and rarely in animals, as well as bacteriophage and eubacteria (31, 32). Group II introns are found in bacterial genomes and in organellar genes of plants, fungi, and protists (33). A recent report shows the presence of group II intron in the mitochondrial genome of a bilaterian worm (34). Group III introns are a special class of introns. They have a conventional group II-type domain VI (dVI) with a bulged adenosine and a degenerated dI, but they have no dII–dV. They also have a relatively relaxed splice site consensus sequence (35). Splicing of group III introns is the same as group II introns. Because of their resemblance of structure and splicing pattern, group III introns are often considered a truncated remnant subgroup of group II introns.

Group I and II introns have very conserved secondary structure and are capable of self-splicing, a process usually aided by protein factors, like the maturase encoded by the open reading frame (ORF) in the case of group II introns. Like the nuclear mRNA introns, the splicing mechanism of group I and group II introns involves two consecutive transesterification steps. The only difference is the nucleophile used in the first step: group I introns use the 3'-OH of an external guanosine whereas group II and nuclear mRNA introns use the 2'-OH of an internal nucleotide, usually an adenosine (36). Even though some group I and group II introns have the capacity of self-splicing *in vitro*, the majority of them require the aid of protein cofactors to aid in their splicing and retrotransposition. The protein cofactor helps to stabilize the intron core structures, thus improving the efficiency of the reactions. A 3D model of the conserved core of group I intron was built in 1990 by Michel and Westhof (37). They aligned 87 sequences with well-defined secondary structures and looked for covariations not involved in secondary structures, which are explained to be involved in tertiary interactions. They built the model based on these deducted tertiary interactions. The crystal structure of group I intron came out in 2005, and confirmed the arrangement of the previous model (38). Dai and Chai *et al.* recently built a model for group II intron *Lactococcus lactis* Ll.LtrB. Photocross-linking method was used to obtain constraints for that model (39). At the same time, Toor *et al.* published a crystal structure of a smaller group II intron from *Oceanobacillus iheyensis* at 3.1-Å resolution (40). Even though Dai and Chai built the model deductively from constraints, the basic arrangements of the resulting model is very close to the crystal structure, similar as in the case of group I intron. These reflect the fact that thorough knowledge of RNA biochemical properties can be good enough to be applied for structure deduction.

The flow line from the primary messenger RNA (pre-mRNA) transcript to the final protein assembly also exemplifies the importance of RNAs and their structures. For the pre-mRNA to be processed into mRNA, introns have to be removed. The splicing of introns is catalyzed by the spliceosome, which is a large RNA–protein complex composed of five small nuclear ribonucleoproteins (snRNPs). There are certain splice signals like the 3′ splice site, 5′ splice site, and branch site to guide the splicing of these introns by the spliceosome. The reaction mechanism of pre-mRNAs excision is exactly the same as group II introns and their splicing depends on various *trans*-acting ribonucleoprotein complexes (snRNPs). It has frequently been hypothesized that the snRNAs in the spliceosome and pre-mRNA introns were initially derived from group II introns (*41*).

Ribosomal RNA (rRNA) is the primary component of the ribosome, the protein-manufacturing organelle of cells in the cytoplasm. rRNA makes up the majority of RNAs found in a typical cell. Even though proteins are also present in the ribosomes, rRNA has catalytic function and is the crucial component (*42*). Crystal structures of ribosome were solved by three independent groups in 2000 (*43–45*). Transfer RNA (tRNA), the transporter that carries the correct amino acid to the ribosomal site of ribosome during protein biosynthesis, is also a highly structured RNA unit, so both the transporter and constructor for proteins are actually RNA molecules. The high-resolution cloverleaf structure of tRNA was resolved in the early 1970s by gradual refinement, most well known in the Alexander Rich group and Aaron Klug group (*46–48*).

In recent years, the study of gene regulation has been attracting more and more attention. MicroRNAs (miRNAs) have been one of the intense focuses. miRNAs are single-stranded RNA molecules of about 22 nt long and serve as gene expression regulators in many organisms (*49–52*). miRNA forms an RNA-induced silencing complex together with some protein cofactors after being processed from hairpin precursor miRNA. The binding between the miRNA and its target mRNA sequence usually requires a "seed" region of perfect and continuous base pairing of 2–8 nt in the 5′ of the miRNA, a bulge in the central region of the miRNA-target duplex, and reasonably good pairing for the 3′ half of miRNA to its target (*53, 54*). However, our knowledge about the pairing and structure of miRNAs is very limited, and it is far from enough to guide us to find the majority population of miRNAs. It is possible that the tertiary structure of the miRNA and the pairing complex play important roles in the recognition and silencing process. So that the understanding of the secondary and tertiary structure of the paired complex would help us make better use of this burgeoning tool.

Overall, RNA molecules are versatile and they function in various reactions in living organisms, with definite 3D forms. The understanding of ribozymes will also help to explain how spliceosomal introns get spliced out from the pre-mRNA. If ribozymes were the first molecular machines used at the origin of life, before the generation and substitution of protein machinery as proposed by Gilbert (19), then the ribozymes would need well-defined tertiary structure for their function. Some studies also reported that even some mRNA molecules have highly structured domains to mediate their gene expression according to the need and environment (55, 56). Even though the biotic world today is mainly a protein world, RNAs play very important roles from information storage to protein synthesizing. Their functions in regulation and enzymatic catalysis are also indispensable. Understanding these properties relies on a basic understanding of RNA structure and has given rise to the field of RNA structure modeling.

II. Chemical and Enzymatic Methods

To characterize the structural properties of RNAs, many experimental methods have been developed, either brand new or adapted from protein research. Many of them have been developed well enough to help us to interpret the structure of RNAs.

A. Enzyme and Chemical Probing

The characterization of RNA structure and structural dynamics is greatly enhanced by the application of chemical and enzymatic probes of RNA structure in solution. These probes enable us to study RNA structures in conditions that are physiologically relevant and to investigate their interactions with partner molecules and to detect enzymatic active sites. Comparatively, the enzymatic analysis offers closer to physiological conditions than chemical reactions, whereas the chemical reactions often require very extreme pH or strong ionic environment. Another difference is that enzymes are more easily blocked sterically because the RNases are more bulky than the chemicals. An advantage in these probing methods is that we have the chance to test if our RNA molecules of interest have the right conformation by checking their activity. The widely used nucleases and base-specific chemical probes are summarized in Tables I and II, respectively.

B. Hydroxyl Radical Mapping

For many years, hydroxyl radical footprinting has been widely used to probe the solvent accessibility of local regions of RNA structure in solution. The most common reaction to generate hydroxyl radicals is the Fenton reaction, in which H_2O_2 reacts with Fe^{2+} to form free hydroxyl molecules. These hydroxyl molecules are extremely active and have no dependence on the RNA sequence. When they touch the DNA backbone, strand breaks result. Because of the

TABLE I
Enzymatic Probes for RNA Structures

Nucleases	Enzyme cut single- or double-stranded region	Sequence specificity
S1 nuclease	Single	None
RNase T1	Single	3'-end of unpaired G
RNase PhyM	Single	3'-end of unpaired A and U
RNase U2	Single	3'-end of unpaired A
RNase V1	Double	None
RNase I	Single	3'-end of ssRNA
Mung bean nuclease	Single	None
RNase T2	Single	None
RNase A	Single	Pyrimidine

Table is taken from (57).

TABLE II
Chemical Probes for RNA Structures

Chemicals	Specificity	Function	Future detection
DMS	Accessible N7G, N1A, and N3C	Methylation	Primer extension
DEPC	Accessible purine bases	Carbethoxylation	Aniline cleavage
Hydrazine	Accessible pyrimidine bases	Removal	Aniline cleavage
Kethoxal	Accessible N1G and N2G	Form a ring in the G	Primer extension
CMCT	Accessible N3U and N1G	Modification	Primer extension
ENU	Accessible phosphate	Alkylation	Polyacrylamide gel electrophoresis

DMS, dimethyl sulfate; DEPC, diethyl pyrocarbonate; kethoxal, α-keto-β-ethoxybutyraldehyde; CMCT, 1-cyclohexyl-3-(2-morpholinoethyl) carbodiimide metho-p-toluenesulfonate; ENU, ethylnitrosourea. Table is taken from (58–60).

small size of those hydroxyl radicals, which is almost the same size as water molecules, the susceptibility of a particular nucleotide to the radical attack is governed by its accessibility to solvent, and the resulting resolution for RNA footprint can be traced as high as single-nucleotide resolution (61). For short radioisotope labeled RNA sequences, denaturing polyacrylamide gels are enough to differentiate the cleavage sites. For larger molecules, primer extension offers a precise solution and the RNA does not need to be labeled.

Fe^{2+} can be tethered to proteins or RNA using 1-(p-bromoacetamidobenzyl)–EDTA. This tethering can cause the cleavages to be directed in a proximity to the bound probe. This method (often called Fe-BABE) can be used to obtain comprehensive structural information around individual nucleic acids (62). Another modification of hydroxyl radical footprinting is time-resolved hydroxyl radical footprinting, in which the hydroxyl radicals are only generated and last for a short period to probe the structure, so interactions at a certain stage of the folding pathway can be captured. The most precise way of achieving this is through radiolysis of water by high-energy synchrotron radiation, in which the hydroxyl radicals are generated on only a millisecond scale.

Bergman *et al.* used hydroxyl radical footprinting data to produce a 3D structural model of the class I ligase ribozyme (63). Their experiment shows that a substantial fraction (17 of 109) of the nucleotides is protected when the ribozyme assumes a compact structure in the presence of Mg^{2+}. Lease *et al.* used time-resolved hydroxyl radical footprinting to find the communication between RNA folding domains (64). Russell *et al.* used the same method to characterize the effect of P5abc peripheral element on the folding kinetics of the tetrahymena group I ribozyme (65).

C. Nucleotide Analogues

To analyze RNA structure and function with nucleotide analogues, we can selectively use a nucleotide analogue with either a functional group deleted or a new functional group added. Among the various methods, nucleotide analogue interference mapping (NAIM) is the most effective and widely used. NAIM utilizes a series 5′-*O*-(1-thio)-nucleoside analogue triphosphates that are randomly incorporated into the RNA molecule of interest in *in vitro* transcription. By cleavage at the phosphorothioate tag with iodine and gel electrophoresis of the cleaved fragments, the location of the analogue substitution and its effect on ribozyme activity can be identified. There are about two dozens nucleotide analogues available for use (66–68). This method and its derivative nucleotide analogue interference suppression (NAIS) have made it possible to probe the contribution of individual functional group at every nucleotide position in an RNA molecule simultaneously, which helps to determine the chemical basis of RNA function and structure.

One example of the application of NAIM and NAIS can be seen in the investigation of Jansen *et al.* on backbone and nucleobase functional groups in glmS riboswitch. They were able to identify essential structural features and potential sites of ligand and metal ion interaction. They also revealed sites that coordinate the recognition of ligand phosphate (69). Recently, Pyle's laboratory used this method extensively on group II intron ai5gamma and identified that a group of atoms within a small section of D1, including the kappa and zeta elements, are crucial for intron folding. This kappa–zeta element controls the sequential collapse of the molecule and forms the docking sites for catalytic D5 in later steps of the folding pathway (70). This element is also shown to form upon the binding of Mg^{2+} to the folding intermediate and its formation triggers the D1 structure to collapse into a pocket-like scaffold with the help of long-range tertiary interactions. Then domain 3, 5, and 6 quickly dock into the ready set position (71). They also found that the linker sequence between domain 2 and 3 has a functional role during the first step of splicing in addition to their known involvement in the second step. This information helps us to understand the group II intron active-site architecture.

D. Cross-Linking

While the hydroxyl radical footprinting and nucleotide analogue modification probe the accessibility of nucleotide, cross-linking method uses photoaffinity agents to infer the structural packing information. Data from cross-linking can be used as distance constraints between specific nucleotides in the RNA structure and be used for structural modeling. To apply cross-linking, short-wave UV light is used to induce the nonspecific cross-linking of unmodified nucleotides. This approach is easy to initiate and not time-consuming, but the data interpretation is very difficult. An alternative method would be to attach photoaffinity cross-linking agents to specific sites in the RNA of interest, and induce the reaction. Then one end of the cross-linking is known, and the other end can be determined by primer extension and polyacrylamide gel electrophoresis.

The photoaffinity agents used in cross-linking are highly active chemical moieties that can form covalent bonds to atoms within their reaching distance upon activation from light of certain wavelength. Azido- or azidophenacyl-substituted nucleotides are the most widely used long-range cross-linking agents and have a reaction radius of about 9 Å. Thionucleotides such as 6-thioguanosine and 4-thiouridine are short-range agents whose active groups are attached to the nucleotide bases. Their sizes are smaller than that of azidophenacyl, so that their incorporation in the RNA will have less perturbation to the whole structure and less damage on the RNA's activities. Because the active groups are on the nucleotide bases, they react only with nucleotides in direct contact (within 1.5 Å) (72). The observation that RNA molecules with individual nicks in the backbone usually maintain their structure and reactivity made the use of circular permuted RNAs (cpRNAs) possible (73). There are well-developed methods to attach cross-linking agents onto the 5' end or 3' end (Fig. 2) (74). Alternatively, the cross-linking agent can be site-specifically incorporated into desired places in the RNA strand either by transcriptional incorporation or by bridging oligonucleotide directed ligation (75, 76).

The cpRNA and photocross-linking approach was first applied by Nolan *et al.* to map the part of bacterial RNase P in vicinity to tRNA (77). Later, Harris *et al.* used the same method to get intra- and intermolecular constraints within the *Escherichia coli* RNase P–pre-tRNA complex. They built a model based on these constraints together with known secondary structure and tertiary interactions (78, 79). The same laboratory used the same technique to prove that

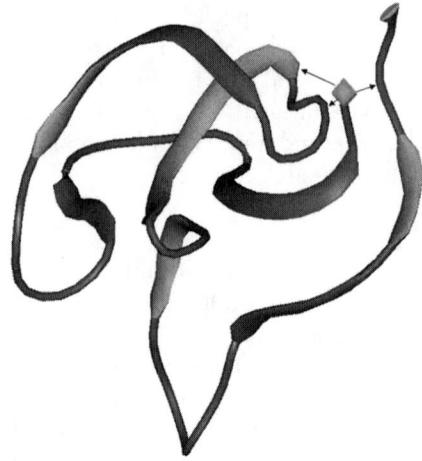

Fig. 2. Cross-linking in CP RNA. Cross-linker is attached to the 5' end of the circular permutated RNA (diamond), and by activation of UV, it can react with nucleic acids close to it (indicated by small black arrows).

eukaryal RNase P folds into functional forms and binds tRNA without aid from protein cofactors. Based on the crystal structure from *Bacillus stearothermophilus* and their previous model of bacterial RNase P, they built another model for eukaryal RNase P (*80*). Lambert *et al.* also built a model of the catalytic core of the hairpin ribozyme using cobalt (III)-induced cross-linking (*81*). Recently, Dai *et al.* modeled the structure of *L. lactis* Ll.LtrB group II intron based on photocross-linking of cpRNA and known interactions (*39*).

III. Physical Approaches to Study RNA Folding and Structure

A. X-Ray and Nuclear Magnetic Resonance

X-ray crystallography is one of the most favored methods for structural characterization. By getting to very high resolution (lower than 1 Å), the atomic lattice detail can be shown (*82, 83*). It usually takes less time to decode and costs less than most other methods, and in theory, it can determine the structure without any additional information. However, this method requires the availability of a crystal, which can be prohibitive for large RNA molecules. Nonetheless, with the development of purification and crystallization techniques and advances in computer software, even difficult structures are being solved using this method (see Introduction section I.C). For example, as of April 2008, there were more than 894 RNA crystal structures in the Nucleic acid DataBase (NDB, website at http://ndbserver.rutgers.edu/index.html).

An additional tool for structural determination of small RNA oligonucleotides is nuclear magnetic resonance (NMR) spectroscopy (84). NMR spectroscopy can be done for RNA in solution, thus giving results closer to physiological conditions. Conventional NMR can determine RNA structures up to 15 kDa with accuracy of 1–1.5 Å (84). With the development of transverse relaxation-optimized spectroscopy, residual dipolar coupling, and labeling strategies, RNA molecules with a molecular mass of up to 35 kDa can be determined (85–87). A strong advantage for NMR is that real-time dynamics in 1D, 2D, and 3D can be determined (88). This makes it a powerful tool for understanding intermediate folding stages, as well as how RNA molecules interact with partners and substrates, such as in the case of ribosomes and spliceosomes (88, 89).

RNA conformational changes occur on a wide range of timescales, and we need different experimental tools to decide with the structures in these different situations. The folding of secondary structure happens within 100 μs, while tertiary structure formation is slower and occurs on a millisecond timescale. For rapid folding events such as hairpin structures, laser temperature-jump spectroscopy can capture kinetic properties (90). For slower reactions on millisecond scale, X-ray synchrotron hydroxyl-radical footprinting can be utilized to capture phases of RNA conformational changes. Its principle is same as that of footprinting method described previously. The difference is that the hydroxyl radicals are generated by a short exposure to an X-ray beam. If the folding event is even slower, then hydroxyl-radical footprinting, chemical base modification, and UV-cross-linking, which are described in previous sections, can be used to analyze the structural transitions. Thermal denaturation profiling and temperature gradient gels provide useful information as well. If the structural transition is very slow (minutes to hours scale), native gel electrophoresis is a straightforward option (88). In recent years, the development of techniques in single-molecule research offers additional methods to measure the structural dynamics of RNA during folding in real time. A modified Mass-Spec technique–ESI-FTICR-MS (ElectroSpray Ionization–Fourier Transform Ion Cyclotron Resonance–Mass Spectrometry) can provide a better readout platform for those chemical probing and cross-linking methods described above. This technique is called mass spectrometric three-dimensional (MS3D) analysis. It can save the trouble in the labeling step, read all the fragments in the solution, and position the modified nucleotides. The resulting protection maps and distance information can be utilized as the constraint to generate 3D models (91–93).

B. Single-Molecule Studies

In their folding pathways, RNAs are shown to go through multiple routes and intermediates (94, 95). In the reactions catalyzed by ribozymes, there are many intermediate complexes formed in multiple steps. The properties of individual pathways and intermediates are typically hard to trace with conventional methods that measure average properties of an ensemble of molecules. The techniques to measure and manipulate single molecules have developed very quickly in recent years, and are promoting a new understanding of molecular interactions and folding energy landscape of ribozymes. Being able to track and measure the transient conformation and interaction of single RNA molecules has allowed researchers to more closely examine RNA folding and molecular dynamics. The extensive application of single-molecule techniques in cells has the potential to reveal *in vivo* folding and functioning dynamics of ribozymes. There are mainly two aspects of utilization in single-molecule studies: force measurement, which is implemented with optical or magnetic tweezers, and optical measurement, which measures single-molecule fluorescence.

To maneuver individual RNA molecules in single-molecule studies, the RNA molecule is usually either adhered to the tip of the microscope using a strong streptavidin-biotin bond or paired with a short sequence on the tip. The tip bound to the RNA molecule can exert and measure forces on the RNA from piconewton to nanonewton scale (96). By applying mechanical forces at a slow rate, folding and unfolding can be controlled in a well-defined environment, and the folding free energy can be readily calculated from the force-extension curves (Fig. 3). The results from this method usually correspond well with

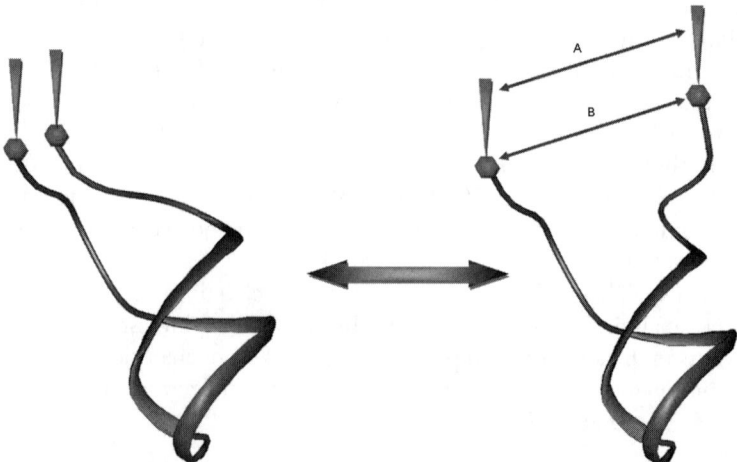

Fig. 3. Maneuvering individual RNA molecules in single-molecule studies. By applying mechanical forces at a slow rate, folding and unfolding can be controlled in a well-defined environment, and the folding free energy can be readily calculated from the forces measured on the tips (A) or distance change between the tips (B).

theoretical predictions. This method has been effectively applied to some large RNA molecules, such as the 1540-base long E. coli 16S rRNA (97). For folding and unfolding processes carried out under nonequilibrium conditions, the free energy change in the reaction can be calculated by averaging Boltzmann-weighted work values obtained from multiple irreversible repeats (98, 99).

A second single-molecule method that can measure molecular motions in real time is single-molecule fluorescence resonance energy transfer (FRET). In this assay, two dyes, fluorescence donor, and acceptor are attached to the ends of the RNA molecule of interest. If the dyes are separated by a large distance, there is little interaction between them and the donor will emit photons by excitation of laser. The close interaction between donor and acceptor leads to the transfer of emission energy from donor to acceptor and the acceptor's emission of photons of a different color (100). The distance law for Förster resonance energy transfer efficiency is given by Eq. (1):

$$E = \frac{1}{\left[1 + (R/R_0)^6\right]} \quad (1)$$

where R is the distance between the donor and acceptor, and R_0 is the Förster radius (typically 3–8 nm). When $R = R_0$, the efficiency E is 50% (96). FRET is very sensitive to conformational changes of the RNA molecule because the energy transfer efficiency between the donor and the acceptor depends on their intermolecular distance and orientation. Thus, FRET can be used to measure distance changes between two definite positions in the RNA molecule in the nanometer scale, a relevant length scale for RNA folding.

A FRET application on a large RNA molecule, the RNase P in *Bacillus subtilis*, detected two more folding intermediates than the two transition states by previous ensemble studies (101). FRET studies on hairpin ribozyme show that cooperative binding of two ions is required to bring the two loops of the ribozyme together, and the folding rate of the four-way junction by FRET studies suggests an unknown intermediate for the large acceleration of the hairpin ribozyme folding. Multiple cycles of cleavage and ligation were also observed in the same molecule, and the rates of these conversions were measured in FRET (102–105).

C. Development of Microscopies

The development of scanning probe microscopy especially in atomic force microscope (AFM) contributed a lot to the realization of single-molecule studies. The AFM is the only instrument that can image samples at subnanometer resolutions and be operated in solution (106). Noah. examined complexes formed between the RNPs and DNA substrates of group II intron *L. lactis* Ll.LtrB. Under AFM, he was able to measure the dimensions of the RNPs and the bend angles of various DNA substrates at different reaction stages (*107*).

Although widely used, AFM has limitations. At room temperature, the force from the probe could partially deform the sample in the solution, as mentioned by Noah (*107*). Also, the thermo motion of flexible molecules makes high-resolution imaging difficult. This limitation was solved by electron cryomicroscopy (cryo-EM). In cryo-EM, the sample is in a fully hydrated state, and the imaging does not cause deformations. When combined with 3D reconstruction, the data can provide molecular fitting and docking information (*108*). Recently, many RNA complexes involved in transcription were resolved by cryo-EM, including the structure of P/E-transition tRNA and its interactions with the ribosome (*109*). Taylor *et al.* used cryo-EM to determine the 3D structures of eukaryotic ribosomes complexed with an elongation factor (eEF2), before and after GTP hydrolysis. Their data provide a detailed two-step translocation model of the mRNA–tRNA complex (*110*). With cryo-EM, Gilbert *et al.* reconstructed the structure of yeast 40S ribosomal subunit in the translation initiation multifactor complex, and showed how the binding of eukaryotic initiation factors induces the scanning mobility of the 40S subunit along the 5′ untranslated region to search for a start codon (*111*). Kaur *et al.* visualized the complex of transfer messenger RNA with protein factor (SmpB) in stalled ribosome. Their data made the understanding of the translocation complex clearer (*112*). The resolution of cryo-EM is usually not high, so sometimes it needs to be combined with other methods like molecular dynamics simulations or crystal structures to build structure models for large molecules whose individual components have been characterized (*113*, *114*).

A combination of cryo-EM and AFM, cryogenic atomic force microscopy (cryo-AFM) can overcome the problem of AFM. cryo-AFM operates in liquid nitrogen vapor so that the thermo motion is suppressed and the frozen sample is not compressed by the probe. However, an obvious drawback of cryo-AFM is that we can not follow dynamic process (*115*). Mat-Arip *et al.* used this technique to visualize the dimerization of pRNAs from bacterial virus phi29 and confirmed the head to head confirmation in the dimer formation (*116*). 3D electron microscopic imaging was used to visualize many of the proteins and nucleic acids in the ribosome complex and was able to detect conformational changes of the ribosome (*117*). Mueller *et al.* constructed an approximate model for the 50S ribosomal subunit from *E. coli* to fit a 7.5 Å resolution electron microscopic map (*118*).

IV. A Molecular Dynamic View of RNA Molecules

As opposed to the experimental methods described above, theoretical approaches are also implemented to solve RNA structures. These approaches' use were grafted from protein structure simulations. They use physical and chemical properties of RNAs as parameters to calculate the interactions between any two particles in the system and sum them up to deduct the outcome of the molecule. The total energy, the forces on each atom, and all the intermediates and final stages of the molecules are calculable in theory.

A. Free Energy

RNA molecules fold into structures with minimal free energy. Different environmental conditions can stabilize or destabilize the conformation of RNA helices. Of these, temperature, pressure, and ionic concentration have the most influence.

In order to predict base pairing in RNA helices, the melting temperature (T_m) (the temperature at which half of the nucleotides are helical and half are single stranded) of RNA first needs to be calculated. In accurate simulations, the semiempirical nearest-neighbor method is widely used to predict melting temperatures of nucleic acid duplexes. This method is obtained through graphical analysis of the Gibbs function [Eq. (2)] and reciprocal melting temperature versus concentration plots. It assumes that the stability of a given base pair depends on the identity and orientation of neighboring base pairs and is generally expressed as Eq. (3) (*119*):

$$G = H - TS = U + pV - TS \qquad (2)$$

$$T_m = \frac{\Delta H}{\Delta S + \ln\left[C_1 - (C_2/2)\right]} - 273.15\,^\circ\text{C} \qquad (3)$$

where G is the Gibbs free energy, H the enthalpy, S the entropy, T the temperature, U the internal energy, p the pressure, and V the volume. ΔH is the standard enthalpy and ΔS is the standard entropy for formation of the duplex from two single strands. C_1 and C_2 are the concentrations of the more concentrated and less concentrated strands, respectively.

From Eq. (2), we can see that the increase in pressure will cause a corresponding increase in the free energy, thus fewer base pairings would form when other parameters remain unchanged. The increase of temperature will lower the change of free energy, so that higher temperature is inhibitory for the formation of helix. From a physical view, because temperature is defined by the ensemble average of kinetic energies of all particles in the system, higher temperature means more atoms will possess higher energy than the hydrogen bond, thus breaking the bond. In molecular dynamics, pressure is a measurement of the kinetic energies of particles in the system. Higher pressure means higher kinetic energy, which is essentially the same effect as an increase in temperature. In the search for the lowest energy for the whole molecule (energy minimization), the total energy will go through an energy surface (a multidimensional surface representing the total energy). The conformation with the lowest free energy is the most probable one in that equilibrium (Fig. 4).

B. Ionic Environment

The second influence on RNA folding is the ionic environment. Most RNAs rely on metal ions to fold, to stabilize tertiary structure, and to aid in catalysis. This reliance on metal ions is because RNA molecules possess high negative charge that acts antagonistically against their folding. Metal ions can promote RNA packing and function in at least four ways (*120*, *121*). First, diffuse ions can nonspecifically screen the charge of the backbone, thus reducing repulsion between RNA strands. There is no contact between the ion and the RNA surface and the ions are not confined to any particular location. Second, water-positioned ions can interact with RNA through contacts mediated by coordinated water molecules. They are so close to the RNA that the sterical

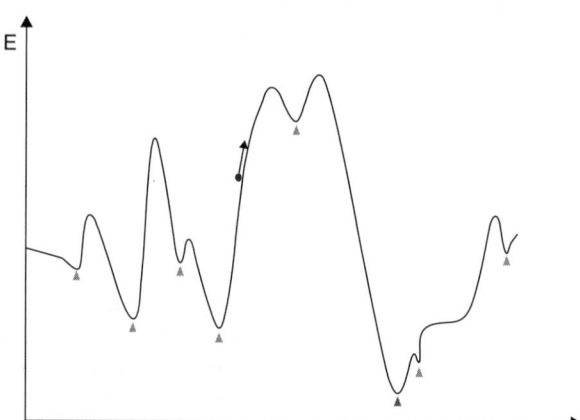

FIG. 4. A two-dimensional view of an energy surface. In the molecular dynamic simulation, a molecule (represented by the climbing ball) goes through all its possible configurations and the changes in its energy of are recorded. The dark triangle indicates the global minima (the state with the lowest energy for that RNA molecule) we are looking for. Because we usually cannot search through the entire energy space, the molecule will most likely get stuck in one of the local minima (light triangles) instead of the global minima (dark triangle), depending on how through we do the simulation.

packing and hydrogen bonding of the water molecule will influence the position of the ion. Third, they can chelate to specific RNA sites with at least two direct contacts, providing local stability to regions with strong negative charge. Finally, they can coordinate directly with RNA functional groups, and help in catalytic reactions (122, 123). Quite often, it is hard to define the exact role of a particular ion, as these mechanisms are cooperative in most circumstances. The crystal structure of the group I intron by Vicens et al. supports the previous two-metal-ion catalytic core model, in which one ion serves as nucleophile activator and the other serves as leaving group stabilizer. The two ions switch roles in the second step of splicing and they also coordinate the scissile phosphate at the splicing sites (124). For the splicing of group II intron *Bacillus halodurans* I1, Mg^{2+} ions are catalytically essential (125). Different concentrations and compositions of the ionic environment also lead to different splicing activity. 100 mM $MgCl_2$ gives the intron more reactivity but lower splicing site fidelity than 10 mM $MgCl_2$. Different divalent ions also result significant different splicing products. 10 mM Mn^{2+} leads to the formation of strong band of lariat, which is very weak in $MgCl_2$ environment (126).

The Hill equation is commonly used to quantitatively calculate the ion–RNA interactions in equilibrium (127). RNA folding is represented by the Eq. (4). The calculation of folding free energy is expressed as Eq. (5):

$$U + nM \Leftrightarrow F \cdot nM \tag{4}$$

$$\Delta G' = \Delta G^{\circ\prime} + RT \ln \frac{F \cdot nM}{U \cdot M^n} \tag{5}$$

$$\Delta G^{\circ\prime} = nRT \ln M_{1/2} \tag{6}$$

where n is the Hill coefficient, R the ideal gas constant, U the unfolded RNA, F the folded RNA, and M the concentration of ions. At equilibrium point, $\Delta G' = 0$. At the midpoint of the folding transition where half the molecules are folded, $U = F \cdot nM$. Therefore, we have Eq. (6). Here, $M_{1/2}$ is the ion concentration for folding of half of the RNAs (124). So an increase in M will result in a larger $\Delta G^{\circ\prime}$, resulting in the RNA forming a more compact folded structure. As ion concentration goes up exponentially, the T_m goes up linearly (128).

V. Computer-Assisted Modeling

With the energy functions and environmental parameters set up, interactions and energies among all the atoms can be calculated. The interactions will "pull" or "push" the atoms to move in relative to each other. After a short interval, the speed, energy, and forces on each atom can be recalculated and motions of the atoms will be adjusted accordingly. This kind of calculation keeps acting until the molecule reaches a minima, where the total energy gets lowest and forces reach balance. In computer-assisted modeling, we can start simulation for all the atoms purely based on the knowledge of quantum mechanics or build the tertiary model from known secondary structures.

A. Ab Initio Tertiary Modeling

There are basically two routes to predict tertiary structure with *ab initio* modeling: either simulating the folding process or searching the entire energy surface for the lowest point. But there are almost endless conformational choices, neither is currently computationally feasible. So the current methods are all based on the secondary structures being first solved (*129*).

B. RNA Secondary Structure Prediction

As stated before, the RNA structure with the lowest free energy is the most probable one at that equilibrium, and we can use the nearest-neighbor method. Still, it is not simple. For a sequence of n nucleotides, the number of possible secondary structures is estimated to be 1.8^n (*130*). It means that for an RNA sequence of 100 nt, there will be 3.367×10^{25} possible solutions. Suppose the calculation is already finished by CPU and each structure takes only 1 bit in the hard drive (impossibly small), outputting the result alone will take more than 100 years and use up all the hard disk space that has ever been manufactured. In reality, the simulation would take much more time than the output, and the data size would be thousands of folds larger. It is obvious that even for a small RNA fragment, using the nearest-neighbor method rigorously is not feasible. There are two common ways in which the calculation task is simplified: one is free energy minimization based on dynamic programming algorithms and the other is multiple sequence alignment. There are also some programs combining these strategies.

The most popular free energy minimizations are based on dynamic programming algorithms due to the calculation problem described above. The advantage of dynamic programming algorithms is that they can implicitly consider all the possibilities without generating the structures first. These algorithms usually divide the whole sequence into fragments, predict one overall free energy for the whole secondary structure and treat it as the sum of those fragments. Then the lowest folding free energy of each individual fragment is calculated and scored together (*131*). In this way, the process is speeded up by the calculation of smaller fragments, consuming exponentially less CPU time (*132, 131*). However, most of these algorithms can not predict pseudoknots or similar structures because they can not track the recursion involving multiple interactions. There are some algorithms developed for the prediction of pseudoknots, like the Sfold and RNAshapes. They, however, only sample the structure based on the Boltzmann probability distribution or narrow the search to some representative structures (*133*, *134*). The general problem of these programs is that they are usually slow and thus, very limited (*131*).

The theoretical reason for multiple sequence alignment method is that the structures of RNA are more conservative than their sequence. By aligning multiple sequences together, sequences constraining their common structures could be summarized. Molecules with similar sequences and similar functions should fold into similar structures. Rigorous mathematical treatments of these alignments are computationally expensive, so some programs use variations of the alignments in the calculation. In Chen's algorithms, random mutations are made on the sequence, and then the free energy of similar conformations is scored as criteria for future calculation (*135*). Dynamic programming can also be implemented in this alignment method (*136*).

C. Tertiary Structure Modeling

There are several different routes for tertiary structure modeling after the secondary structural information is obtained. In the first method, we can set up straightforward calculation of all possible conformations on the basis of known secondary structures. This idea is used in Macromolecular conformations by symbolic programming (MC-Sym) (*137*). In the program, the RNA structure properties and constraints are entered in a script or interactively within the MC-Sym interpreter. The program backtracks each nucleotide based on a Monte Carlo algorithm to produce 3D structures. It calculates every possible position and combination for each nucleotide, so theoretically this program can calculate all the possible solutions for a structure. But, it can only deal with small or simple structures. For large RNA molecules, having too many conformational choices on single-stranded regions makes the calculation too expensive.

The second route is an extended comparative sequence analysis. In the alignment of closely related RNA sequences, nucleotides that covary with statistical significance but not are involved in direct base pairing are considered to be tertiary interactions. This idea was successfully used on the prediction of the catalytic core of a group I intron and later generated a modular building program of RNA (MANIP) (*138*, *139*). However, the resulting structure is quite coarse and many details are lost using this method.

With the availability of solved crystal structures and models, homologous modeling can be used to infer structures for close relatives. This method is already widely used in protein modeling due to the availability of multiple crystal structures. For regions with good sequence alignment, direct substitution can be used, whereas for insertions and deletions, similar structures can be inferred from other crystal structures and assembled together. An application of this method can be seen in the model of 30S rRNA subunit. Based on the notion that molecules with similar sequences and similar functions should have similar 3D folding, Tung *et al.* built a model for the *E. coli* 30S ribosomal subunit with the crystal structure of the *Thermus thermophilus* 30S ribosomal subunit as the template. For regions with good alignment, nucleotides are directly substituted. For regions with insertions and deletions, the structure is dissected, and individual motifs are modeled and aligned based on the overlapping sequences (*140*). The Gutell Lab developed a database website for RNA comparative analysis that provides a comprehensive collection of RNA sequence comparison, known motifs, and solved structures including models and crystal structures. By sequence alignment, comparative and phylogenetic analysis, secondary and higher-order structure can be inferred (*141*).

The third one is called coarse-grained simulation. This method simplifies a certain structural element into one solid unit and treats it like one atom. All the forces on this element or motif are exerted in the center of this simplified object. In the modeling of large RNA molecules, known or solved secondary structures can be represented in this way, so the calculation of interactions among thousands of atoms are lowered several grades to interactions among dozens of elements, and thus is much faster (Fig. 5). The *E.coli* 16S RNA in the 30S ribosomal subunit was modeled this way (*142*).

Each method above is not limited to be used alone. The combination of different methods is very helpful in modeling. For example, Dai *et al.* built a complete 3D structure for *L. lactis Ll*.LtrB-ΔORF intron recently (*39*). The secondary structure was achieved before using sequence alignment. Before the computer-assisted modeling, circular permutation and cross-linking experiments (*143*) were used to map tertiary interactions in the RNA. At the same time, RNA helices and certain simple structures were generated in MC-sym. The data we obtained through bench work and some other known interactions were used together in the building process. Crystal structures were borrowed

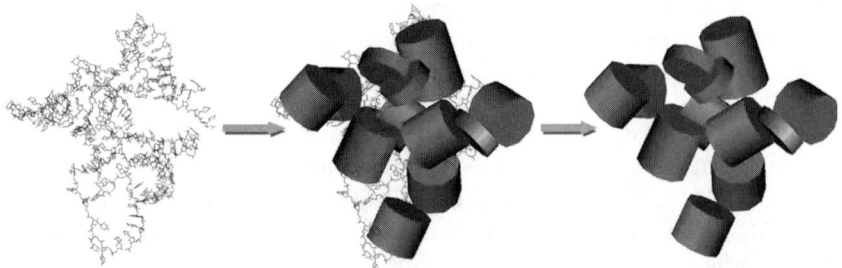

FIG. 5. Coarse-grained simulation. Instead of doing calculation on interactions among thousands of atoms in the 120 nucleic acids, coarse-grained simulation can be utilized to summarize the representative structures. The domains can be simplified into only a dozen of solid units, and the difficulty of calculation is reduced exponentially.

for the existing known interactions. All the helices and known interactions were treated as solid elements when the model was built in ERNA3D. The model was finally optimized in CharMM to eliminate sterical clashes and fix bond discontinuities (39).

VI. Conclusion

Although a great deal of effort has been put into structure prediction research, successful structures or models especially for large molecules are rare today. Of the many reasons, the most important ones are that the thermodynamic properties of RNA folding are not well understood, and few solved structures can be referred to as examples (especially crystal ones). The structures are influenced by folding kinetics and environmental conditions such as temperature and ion concentration. Experimental methods all have their limitation. In the simulation algorithms, approximation or sampling is used, so not all configurations are explored. When RNAs fold in the cellular environment, they are aided by chaperon molecules. This, however, is poorly characterized and therefore can not be taken into consideration in computer modeling. One more issue in all the computer modeling is that the structure with the lowest free energy is only "most likely" to be the right one. All near-optimal structures can be presented, but there is no indication which one is correct. Building 3D structures from input data automatically is still beyond reach. There is more work needed to achieve the ultimate goal of atomic-level RNA 3D structure.

Acknowledgment

I thank Dr. Dawn Simon and Bonnie McNeil for proofreading and useful suggestions.

References

1. Kazantsev, A. V., Krivenko, A. A., Harrington, D. J., Holbrook, S. R., Adams, P. D., and Pace, N. R. (2005). Crystal structure of a bacterial ribonuclease P RNA. *Proc. Natl. Acad. Sci. USA* **102**, 13392–13397.
2. Pyle, A. M., and Lambowitz, A. M. (2006). Group II introns: Ribozymes that splice RNA and invade DNA. *In* "The RNA World" (R. F. Gesteland, T. R. Cech, and J. F. Atkins, Eds.), 3rd edn. pp. 469–505. Cold Spring Harbor Laboratory Press, Cold Spring Harbor, New York.
3. Leontis, N. B., and Westhof, E. (2003). Analysis of RNA motifs. *Curr. Opin. Struct. Biol.* **13**(3), 300–308.
4. Leontis, N. B., Stombaugh, J., and Westhof, E. (2002). The non-Watson-Crick base pairs and their associated isostericity matrices. *Nucleic Acids Res.* **30**(16), 3497–3531.

5. Conn, G. L., Gutell, R. R., and Draper, D. E. (1998). A functional ribosomal RNA tertiary structure involves a base triple interaction. *Biochemistry* **37**(34), 11980–11988.
6. Dinman, J. D., Richter, S., Plant, E. P., Taylor, R. C., Hammell, A. B., and Rana, T. M. (2002). The frameshift signal of HIV-1 involves a potential intramolecular triplex RNA structure. *Proc. Natl. Acad. Sci. USA* **99**(8), 5331–5336.
7. Nissen, P., Ippolito, J. A., Ban, N., Moore, P. B., and Steitz, T. A. (2001). RNA tertiary interactions in the large ribosomal subunit: The A-minor motif. *Proc. Natl. Acad. Sci. USA* **98**(9), 4899–4903.
8. Doherty, E. A., Batey, R. T., Masquida, B., and Doudna, J. A. (2001). A universal mode of helix packing in RNA. *Nat. Struct. Biol.* **8**(4), 339–343.
9. Staple, D. W., and Butcher, S. E. (2005). Pseudoknots: RNA structures with diverse functions. *PLOS Biol.* **3**(6), e213.
10. Ferré-D'Amaré, A. R., Zhou, K., and Doudna, J. A. (1998). Crystal structure of a hepatitis delta virus ribozyme. *Nature* **395**(6702), 567–574.
11. Cate, J. H., Gooding, A. R., Podell, E., Zhou, K., Golden, B. L., Kundrot, C. E., Cech, T. R., and Doudna, J. A. (1996). Crystal structure of a group I ribozyme domain: Principles of RNA packing. *Science* **273**(5282), 1678–1685.
12. Duckett, D. R., Murchie, A. I., Diekmann, S., von Kitzing, E., Kemper, B., and Lilley, D. M. (1988). The structure of the Holliday junction, and its resolution. *Cell* **55**(1), 79–89.
13. Holland, J. A., Hansen, M. R., Du, Z., and Hoffman, D. W. (1999). An examination of coaxial stacking of helical stems in a pseudoknot motif: The gene 32 messenger RNA pseudoknot of bacteriophage T2. *RNA* **5**(2), 257–271.
14. Walter, A. E., Turner, D. H., Kim, J., Lyttle, M. H., Müller, P., Mathews, D. H., and Zuker, M. (1994). Coaxial stacking of helices enhances binding of oligoribonucleotides and improves predictions of RNA folding. *Proc. Natl. Acad. Sci. USA* **91**(20), 9218–9222.
15. Schneider, C., and Sühnel, J. (2000). A molecular dynamics simulation study of coaxial stacking in RNA. *J. Biomol. Struct. Dyn.* **18**(3), 345–352.
16. Qin, P. Z., and Pyle, A. M. (1998). The architectural organization and mechanistic function of group II intron structural elements. *Curr. Opin. Struct. Biol.* **8**(3), 301–308.
17. Kolb, V. A., Makeyev, E. V., and Spirin, A. S. (2000). Co-translational folding of an eukaryotic multidomain protein in a prokaryotic translation system. *J. Biol. Chem.* **275**(22), 16597–16601.
18. Guerrier-Takada, C., Gardiner, K., Marsh, T., Pace, N., and Altman, S. (1983). The RNA moiety of ribonuclease P is the catalytic subunit of the enzyme. *Cell* **35**, 849–857.
19. Gilbert, W. (1986). The RNA world. *Nature* **319**, 618.
20. Baird, N. J., Fang, X. W., Srividya, N., Pan, T., and Sosnick, T. R. (2007). Folding of a universal ribozyme: The ribonuclease P RNA. *Q. Rev. Biophys.* **40**(2), 113–161.
21. Krasilnikov, A. S., Yang, X., Pan, T., and Mondragón, A. (2003). Crystal structure of the specificity domain of ribonuclease P. *Nature* **421**, 760–764.
22. Krasilnikov, A. S., Xiao, Y., Pan, T., and Mondragon, A. (2004). Basis for structural diversity in homologous RNAs. *Science* **306**, 104–107.
23. Torres-Larios, A., Swinger, K. K., Krasilnikov, A. S., Pan, T., and Mondragon, A. (2005). Crystal structure of the RNA component of bacterial ribonuclease P. *Nature* **437**, 584–587.
24. Pley, H. W., Flaherty, K. M., and McKay, D. B. (1994). Three-dimensional structure of a hammerhead ribozyme. *Nature* **372**, 68–74.
25. Scott, W. G., Finch, J. T., and Klug, A. (1995). The crystal structure of an all-RNA hammerhead ribozyme: A proposed mechanism for RNA catalytic cleavage. *Cell* **81**(7), 991–1002.
26. Martick, M., and Scott, W. G. (2006). Tertiary contacts distant from the active site prime a ribozyme for catalysis. *Cell* **126**(2), 309–320.

27. Ferré-D'Amaré, A. R., Zhou, K., and Doudna, J. A. (1998). Crystal structure of a hepatitis delta virus ribozyme. *Nature* **395**(6702), 567–574.
28. Ferré-D'amaré, A. R., and Rupert, P. B. (2002). The hairpin ribozyme: From crystal structure to function. *Biochem. Soc. Trans.* **30**(Pt 6), 1105–1109.
29. Klein, D. J., and Ferré-D'Amaré, A. R. (2006). Structural basis of glmS ribozyme activation by glucosamine-6-phosphate. *Science* **313**(5794), 1752–1756.
30. Cochrane, J. C., Lipchock, S. V., and Strobel, S. A. (2007). Structural investigation of the GlmS ribozyme bound to its catalytic cofactor. *Chem. Biol.* **14**(1), 97–105.
31. Cannone, J. J., Subramanian, S., Schnare, M. N., Collett, J. R., D'Souza, L. M., Du, Y., Feng, B., Lin, N., Madabusi, L. V., Müller, K. M., Pande, N., Shang, Z. *et al.* (2002). The comparative RNA web (CRW) site: An online database of comparative sequence and structure information for ribosomal, intron, and other RNAs. *BMC Bioinformatics* **3**, 2.
32. Edgell, D. R., Belfort, M., and Shub, D. A. (2000). Barriers to intron promiscuity in bacteria. *J Bacteriol.* **182**(19), 5281–5289.
33. Toro, N., Jiménez-Zurdo, J. I., and García-Rodríguez, F. M. (2007). Bacterial group II introns: Not just splicing. *FEMS Microbiol. Rev.* **31**(3), 342–358.
34. Vallès, Y., Halanych, K. M., and Boore, J. L. (2008). Group II introns break new boundaries: Presence in a bilaterian's genome. *PLoS ONE* **3**(1), e1488.
35. Christopher, D. A., and Hallick, R. B. (1989). Euglena gracilis chloroplast ribosomal protein operon: A new chloroplast gene for ribosomal protein L5 and description of a novel organelle intron category designated group III. *Nucleic Acids Res.* **17**(19), 7591–7608.
36. Lykke-Andersen, J., Aagaard, C., Semionenkov, M., and Garrett, R. A. (1997). Archaeal introns splicing, intercellular mobility and evolution. *Trends Biochem. Sci.* **22**(9), 326–331.
37. Michel, F., and Westhof, E. (1990). Modelling of the three-dimensional architecture of group I catalytic introns based on comparative sequence analysis. *J. Mol. Biol.* **216**(3), 585–610.
38. Vicens, Q., and Cech, T. R. (2006). Atomic level architecture of group I introns revealed. *Trends Biochem. Sci.* **31**(1), 41–51.
39. Dai, L., Chai, D., Gu, S. Q., Gabel, J., Noskov, S. Y., Blocker, F. J. H., Lambowitz, A. M., and Zimmerly, S. (2008). A three-dimensional model of a group II intron RNA and its interaction with the intron-encoded reverse transcriptase. *Mol. Cell* **30**, 1–14.
40. Toor, N., Keating, K. S., Taylor, S. D., and Pyle, A. M. (2008). Crystal structure of a self-spliced group II intron. *Science* **320**(5872), 77–82.
41. Valadkhan, S. (2005). snRNAs as the catalysts of pre-mRNA splicing. *Curr. Opin. Chem. Biol.* **9**(6), 603–608.
42. Spirin, A. S. (2004). The ribosome as an RNA-based molecular machine. *RNA Biol.* **1**(1), 3–9.
43. Ban, N., Nissen, P., Hansen, J., Moore, P. B., and Steitz, T. A. (2000). The complete atomic structure of the large ribosomal subunit at 2.4 A resolution. *Science* **289**(5481), 905–920.
44. Schluenzen, F., Tocilj, A., Zarivach, R., Harms, J., Gluehmann, M., Janell, D., Bashan, A., Bartels, H., Agmon, I., Franceschi, F., and Yonath, A. (2000). Structure of functionally activated small ribosomal subunit at 3.3 angstroms resolution. *Cell* **102**(5), 615–623.
45. Wimberly, B. T., Brodersen, D. E., Clemons, W. M., Jr., Morgan-Warren, R. J., Carter, A. P., Vonrhein, C., Hartsch, T., and Ramakrishnan, V. (2000). Structure of the 30S ribosomal subunit. *Nature* **407**(6802), 327–339.
46. Kim, S. H., Quigley, G., Suddath, F. L., McPherson, A., Sneden, D., Kim, J. J., Weinzierl, J., Blattmann, P., and Rich, A. (1972). The three-dimensional structure of yeast phenylalanine transfer RNA: Shape of the molecule at 5.5-A resolution. *Proc. Natl. Acad. Sci. USA* **69**(12), 3746–3750.
47. Robertus, J. D., Ladner, J. E., Finch, J. T., Rhodes, D., Brown, R. S., Clark, B. F., and Klug, A. (1974). Structure of yeast phenylalanine tRNA at 3 A resolution. *Nature* **250**(467), 546–551.
48. Clark, B. F. (2006). The crystal structure of tRNA. *J. Biosci.* **31**(4), 453–457.

49. Reinhart, B. J., and Bartel, D. P. (2002). Small RNAs correspond to centromere heterochromatic repeats. *Science* **297,** 1831.
50. Llave, C., Kasschau, K. D., Rector, M. A., and Carrington, J. C. (2002). Endogenous and silencing-associated small RNAs in plants. *Plant Cell* **14,** 1605–1619.
51. Djikeng, A., Shi, H. F., Tschudi, C., and Ullu, E. (2001). RNA interference in *Trypanosoma brucei*: Cloning of small interfering RNAs provides evidence for retroposon-derived 24–26-nucleotide RNAs. *RNA* **7,** 1522–1530.
52. Ambros, V., Lee, R. C., Lavanway, A., Williams, P. T., and Jewell, D. (2003). MicroRNAs and other tiny endogenous RNAs in *C. elegans*. *Curr. Biol.* **13,** 807–818.
53. Grimson, A., Farh, K. K., Johnston, W. K., Garrett-Engele, P., Lim, L. P., and Bartel, D. P. (2007). MicroRNA targeting specificity in mammals: Determinants beyond seed pairing. *Mol. Cell* **27,** 91–105.
54. Nielsen, C. B., Shomron, N., Sandberg, R., Hornstein, E., Kitzman, J., and Burge, C. B. (2007). Determinants of targeting by endogenous and exogenous microRNAs and siRNAs. *RNA* **13,** 1894–1910.
55. Stormo, G. D. (2003). New tricks for an old dogma: Riboswitches as cis-only regulatory systems. *Mol. Cell* **11**(6), 1419–1420.
56. Pandey, N. B., Williams, A. S., Sun, J. H., Brown, V. D., Bond, U., and Marzluff, W. F. (1994). Point mutations in the stem-loop at the 3′ end of mouse histone mRNA reduce expression by reducing the efficiency of 3′ end formation. *Mol. Cell Biol.* **14**(3), 1709–1720.
57. Knapp, G. (1989). Enzymatic approaches to probing of RNA secondary and tertiary structure. *In* "Methods in Enzymology" (J. E. Dahlberg and J. N. Abelson, Eds.), pp. 192–211. Academic Press, Inc., San Diego.
58. Romby, P., Moras, D., Bergdoll, M., Dumas, P., Vlassov, V. V., Westhof, E., Ebel, J. P., and Giegé, R. (1985). Yeast tRNAAsp tertiary structure in solution and areas of interaction of the tRNA with aspartyl-tRNA synthetase. A comparative study of the yeast phenylalanine system by phosphate alkylation experiments with ethylnitrosourea. *J. Mol. Biol.* **184**(3), 455–471.
59. Conway, L., and Wickens, M. (1989). Modification interference analysis of reactions using RNA substrates. *In* "Methods in Enzymology" (J. E. Dahlberg and J. N. Abelson, Eds.), pp. 369–379. Academic Press, Inc., San Diego.
60. Balzer, M., and Wagner, R. (1998). A chemical modification method for the structural analysis of RNA and RNA–protein complexes within living cells. *Anal. Biochem.* **256**(2), 240–242.
61. Tullius, T. D., and Greenbaum, J. A. (2005). Mapping nucleic acid structure by hydroxyl radical cleavage. *Curr. Opin. Chem. Biol.* **9**(2), 127–134.
62. Joseph, S., and Noller, H. F. (2000). Directed hydroxyl radical probing using iron(II) tethered to RNA. *Methods Enzymol.* **318,** 175–190.
63. Bergman, N. H., Lau, N. C., Lehnert, V., Westhof, E., and Bartel, D. P. (2004). The three-dimensional architecture of the class I ligase ribozyme. *RNA* **10**(2), 176–184.
64. Lease, R. A., Adilakshmi, T., Heilman-Miller, S., and Woodson, S. A. (2007). Communication between RNA folding domains revealed by folding of circularly permuted ribozymes. *J. Mol. Biol.* **373**(1), 197–210.
65. Russell, R., Tijerina, P., Chadee, A. B., and Bhaskaran, H. (2007). Deletion of the P5abc peripheral element accelerates early and late folding steps of the Tetrahymena group I ribozyme. *Biochemistry* **46**(17), 4951–4961.
66. Strobel, S. A. (1999). A chemogenetic approach to RNA function/structure analysis. *Curr. Opin. Struct. Biol.* **9**(3), 346–352.
67. Ryder, S. P., and Strobel, S. A. (1999). Nucleotide analog interference mapping. *Methods* **18**(1), 38–50.
68. Ryder, S. P., Ortoleva-Donnelly, L., Kosek, A. B., and Strobel, S. A. (2000). Chemical probing of RNA by nucleotide analog interference mapping. *Methods Enzymol.* **317,** 92–109.

69. Jansen, J. A., McCarthy, T. J., Soukup, G. A., and Soukup, J. K. (2006). Backbone and nucleobase contacts to glucosamine-6-phosphate in the glmS ribozyme. *Nat. Struct. Mol. Biol.* **13**(6), 517–523.
70. Waldsich, C., and Pyle, A. M. (2007). A folding control element for tertiary collapse of a group II intron ribozyme. *Nat. Struct. Mol. Biol.* **14**(1), 37–44.
71. Waldsich, C., and Pyle, A. M. (2008). A kinetic intermediate that regulates proper folding of a group II intron RNA. *J. Mol. Biol.* **375**(2), 572–580.
72. Favre, A., Saintomé, C., Fourrey, J. L., Clivio, P., and Laugâa, P. (1998). Thionucleobases as intrinsic photoaffinity probes of nucleic acid structure and nucleic acid-protein interactions. *J. Photochem. Photobiol. B.* **42**(2), 109–124.
73. Harris, M. E., and Christian, E. L. (1999). Use of circular permutation and end modification to position photoaffinity probes for analysis of RNA structure. *Methods* **18**(1), 51–59.
74. Thomas, B. C., Kazantsev, A. V., Chen, J. L., and Pace, N. R. (2000). Photoaffinity cross-linking and RNA structure analysis. *Methods Enzymol.* **318**, 136–147.
75. Sontheimer, E. J. (1994). Site-specific RNA crosslinking with 4-thiouridine. *Mol. Biol. Rep.* **20**(1), 35–44.
76. Moore, M. J., and Sharp, P. A. (1992). Site-specific modification of pre-mRNA: The 2′-hydroxyl groups at the splice sites. *Science* **256**(5059), 992–997.
77. Nolan, J. M., Burke, D. H., and Pace, N. R. (1993). Circularly permuted tRNAs as specific photoaffinity probes of ribonuclease P RNA structure. *Science* **261**(5122), 762–765.
78. Harris, M. E., Nolan, J. M., Malhotra, A., Brown, J. W., Harvey, S. C., and Pace, N. R. (1994). Use of photoaffinity crosslinking and molecular modeling to analyze the global architecture of ribonuclease P RNA. *EMBO J.* **13**(17), 3953–3963.
79. Harris, M. E., Kazantsev, A. V., Chen, J. L., and Pace, N. R. (1997). Analysis of the tertiary structure of the ribonuclease P ribozyme-substrate complex by site-specific photoaffinity crosslinking. *RNA* **3**(6), 561–576.
80. Marquez, S. M., Chen, J. L., Evans, D., and Pace, N. R. (2006). Structure and function of eukaryotic Ribonuclease P RNA. *Mol. Cell.* **24**(3), 445–456.
81. Lambert, D., Heckman, J. E., and Burke, J. M. (2006). Cation-specific structural accommodation within a catalytic RNA. *Biochemistry* **45**(3), 829–838.
82. Podjarny, A., Howard, E., Mitschler, A., and Chevrier, B. (2002). X-ray crystallography at subatomic resolution. *Europhys. News* **33**, 4.
83. Smyth, M. S., and Martin, J. H. J. (2000). X-ray crystallography. *Mol. Pathol.* **53**(1), 8–14.
84. Lukavsky, P. J., and Puglisi, J. D. (2001). RNAPack: An integrated NMR approach to RNA structure determination. *Methods* **25**(3), 316–332.
85. Furtig, B., Richter, C., Wohnert, J., and Schwalbe, H. (2003). NMR spectroscopy of RNA. *Chembiochemistry* **4**, 936–962.
86. Lukavsky, P. J., and Puglisi, J. D. (2005). Structure determination of large biological RNAs. *Methods Enzymol.* **394**, 399–416.
87. Tzakos, A. G., Grace, C. R., Lukavsky, P. J., and Riek, R. (2006). NMR techniques for very large proteins and rnas in solution. *Annu. Rev. Biophys. Biomol. Struct.* **35**, 319–342.
88. Furtig, B., Buck, J., Manoharan, V., Bermel, W., Jäschke, A., Wenter, P., Pitsch, S., and Schwalbe, H. (2007). Time-resolved NMR studies of RNA folding. *Biopolymers* **86**(5–6), 360–383.
89. Getz, M., Sun, X., Casiano-Negroni, A., Zhang, Q., and Al-Hashimi, H. M. (2007). NMR studies of RNA dynamics and structural plasticity using NMR residual dipolar couplings. *Biopolymers* **86**(5–6), 384–402.
90. Ma, H., Proctor, D. J., Kierzek, E., Kierzek, R., Bevilacqua, P. C., and Gruebele, M. (2006). Exploring the energy landscape of a small RNA hairpin. *J. Am. Chem. Soc.* **128**(5), 1523–1530.

91. Meng, Z., and Limbach, P. A. (2006). Mass spectrometry of RNA: Linking the genome to the proteome. *Brief. Funct. Genomic. Proteomic.* **5**(1), 87–95.
92. Yu, E., and Fabris, D. (2004). Toward multiplexing the application of solvent accessibility probes for the investigation of RNA three-dimensional structures by electrospray ionization-Fourier transform mass spectrometry. *Anal. Biochem.* **334**(2), 356–366.
93. Yu, E. T., Zhang, Q., and Fabris, D. (2005). Untying the FIV frameshifting pseudoknot structure by MS3D. *J. Mol. Biol.* **345**(1), 69–80.
94. Pan, T., and Sosnick, T. R. (1997). Intermediates and kinetic traps in the folding of a large ribozyme revealed by circular dichroism and UV absorbance spectroscopies and catalytic activity. *Nat. Struct. Biol.* **4**(11), 931–938.
95. Treiber, D. K., Rook, M. S., Zarrinkar, P. P., and Williamson, J. R. (1998). Kinetic intermediates trapped by native interactions in RNA folding. *Science* **279**(5358), 1943–1946.
96. Greulich, K. O. (2005). Single-molecule studies on DNA and RNA. *Chem. Phys. Chem.* **6**, 2458–2471.
97. Harlepp, S., Marchal, T., Robert, J., Léger, J. F., Xayaphoummine, A., Isambert, H., and Chatenay, D. (2003). Probing complex RNA structures by mechanical force. *Eur. Phys. J. E. Soft. Matter.* **12**(4), 605–615.
98. Jarzynski, C. (1997). Nonequilibrium equality for free energy differences. *Phys. Rev. Lett.* **78**, 2690–2693.
99. Liphardt, J., Dumont, S., Smith, S. B., Tinoco, I., Jr., and Bustamante, C. (2002). Equilibrium information from nonequilibrium measurements in an experimental test of Jarzynski's equality. *Science* **296**(5574), 1832–1835.
100. Zhuang, X. (2005). Single-molecule RNA science. *Annu. Rev. Biophys. Biomol. Struct.* **34**, 399–414.
101. Xie, Z., Srividya, N., Sosnick, T. R., Pan, T., and Scherer, N. F. (2004). Single-molecule studies highlight conformational heterogeneity in the early folding steps of a large ribozyme. *Proc. Natl. Acad. Sci. USA* **101**(2), 534–539.
102. Murchie, A. I. H., Thomson, J. B., Walter, F., and Lilley, D. M. J. (1998). Folding of the hairpin ribozyme in its natural conformation achieves close physical proximity of the loops. *Mol. Cell* **1**(6), 873–881.
103. Walter, F., Murchie, A. I. H., Thomson, J. B., and Lilley, D. M. J. (1998). Structure and activity of the hairpin ribozyme in its natural junction conformation; effect of metal ions. *Biochemistry* **37**(40), 14195–14203.
104. Tan, E., Wilson, T. J., Nahas, M. K., Clegg, R. M., Lilley, D. M. J., and Ha, T. (2003). A four-way junction accelerates hairpin ribozyme folding via a discrete intermediate. *PNAS* **100**(16), 9308–9313.
105. Nahas, M. K., Wilson, T. J., Hohng, S., Jarvie, K., Lilley, D. M. J., and Ha, T. (2004). Observation of internal cleavage and ligation reactions of a ribozyme. *Nat. Struct. Mol. Biol.* **11**(11), 1107–1113.
106. Engel, A., and Müller, D. J. (2000). Observing single biomolecules at work with the atomic force microscope. *Nat. Struct. Biol.* **7**(9), 715–718.
107. Noah, J. W. (2006). Atomic force microscopy reveals DNA bending during group II intron ribonucleoprotein particle integration into double-stranded DNA. *Biochemistry* **45**, 12424–12435.
108. Frank, J. (2001). Cryo-electron microscopy as an investigative tool: The ribosome as an example. *Bioessays* **23**(8), 725–732.
109. Li, W., and Frank, J. (2007). Transfer RNA in the hybrid P/E state: Correlating molecular dynamics simulations with cryo-EM data. *Proc. Natl. Acad. Sci. USA* **104**(42), 16540–16545.
110. Taylor, D. J., Nilsson, J., Merrill, A. R., Andersen, G. R., Nissen, P., and Frank, J. (2007). Structures of modified eEF2 80S ribosome complexes reveal the role of GTP hydrolysis in translocation. *EMBO J.* **26**(9), 2421–2431.

111. Gilbert, R. J., Gordiyenko, Y., von der Haar, T., Sonnen, A. F., Hofmann, G., Nardelli, M., Stuart, D. I., and McCarthy, J. E. (2007). Reconfiguration of yeast 40S ribosomal subunit domains by the translation initiation multifactor complex. *Proc. Natl. Acad. Sci. USA* **104**(14), 5788–5793.
112. Kaur, S., Gillet, R., Li, W., Gursky, R., and Frank, J. (2006). Cryo-EM visualization of transfer messenger RNA with two SmpBs in a stalled ribosome. *Proc. Natl. Acad. Sci. USA* **103**(44), 16484–16489.
113. Fabiola, F., and Chapman, M. S. (2005). Fitting of high-resolution structures into electron microscopy reconstruction images. *Structure* **13**(3), 389–400.
114. Jolley, C. C., Wells, S. A., Fromme, P., and Thorpe, M. F. (2008). Fitting low-resolution cryo-EM maps of proteins using constrained geometric simulations. *Biophys. J.* **94**(5), 1613–1621.
115. Shao, Z., and Zhang, Y. (1996). Biological cryo atomic force microscopy: A brief review. *Ultramicroscopy* **66**(3–4), 141–152.
116. Mat-Arip, Y., Garver, K., Chen, C., Sheng, S., Shao, Z., and Guo, P. (2001). Three-dimensional interaction of Phi29 pRNA dimer probed by chemical modification interference, cryo-AFM, and cross-linking. *J. Biol. Chem.* **276**(35), 32575–32584.
117. Stark, H., Rodnina, M. V., Wieden, H. J., van Heel, M., and Wintermeyer, W. (2000). Large-scale movement of elongation factor G and extensive conformational change of the ribosome during translocation. *Cell* **100**(3), 301–309.
118. Mueller, F., Sommer, I., Baranov, P., Matadeen, R., Stoldt, M., Wöhnert, J., Görlach, M., van Heel, M., and Brimacombe, R. (2000). The 3D arrangement of the 23 S and 5 S rRNA in the *Escherichia coli* 50 S ribosomal subunit based on a cryo-electron microscopic reconstruction at 7.5 A resolution. *J. Mol. Biol.* **298**(1), 35–59.
119. SantaLucia, J. (1998). A unified view of polymer, dumbbell, and oligonucleotide DNA nearest-neighbor thermodynamics. *Proc. Natl. Acad. Sci. USA* **95**(4), 1460–1465.
120. Misra, V. K., and Draper, D. E. (1998). On the role of magnesium ions in RNA stability. *Biopolymers* **48**(2–3), 113–135.
121. Shiman, R., and Draper, D. E. (2000). Stabilization of RNA tertiary structure by monovalent cations. *J. Mol. Biol.* **302**(1), 79–91.
122. Draper, D. E., Grilley, D., and Soto, A. M. (2005). Ions and RNA folding. *Annu. Rev. Biophys. Biomol. Struct.* **34**, 221–243.
123. Draper, D. E. (2004). A guide to ions and RNA structure. *RNA* **10**, 335–343.
124. Vicens, Q., and Cech, T. R. (2006). Atomic level architecture of group I introns revealed. *Trends Biochem. Sci.* **31**(1), 41–51.
125. Lambowitz, A. M., and Zimmerly, S. (2004). Mobile group II introns. *Annu. Rev. Genet.* **38**, 1–35.
126. Toor, N., Robart, A. R., Christianson, J., and Zimmerly, S. (2006). Self-splicing of a group IIC intron: 5′ exon recognition and alternative 5′ splicing events implicate the stem-loop motif of a transcriptional terminator. *Nucleic Acids Res.* **34**(22), 6461–6471.
127. Silverman, S. K., and Cech, T. R. (1999). Energetics and cooperativity of tertiary hydrogen bonds in RNA structure. *Biochemistry* **38**(27), 8691–8702.
128. Tinoco, I., Jr., and Bustamante, C. (1999). How RNA folds. *J. Mol. Biol.* **293**(2), 271–281.
129. Shapiro, B. A., Yingling, Y. G., Kasprzak, W., and Bindewald, E. (2007). Bridging the gap in RNA structure prediction. *Curr. Opin. Struct. Biol.* **17**(2), 157–165.
130. Zuker, M. (1984). RNA secondary structures and their prediction. *Bull. Math. Biol.* **46**(4), 591–621.
131. Eddy, S. R. (2004). How do RNA folding algorithms work? *Nat. Biotechnol.* **22**(11), 1457–1458.
132. Mathews, D. H., and Turner, D. H. (2006). Prediction of RNA secondary structure by free energy minimization. *Curr. Opin. Struct. Biol.* **16**(3), 270–278.

133. Ding, Y., Chan, C. Y., and Lawrence, C. E. (2005). RNA secondary structure prediction by centroids in a Boltzmann weighted ensemble. *RNA* **11**(8), 1157–1166.
134. Steffen, P., Voss, B., Rehmsmeier, M., Reeder, J., and Giegerich, R. (2006). RNAshapes: An integrated RNA analysis package based on abstract shapes. *Bioinformatics* **22**(4), 500–503.
135. Chen, J. H., Le, S. Y., and Maizel, J. V. (2000). Prediction of common secondary structures of RNAs: A genetic algorithm approach. *Nucleic Acids Res.* **28**(4), 991–999.
136. Mathews, D. H., and Turner, D. H. (2002). Dynalign: An algorithm for finding the secondary structure common to two RNA sequences. *J. Mol. Biol.* **317**(2), 191–220.
137. Major, F., Turcotte, M., Gautheret, D., Lapalme, G., Fillion, E., and Cedergren, R. (1991). The combination of symbolic and numerical computation for three-dimensional modeling of RNA. *Science* **253**(5025), 1255–1260.
138. Michel, F., and Westhof, E. (1990). Modelling of the three-dimensional architecture of group I catalytic introns based on comparative sequence analysis. *J. Mol. Biol.* **216**(3), 585–610.
139. Massire, C., and Westhof, E. (1998). MANIP: An interactive tool for modelling RNA. *J. Mol. Graph. Model.* **16**(4–6), 197–205, 255–257.
140. Tung, C. S., Joseph, S., and Sanbonmatsu, K. Y. (2002). All-atom homology model of the *Escherichia coli* 30S ribosomal subunit. *Nat. Struct. Biol.* **9**(10), 750–755.
141. Cannone, J. J., Subramanian, S., Schnare, M. N., Collett, J. R., D'Souza, L. M., Du, Y., Feng, B., Lin, N., Madabusi, L. V., Müller, K. M., Pande, N., Shang, Z. *et al.* (2002). The comparative RNA web (CRW) site: An online database of comparative sequence and structure information for ribosomal, intron, and other RNAs. *BMC Bioinformatics* **3**, 2.
142. Malhotra, A., and Harvey, S. C. (1994). A quantitative model of the *Escherichia coli* 16 S RNA in the 30 S ribosomal subunit. *J. Mol. Biol.* **240**(4), 308–340.
143. Thomas, B. C., Kazantsev, A. V., Chen, J. L., and Pace, N. R. (2000). Photoaffinity cross-linking and RNA structure analysis. *Methods Enzymol.* **318**, 136–147.

DNA Polymerase ε: A Polymerase of Unusual Size (and Complexity)

Zachary F. Pursell and
Thomas A. Kunkel

Department of Health and Human Services, Laboratory of Molecular Genetics and Laboratory of Structural Biology, National Institute of Environmental Health Sciences, National Institutes of Health, Research Triangle Park, North Carolina 27709

I. Introduction	102
II. Pol ε Structure	104
A. The Catalytic Subunit	104
B. Pol ε Holoenzyme	106
III. Physical and Functional Interactions of Pol ε	109
IV. Biochemical Properties of Pol ε	111
A. Polymerization	111
B. Exonuclease	112
C. DNA Binding	113
D. Fidelity	113
V. Pol ε in DNA Replication	114
A. Evidence That Pol ε Is a Major Replicative Polymerase	115
B. Division of Labor at the Replication Fork	116
VI. The Role of Pol ε in Checkpoint Control	119
A. Cell Cycle Progression and Replication	119
B. Pol ε Linked to Cell Cycle Progression and Replication	119
C. Dpb2 and the Cell Cycle	120
VII. Pol ε Involvement in Regulating Chromatin States	121
A. Pol ε Influence on Gene Silencing	121
B. Pol ε and Telomeres	122
C. Telomere-Proximal Effect	122
VIII. Pol ε Relationship with Chromatin Remodeling Complexes	123
A. Chromatin Remodeling	123
B. Dpb4/p17: The Shared Subunit	124
C. Centromeres and Sister Chromatid Cohesion	126
IX. The Roles of Pol ε in Excision Repair of DNA Damage	126
A. Base Excision Repair	127
B. Nucleotide Excision Repair	127
C. Does Pol ε Function in Mismatch Repair?	129
X. Pol ε in Recombination	129
A. Double-Strand Break Repair	129
B. Gene Conversion	129
C. Break-Induced Replication	130

D. Mammalian Recombination Complex.	130
E. rDNA Recombination	131
XI. *Schizosaccharomyces pombe* Pol ε	131
A. Catalytic Subunit	131
B. Accessory Subunits	132
XII. Xenopus Pol ε	133
XIII. Concluding Remarks	134
References	134

DNA polymerase epsilon (Pol epsilon) is a large, multi-subunit polymerase that is conserved throughout all eukaryotes. In addition to its role as one of the three DNA polymerases responsible for bulk chromosomal replication, Pol epsilon is implicated in a wide variety of important cellular processes, including the repair of damaged DNA, DNA recombination and the regulation of proper cell cycle progression. Additionally, recent work has suggested that Pol epsilon is linked to chromatin remodeling and the regulation of epigenetic inheritance. Though much has been learned in the past two decades about the various functions of Pol epsilon, a great deal remains unknown due in large part to the complexities of both the various implicated pathways and the Pol epsilon holoenzyme itself. While most of the progress in understanding Pol epsilon has come from work using the human and baker's yeast systems, the Xenopus and fission yeast systems have provided valuable insights. Here we discuss what is currently known about this unique DNA polymerase, from its initial identification through the present.

I. Introduction

Shortly after Kornberg *et al.* identified the first DNA polymerase in the mid-twentieth century (1), the first eukaryotic DNA polymerase was discovered (2), and was eventually named DNA Pol α. Pol α was initially believed to be the sole polymerase responsible for eukaryotic DNA replication, but that view changed two decades later with the discovery of a second replicative polymerase, Pol δ (3). Although the catalytic subunit of Pol δ is 125 kDa, soon after its discovery even larger polymerases were purified that had somewhat similar properties and were therefore variously called Pol δII, Pol δ*, and big Pol δ (4–6). However, biochemical studies eventually led to the realization that these larger enzyme forms were actually a distinct polymerase, Pol ε (7). Although much of the early work on Pol ε centered on its initially defined role in mammalian DNA repair, by the early 1990s the gene encoding Pol ε was cloned (8, 9), making possible the

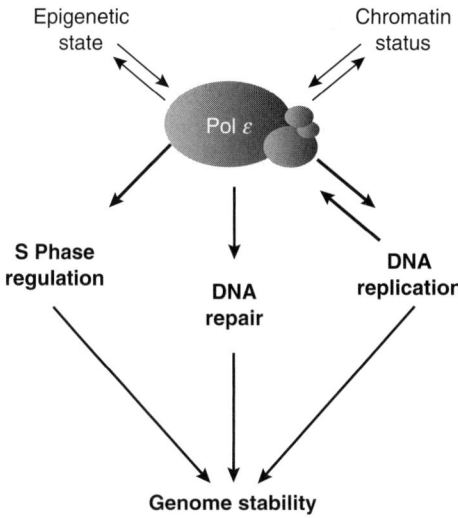

FIG. 1. Pol ε: At the intersection of complex and diverse cellular processes. Pol ε is involved in a wide array of diverse cellular processes. Its involvement is depicted here with the direction and size of the arrows meant to reflect the direction of influence and relative importance of its involvement in each process, respectively. By functioning as the leading strand polymerase in normal replication, a large arrow points from DNA replication toward Pol ε (see text in Section V). However, some data suggest that, while important for normal replication, the essential function of Pol ε lie s in its role as a checkpoint sensor during replication (see Section VI), influencing both replication as well as cell cycle progression. Thus, large arrows point away from Pol ε toward these processes. Pol ε is also implicated in the repair of damaged DNA, though the degree to which it operates in specialized repair pathways remains unclear (see sections IX and X). Pol ε contributes to altering chromatin status, typically by promoting a silenced state either transiently or in a heritable, epigenetic manner (see sections VII and VIII), thus arrows point from Pol ε toward these processes. Additionally, alterations to chromatin or epigenetic modifications may target Pol ε to these regions, thus arrows point from these processes toward Pol ε. The interaction of Pol ε with other factors (see section III) as well as the intrinsic fidelity and other biochemical properties of Pol ε (see section IV) play important roles throughout each of these processes. Proper cell cycle progression, DNA damage repair, and DNA replication, and possibly chromatin and epigenetic states as well, serve the overall goal of maintaining genome stability.

many genetic and biochemical studies that are the subject of this chapter. From these studies, we now know that Pol ε is involved in several processes that are central to maintaining the stability of the eukaryotic nuclear genome (Fig. 1). These include DNA replication, repair of DNA damage, control of cell cycle progression, chromatin remodeling, and epigenetic regulation of the stable transfer of information from mother to daughter cells. Here we consider the evidence supporting these many functions for Pol ε, and in doing so point out that a great deal remains to be learned about this large and complex polymerase.

II. Pol ε Structure

A. The Catalytic Subunit

DNA pol ε is a member of the B family of DNA polymerases that share sequence homology with the catalytic subunit of bacterial Pol II, the product of the *Escherichia coli polB* gene. The open reading frame for the *POLE1* gene encoding the Pol ε catalytic subunit (Fig. 2A) is among the longest of the many known eukaryotic polymerases (*10*), rivaled only by those of its B family sibling Pol ζ and the A family enzyme, Pol θ. The catalytic subunits of human (Fig. 2A) and yeast Pol ε contain 2286 and 2222 amino acids, respectively. The 140 kDa N-terminal half of the protein is fairly well conserved across different species, with 63% sequence identity shared between the yeast and human enzymes (*9*). This conservation reflects the fact that the amino terminal residues of Pol ε harbor the polymerase and exonuclease activities.

The X ray crystal structure of Pol ε has not yet been reported. However, based on homology to two B family siblings for which X ray crystal structures are available [RB69 Pol (*11*) and φ29 Pol (*12*)], the polymerase active site is within the palm of a polymerase domain composed of the palm, fingers, and thumb subdomains (Fig. 2B) that are characteristics of all polymerases regardless of family (*13*). The structural framework for catalysis of the polymerization reaction is composed of highly conserved motifs A, B, and C (Fig. 2C and D) that are characteristics of the "right-handed" polymerases, for example, those

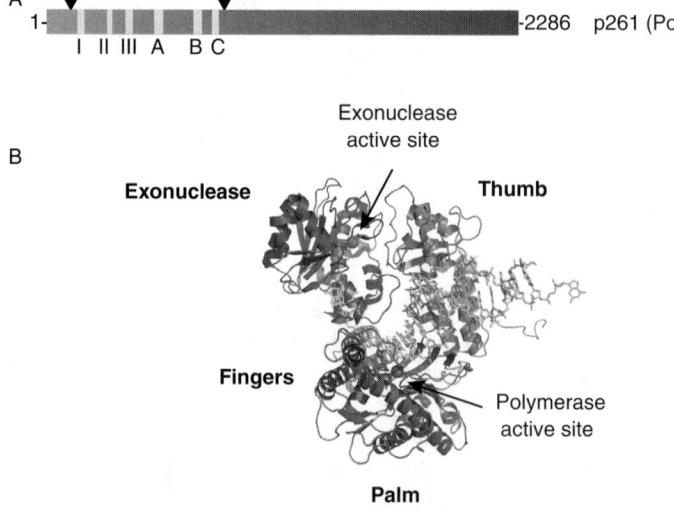

Fig. 2. (Continued)

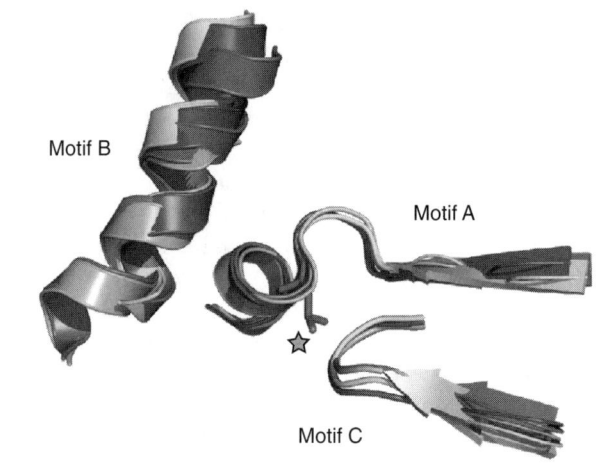

FIG. 2. Pol ε catalytic subunit. (A) A schematic of the Pol ε catalytic subunit. Conserved motifs in the exonuclease and polymerase domains are shown, with the C-terminal protein–protein interaction region at the extreme C-terminus. DEAD-box cleavage sites in human Pol ε are shown as black arrows. (B) The structure of the Pol ε homologue RB69 DNA polymerase complexed with an incoming (correct) dTTP and primer-template DNA is shown using coordinates from PDB accession number 1IG9 (11). The fingers, palm, thumb, and exonuclease domains are each labelled. The duplex DNA is yellow and dTTP shown at the polymerase active site is red. The gray spheres represent the divalent metal ions in the polymerase and exonuclease active sites. (C) Alignment of the amino acid sequences of conserved polymerase motifs A, B, and C from Pol ε and other representative B family polymerases. Conserved catalytic aspartate residues are shown in the black boxes. The conserved motif A methionine that differs between Pol ε and the other B family polymerases is shown in a gray box with a star. Pol2, p261, and cdc20 are Pol ε from *Saccharomyces cerevisiae*, *Homo sapiens*, and *Schizosaccharomyces pombe*, respectively. Pol1 and Pol3 are

in families A, B, and Y. Within this framework are three carboxylates thought to coordinate two metal ions (blue spheres in Fig. 2B) required for catalysis, two of which are in the invariant DTD sequence in motif C and a third in motif A (black boxes in the alignments in Fig. 2C).

The amino terminus of Pol ε also includes residues critical for its intrinsic 3′ exonuclease activity (Fig. 2A). This activity is contained in a domain (Fig. 2B) that is physically separated from the polymerase domain by many angstroms (Fig. 2B). The exonuclease active site is composed of amino acids in conserved motifs designated Exo I, II, and III (Fig. 2A and E), again including three catalytic carboxylates. While a similar domain is present within all B family polymerases, the catalytic carboxylates are not present in some other B family members, including eukaryotic Pols α and ζ. These latter two polymerases naturally lack intrinsic exonuclease activity, such that they cannot proofread replication errors and synthesize DNA less accurately than does Pol ε, or Pol δ, which also has an intrinsic 3′ exonuclease activity.

The 120 kDa C-terminal half of the catalytic subunit of Pol ε does not harbor a known catalytic activity but nonetheless shares 25% sequence homology between yeast and humans. The last one hundred residues at the C-terminus of yeast Pol ε contain two C4 zinc fingers and a spacer region that contains an essential function of Pol ε (see below), and this C-terminal half also mediates interactions with the smaller subunits of the holoenzyme (*14*, *15*). Also present is a putative PCNA interaction motif that is not necessary for normal DNA replication, but which when mutated leads to increased sensitivity to MMS (*16*).

B. Pol ε Holoenzyme

In all species studied to date, the Pol ε holoenzyme is composed of four subunits (Fig. 3A and B and Table I). The second subunit of Pol ε is Dpb2 in budding yeast and p59 in humans, the latter based on its predicted molecular weight. This subunit has no known catalytic activity but it interacts with the catalytic subunit so tightly that it dissociates only under denaturing conditions (*5*). The absence of Dpb2 does not reduce Pol ε catalytic activity but does reduce Pol ε stability during purification (*9*), leading to the idea that it has an

---

S. cerevisiae pols α and δ, respectively. RB69 and φ29 are bacteriophage DNA polymerases. (D) Ribbon diagram depicting an overlay of the structures of polymerase motifs A, B, and C from three B family DNA polymerases. Coordinates from PDB accession numbers 2PYL (φ29 pol), 1IG9 (RB69 pol), 1TGO (Tgo pol), and 1QQC (D.tok pol) were used to align the structures with PyMol. The conserved Leu/Met that was altered to generate the mutator alleles described in the text is shown as a star in the RB69 Pol structure. (E) Alignment of the amino acid sequences of conserved exonuclease motifs I, II, and III from Pol ε and other B family polymerases. Conserved catalytic carboxylates are shown in black boxes. DNA polymerases are as in part (C).

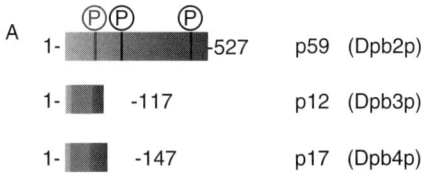
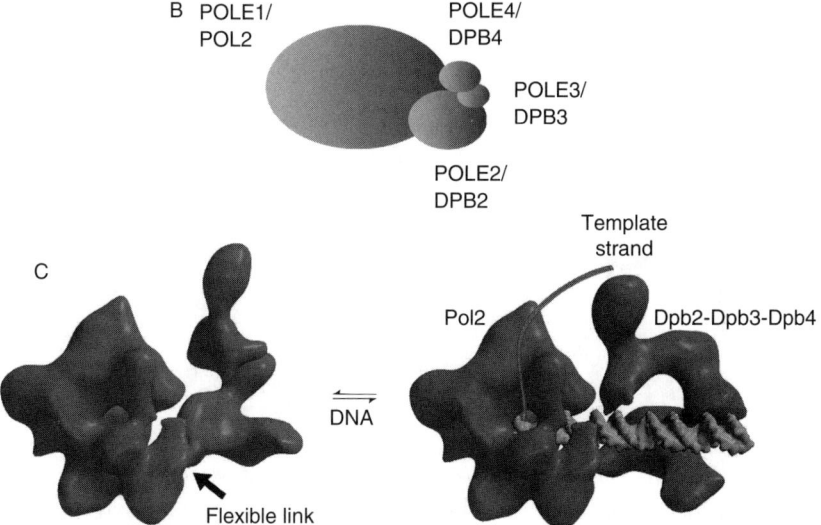

Fig. 3. Pol ε holoenzyme. (A) Schematic representation of each of the three Pol ε accessory subunits. Sites of known *in vitro* and potential *in vivo* phosphorylation (*17*) are shown as gray and black circles, respectively. Histone-fold motifs are shown as gray boxes. (B) Cartoon of four-subunit Pol ε holoenzyme. Each subunit is drawn approximately to scale, based on its predicted molecular weight. Human (and yeast) gene names are indicated next to each subunit. (C) Cryo-EM structure of four-subunit yeast Pol ε (*23*). The open conformation is shown on the left, while the right depicts a model of a closed conformation with duplex DNA bound. This image is reprinted from (*23*) with permission from the authors.

important role in stabilizing the catalytic subunit. Dpb2 is essential in yeast. Moreover, it is phosphorylated in S phase in a Cdc28 dependent manner, suggesting a potential regulatory role (*17*). Based on extensive yeast two-hybrid analysis, Dpb2 interaction with the C-terminus of Pol2 was disrupted with Pol2 mutants that disrupt C-terminus self-interaction in the yeast two-hybrid assay and disrupt normal replication *in vivo*, indicating that the stabilizing effects of Dpb2 on the holoenzyme may allow for an active, possibly dimeric form of Pol ε (*14*). This possibility is somewhat tempered by the extensive biochemical characterization, including analytical ultracentrifugation and sedimentation

TABLE I
Pol ε Gene Names

Organism	Gene name	Number of amino acid residues	Predicted molecular weight (kDa)	Also known as
Saccharomyces cerevisiae	POL2	2222	255	
	DPB2	689	78	
	DPB3	201	23	
	DPB4	196	22	
Homo sapiens	POLE1	2286	261	p261
	POLE2	527	59	p59
	POLE3	117	12	p12
	POLE4	147	17	p17/CHRAC-17
Schizosaccharomyces pombe	cdc20	2199	253	
	dpb2	594	67	
	dpb3	199	22	
	dpb4	210	24	

analysis of the native holoenzyme purified from yeast, showing a 1:1:1:1 stoichiometry for each of the four subunits (18). Additionally, it is likely the Pol2–Dpb2 interaction itself, and therefore likely the presence of all four subunits, that enables normal replication because mutant alleles of Dpb2 that disrupt this interaction show reduced survival (19). These Dpb2 mutant alleles are also mutators, making errors during replication, though it is unclear if this is due to affecting holoenzyme stability, processivity, or some other process (19). Both Dpb2 and its human homologue, p59, contain consensus PCNA interaction motifs, though only the human protein has been shown to physically interact with PCNA (20).

Like Dpb2, the two smallest subunits of human Pol ε, p12 and p17, also lack catalytic activity and also physically interact with the C-terminal half of the catalytic subunit. They also interact with each other through histone-fold motifs (HFMs) similar to those found in histone H2A and H2B (15). Yeast strains with deletions in Dpb3 or Dpb4, the yeast homologues of human p12 and p17, are viable (21, 22). Thus, neither subunit is essential for growth or chromosomal replication. However, dpb3Δ strains have slightly elevated spontaneous mutation rates (21), indicating that Dpb3 is important for genome stability and may possibly modulate the fidelity of DNA synthesis conducted by Pol ε in vivo. While this phenotype could result from reduced fidelity during replication, Dpb3 and Dpb4 both have roles outside of normal DNA replication, including chromatin remodeling and DNA transactions at the ribosomal DNA repeats (see below).

In an exciting recent development (23), the structures of the 4-subunit yeast Pol ε holoenzyme (Fig. 3C) and of the 2-subunit Pol2–Dpb2 enzyme have been solved by cryo-EM. These structures suggest that the three accessory subunits are connected to the catalytic subunit in such a manner that they contact the double-stranded DNA duplex upstream of the polymerase active site (Fig. 3C), possibly reducing polymerase dissociation and increasing processivity. However, this linkage is flexible, indicating that the multisubunit tail may adopt different conformations, perhaps making additional contacts with factors involved in replication or checkpoint activation. A groove is present that is large enough to accommodate approximately 40 nucleotides of double-stranded DNA upstream of the primer terminus. This is the same length of dsDNA that maximizes the intrinsic processivity of Pol ε when copying long primer-templates *in vitro* (23). The position of Dpb2 in the structure suggests the potential to interact with PCNA upstream of the active site, which is likely where RB69 Pol interacts with its sliding clamp (24).

III. Physical and Functional Interactions of Pol ε

In addition to interactions among the four subunits of the holoenzyme, Pol ε has also been shown to interact with other proteins (Table II). One interacting partner is yeast Dpb11 (25), known as TopBP1 in humans and variously known as Rad4, Cut5, and Mus101 in other species. Dpb11 is a BRCT repeat-containing protein initially identified as a multicopy suppressor of temperature-sensitive mutants in the C-terminus of Pol2 and of mutants in Dpb2 (26) that is loaded onto origins after pre-replication complex (RC) formation, and is required for loading Pol α and Pol ε at origins (25). Dpb11 interacts with phosphorylated Sld2 and Sld3, an interaction that is necessary to initiate replication (27, 28). Dpb11 interacts genetically with components of the four-subunit GINS complex that is important in replication initiation (29, 30) and fork progression (31). In addition to Dpb11, the GCM complex contains Cdc45 and a heterohexamer of MCM, ATP-dependent helicase activity, and may be the replicative helicase (32).

TopBP1, the human homologue of Dpb11, interacts with the full-length Pol ε catalytic subunit and a 180 kDa variant of pol ε in human cells that overexpress TopBP1. The functional significance of this interaction has not been explored (33). TopBP1 deficiency is not lethal in human cells but does perturb cell cycle progression and leads to genome instability due to DNA strand breaks accumulated in S phase (34). Human Pol ε is part of a complex

TABLE II
PHYSICAL INTERACTIONS WITH POL ε

Process	Name	Organism[a]	References
Replication			
	Dpb11p	Sc	25
	TopBP1	Hs	33
	PCNA	Hs	20
	Claspin	Xl	206
Repair			
	PCNA	Hs	20
	LigI	Hs	159
Other			
Sister chromatid cohesion	Trf4	Sc	140
Chromatin remodeling	CHRAC (complex)[b]	Sc/Hs/Dm	124, 125, 130
Transcription/Repair?	RNAPII (complex)	Hs	35
Cell cycle	Mdm2	Hs	39
Recombination	LigIII	Bt	178

[a] Bt, *Bos taurus*; Dm, *Drosophila melanogaster*; Hs, *Homo sapiens*; Sc, *Saccharomyces cerevisiae*; Xl, *Xenopus laevis*.

[b] Dpb4 (and its homologues) is a subunit of CHRAC. There is no evidence that Pol ε interacts with CHRAC directly.

with RNA Pol II that increases transcription activation (35). This complex contains a number of repair factors. The Pol ε interaction was later shown to be with the hyperphosphorylated elongation form of RNA polII and occurred throughout the cell cycle (36), possibly pointing to a link between transcription and DNA repair synthesis.

Mdm2 is an oncoprotein that is upregulated in many human tumors (37) and is a major negative regulator of p53, acting as an E3 ubiquitin ligase and targeting p53 for degradation (38). An N-terminal 166 amino acid region of Mdm2 physically interacts with the C-terminal, noncatalytic half of Pol ε (39) and stimulates Pol ε activity (40). It was suggested that Mdm2, and perhaps other factors, may modulate Pol ε functions in response to changing conditions. For example, binding of Mdm2 may displace Pol ε-bound factors in response to stress, thus reconfiguring Pol ε to perform a role in DNA repair and/or checkpoint control. Pol ε interactions with other proteins (Table II) are discussed below in relationship to the many functions proposed for Pol ε (Fig. 1). Given the large size of the essential noncatalytic portion of the Pol ε catalytic subunit (Fig. 2A), the fact that the holoenzyme contains four subunits, and has been understudied due to past challenges in obtaining and working with Pol ε holoenzyme, the number of interacting partners can be expected to increase in the future, especially because some interactions may be condition-specific and/or transient.

IV. Biochemical Properties of Pol ε

A. Polymerization

Pol ε catalyzes DNA template-dependent DNA synthesis by a phosphoryl transfer reaction involving nucleophilic attack by the 3′ hydroxyl of the primer terminus on the α-phosphate of the incoming deoxynucleoside triphosphate (dNTP). The products of this reaction are pyrophosphate and a DNA chain increased in length by one nucleotide. The catalytic mechanism is conserved among DNA polymerases (41). It begins with binding of a primer template to the polymerase. The primer terminus is bound at the polymerase active site, which is largely composed of the A, B, and C sequence motifs mentioned above (Fig. 2C), which harbor the three carboxylate residues that coordinate two divalent metal ions, usually Mg^{2+}. Binding of a correct dNTP results in conformational changes in both the polymerase and the DNA. While the specific nature of these changes depends on the polymerase [reviewed in (42)], the dNTP-induced conformational changes result in assembly of an active site (Fig. 4) with geometry appropriate for the phosphoryl transfer reaction, which proceeds via in-line displacement and results in inversion of the stereochemical configuration of the α-phosphorous atom, as structurally observed for the family X member Pol λ (43). Pyrophosphate is released and translocation of the polymerase allows the newly incorporated base pair to serve as the primer terminus for the next cycle of catalysis.

Pol ε usually translocates following nucleotide incorporation without dissociation from DNA, that is, it synthesizes DNA processively. Indeed, Pol ε was initially described as a larger variant of Pol δ (4, 5) that was highly processive (>5 kb) in the absence of PCNA (6). Later work demonstrated that Pol ε processivity can in fact be stimulated by PCNA (44), but not to the extent that PCNA stimulates Pol δ (7). Another early biochemical distinction between these two polymerases is that poly(dA)·oligo(dT) is a preferred substrate for Pol ε, while Pol δ prefers an alternating poly(dA-dT) substrate (7, 45). Interestingly, >100 mM KCl and NaCl strongly inhibit Pol ε synthesis (44, 46), while potassium glutamate actually slightly stimulates Pol ε activity (44). Glutamate is the physiologically relevant anion in bacteria (47) and is less disruptive to *E. coli* DNA polymerase III complex stability than is chloride (48). A recent study provides evidence that in the presence of PCNA, the processivities of Pol ε and Pol δ are actually quite similar (49). Under reaction conditions where DNA

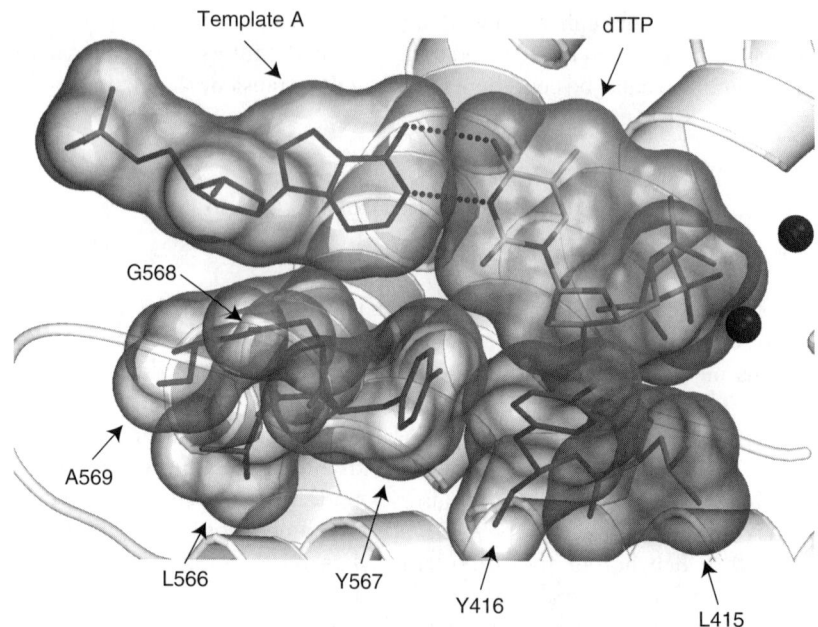

FIG. 4. B family polymerase nascent base pair binding pocket. Surface representation of the nascent base pair and several amino acids in RB69 Pol that form the DNA minor groove edge of the binding pocket at the polymerase active site (Adapted from (212) with the author's permission). Met644 in yeast Pol ε aligns with Leu415 in RB69 Pol. The adjacent Tyr416 in RB69 Pol aligns with Tyr869 in yeast Pol α, which, when substituted with alanine, results in a mutator phenotype.

synthesis results from a single polymerase-DNA binding event, the processivities of both pol ε and pol δ are low (<600 nt). This is likely due to more efficient loading of Pol δ on the RFC-PCNA-containing primer terminus via its much stronger interactions with PCNA. In the absence of PCNA, Pol ε is remarkably only able to synthesize up to ~50 nt in one binding event. These assays were performed in sodium acetate, and while yeast Pol ε holoenzyme stability was enhanced in sodium acetate (18), its optimal concentration for Pol ε activity has not been extensively investigated.

B. Exonuclease

Pol ε and Pol δ are the only two eukaryotic nuclear DNA polymerases with an intrinsic 3′-5′ exonucleolytic activity with which to excise primer terminal nucleotides (10). This activity is especially useful for proofreading of errors made by these polymerases. Proofreading of polymerization errors occurs when the presence of a mismatch at or within a few nucleotides upstream of

DNA POLYMERASE ε   113

the primer terminus in the polymerase active site slows further polymerization. This increases the time available for the primer terminus to fray, generating a single-stranded primer that can move to the exonuclease active site for excision of the nascent error. The exonuclease active site is thought to contain two divalent metal ions coordinated by the carboxylate residues in Exo motifs I and III. These metals catalyze an in-line displacement reaction similar to that described above for polymerization, thereby liberating a dNMP from the primer terminus. The exonuclease activity of Pol ε is robust, such that it substantially increases replication fidelity (see below), and can also excise correctly paired bases. The latter may occur during bypass of lesions in DNA (50) and possibly during DNA mismatch repair, where genetic studies have implicated the exonuclease activity of Pol ε in excision of the nascent strand containing the mismatch (51).

C. DNA Binding

An additional biochemical property of Pol ε that likely has a direct bearing on *in vivo* function is its ability to bind dsDNA through its small subunits (52) (Fig. 3 and see below). Pol ε can also bind ssDNA, the presence of which actually inhibits Pol ε activity in the absence of RPA (52). While ssDNA binding is mediated through the N-terminal catalytic half, the four-subunit holoenzyme dissociates from primer-template DNA almost two orders of magnitude faster in the presence of ssDNA than does the catalytic half (53). This ssDNA binding is proposed to play an integral role in the Pol ε checkpoint function.

D. Fidelity

Initial measurements of the fidelity of DNA synthesis by Pol ε were made with enzyme purified from calf thymus tissue at a time when Pol ε was still referred to as Pol δII (54, 55). More recent measurements have been performed with recombinant yeast Pol ε holoenzyme, as well as with a mutant holoenzyme that lacks 3′ exonuclease activity due to the alanine replacements for the catalytic carboxylates in the exonuclease active site (56, 57). These studies all reveal that, as expected for a polymerase that has a major role in DNA replication, mammalian and yeast Pol ε synthesize DNA with high fidelity. This high fidelity sharply contrasts with the much lower fidelity of other polymerases, for example, the translesion synthesis enzyme Pol η and the DNA repair polymerase yeast Pol IV (Table III). The major contribution to the fidelity of DNA synthesis by Pol ε is the high nucleotide selectivity of the polymerase itself, as illustrated by the low rates at which exonuclease-deficient Pol ε generates the two most common polymerization errors, single base substitution and single base deletion errors (Table III). Such high fidelity is

TABLE III
Fidelity of Yeast DNA Pol ε Compared with other Yeast DNA Polymerases

Enzyme	Family	Base substitution error rate (x10^{-5})	Single-base deletion error rate (x10^{-5})	References
Pol ε (exo−)	B	24	5.6	57
Pol ε (exo+)	B	≤2	≤0.05	57
Pol δ (exo−)	B	13	5.7	207
Pol δ (exo+)	B	≤1.3	1.3	208
Pol α	B	9.6	3.1	209
Pol IV	X	320	360	210
Pol η	Y	950	93	211

thought to partly depend on rigorous selection for correct Watson-Crick base pairing geometry in the binding pocket for the nascent base pair [see (58–63) for recent reviews on fidelity mechanisms].

An additional contribution to Pol ε fidelity comes from proofreading by the 3′ exonuclease. This is revealed by the lower error rates for the wild-type holoenzyme as compared to its exonuclease-deficient derivative (Table III). The contribution of proofreading to fidelity varies depending on the composition of the mismatch and the local sequence environment. Such variations are expected because the degree to which different Pol ε errors slow polymerization and/or sensitize the helix to fraying can vary. On average, proofreading improves Pol ε fidelity about 10- to 100-fold, consistent with the mutator effects seen in yeast when the exonuclease activity of Pol ε is inactivated (57, 64, 65).

Like all polymerases, Pol ε ultimately does not generate all types of errors at equal rates, but rather has a distinctive error specificity (57). Two features of Pol ε error specificity are particularly interesting in light of its proposed biological roles in DNA replication. One is that Pol ε is among the most accurate of DNA polymerases for single base deletion/insertion errors. Because indels are typically generated more frequently within repetitive sequences (66), this property may be relevant to the proposal that Pol ε has a particularly important role in replicating heterochromatic DNA, which is enriched in repetitive sequences (67). Another is that a mutant derivative of Pol ε has a unique base substitution error specificity that has been useful for inferring its role in replication of the leading strand template (see below).

V. Pol ε in DNA Replication

In order for eukaryotic cells to divide and pass along their genetic complement to each daughter cell, the entire genome must be replicated accurately once, and only once, per division. This is an ordered process whereby origins of replication, ranging from well-defined 125 bp ARS sequences in *Saccharomyces cerevisiae* to relatively poorly understood zones of replication of many kb in mammalian cells, are licensed by the binding of pre-RC components, including the ORC complex, during the G1 phase of the cell cycle (68, 69). These origins are then activated for replication during S phase by the binding and phosphorylation of a number of factors including Cdc45, Sld2, Sld3, and Dpb11 (27) that enable the replicative helicase, likely the CMG complex consisting of Cdc45/Mcm2–7/GINS, to unwind the DNA in a bidirectional manner in order to establish replication forks (32, 70). Once the origins have been properly licensed and activated, replication can proceed. Because the two DNA strands are antiparallel and DNA polymerases synthesize DNA in only one direction (5′ to 3′ on each nascent strand), replication is inherently asymmetric, with synthesis thought to be highly processive on the leading strand and discontinuous on the lagging strand (71). Because no DNA polymerase can conduct *de novo* DNA synthesis, a two-subunit primase that is tightly associated with Pol α initiates replication by synthesizing short RNA chains. The primase initiates synthesis both at the origins of replication and at the beginning of every Okazaki fragment during lagging strand replication. These primers are then extended by Pol α, which synthesizes DNA chains of up to 20 nucleotides. This synthesis is followed by a switch to Pol δ and/or Pol ε, which then perform the bulk of chain elongation.

A. Evidence That Pol ε Is a Major Replicative Polymerase

That yeast Pol ε is involved in replication is indicated by a number of studies. Early evidence came from studies revealing phenotypic similarities between the initial pol2 yeast mutants, primarily disruptions of the newly cloned ORF (72), and those from mutants of pol1 and pol3, as well as other mutants known to be essential for replication (73, 74). Later screens for single amino acid residue changes leading to temperature-sensitive alleles identified mutations in two different regions of the catalytic subunit, the polymerase active site and the noncatalytic C-terminus (75, 76). Both types of alleles had similar phenotypes but later studies also indicated a role for Pol ε in checkpoint activation [(77) and see below]. At the restrictive temperature, these temperature-sensitive mutants ceased DNA synthesis, accumulated subchromosomal size DNA fragments, and arrested with unreplicated chromosomes, all similar to Pol α and Pol δ mutants, indicating that, in addition to an essential role for Pol δ in replication, faithful chromosomal replication was also Pol ε-dependent (75, 76). Peak transcript expression during the cell cycle was also coincident with Pol α (75) and subsequent microarray analysis found that mRNA for catalytic and noncatalytic subunits of the replicative DNA polymerases peak just prior to S phase (78). That

Pol ε participates in replication is also indicated by the fact that it is loaded onto DNA at origins of replication. Pol ε in yeast is found associated with replication origins prior to Pol α and along with Dpb11 (25), which is essential for chromosomal replication (79, 80).

Observations derived from inactivation of its intrinsic 3′–5′ exonuclease proofreading activity provide further support for a role of Pol ε in DNA replication. When the exonuclease activity of Pol ε was first identified and inactivated (64), the resulting yeast *pol2-4* mutant had an elevated mutation rate that was further increased by inactivating DNA mismatch repair, indicating that the mutations observed in the *pol2-4* strain are indeed replication errors (51, 57, 64). Additionally, depletion of Pol ε from a Xenopus extract results in a chromosomal replication defect, and this defect is restored upon addition of purified Xenopus Pol ε (81). Further evidence that Pol ε is involved in replication came from immunohistochemical studies showing Pol ε colocalization with actively replicating foci of DNA in normal human fibroblasts (67). This colocalization was observed late in S phase, leading to the suggestion that Pol ε is particularly important for replicating heterochromatin.

B. Division of Labor at the Replication Fork

A two-polymerase model of replication fork progression derives from extensive studies of the *E. coli* replication fork that consists of two molecules of DNA polymerase III coordinately synthesizing DNA, with one operating on the continuous leading strand and the other operating on the discontinuous lagging strand [reviewed in (82)]. Coordination of these two polymerases at the fork is mediated by a multisubunit complex with the τ subunit at the functional center. Following the identification of Pol ε and Pol δ as eukaryotic replicative polymerases, three different models have been put forth regarding the division of labor between Pol ε and Pol δ at the fork. One suggests that Pol δ is primarily responsible for replicating the leading strand and that Pol ε synthesizes the lagging strand. This model arose from early biochemical studies of the two polymerases [(83, 84) and reviewed in (82)]. Based on substantial evidence now suggesting an important role for Pol δ in lagging-strand synthesis (85–87), this model is currently disfavored.

A second model posits that Pol δ is responsible for the majority of synthesis on both the leading and lagging strands. This model is supported by the fact that SV40 origin-dependent replication of double-stranded DNA by primate cells requires Pols α and δ, but not Pol ε (88, 89). A caveat here is that, unlike chromosomal replication, replication from the SV40 origin relies on SV40 T antigen (Tag) for initiation and helicase activities (90). Moreover, SV40 is a polyomavirus and its replication is not subject to the same cell cycle controls as is normal, chromosomal DNA (91). The latter may be particularly relevant, given the proposed role of Pol ε in checkpoint functions (see below). That Pol ε is dispensable for *in vitro* SV40 replication is belied by evidence that Pol ε can be cross-linked to nascent chromosomal DNA in mammalian cells (89), suggesting that, as in yeast cells (75), mammalian Pol ε is involved in chromosomal replication.

Additional support for the model suggesting that Pol δ is responsible for the majority of synthesis on both the leading and lagging strands comes from experiments in budding and fission yeast using Pol2 mutants lacking the open reading frame for the polymerase catalytic activity (92–94). While deleting the entire Pol2 gene is lethal, deleting only the N-terminus encoding the polymerase activity is not. This indicates that Pol ε activity per se is not absolutely required for chromosomal replication and that the essential function of the Pol2 gene is contained within its C-terminal residues. Nonetheless, cells lacking the N-terminus of *Pol2* encoding Pol ε activity are not completely healthy and they have a prolonged S phase. In contrast to the viability of an in frame deletion of the N-terminus, mutation of two of the catalytic aspartic acid residues of yeast Pol ε to alanines is lethal (92). This is likely to encode a dominant negative polymerase that binds to the replication fork and inhibits replication.

A third model (72) posits that Pol ε is the primary leading strand DNA polymerase, whereas Pol δ is primarily responsible for lagging-strand synthesis. This model is supported by the fact that Pol ε is intrinsically more processive than Pol δ and therefore is better suited for continuous leading strand replication (16). In addition, Pol δ is clearly implicated in lagging strand replication because its 3′ exonuclease cooperates with FEN1 to generate ligatable nicks at the 5′ ends of Okazaki fragments (86) and because it participates in completing the replication of telomeres (95), a lagging strand replication process. Genetic evidence from yeast indicating that the exonuclease activities of Pol ε and Pol δ proofread replication errors on opposite strands during replication (65, 96) leads to the further inference that Pol ε participates in leading strand replication.

Strong additional evidence for Pol ε replicating the leading strand came from a combination of biochemical and genetic evidence in yeast (97). An M664G mutant allele of yeast Pol ε with reduced fidelity was identified (Fig. 2C and D) that possessed a unique *in vitro* mutational signature, while otherwise retaining normal biochemical properties. When this allele was placed in a haploid yeast strain, an *in vivo* mutational signature was observed that was similar to that seen *in vitro* and was dependent on both the position and the orientation of a reporter gene relative to efficient replication origins. The mutational signature was consistent with participation of Pol ε in leading strand replication. This mutational signature pattern consistent with leading strand replication occurred on two different chromosomes and at both early- and late-firing origins.

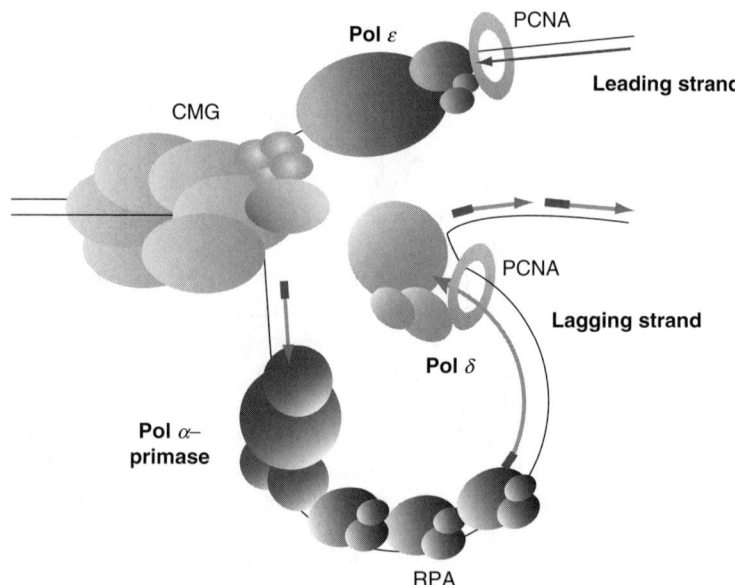

Fig. 5. Model of a eukaryotic replication fork. This model is based on the currently favored hypothesis that Pol ε is primarily responsible for leading strand synthesis, and Pol δ is primarily responsible for lagging strand synthesis. Pol α-primase cooperates with Pol δ to conduct lagging strand synthesis, with the initiating RNA primers shown as dark gray. RPA heterotrimers are shown as boxes. The CMG replicative helicase is shown as a heterohexameric MCM complex associated with the GINS complex and Cdc45.

Recent evidence using a similar approach has now firmly implicated Pol δ in replicating the lagging strand (98). A mutant allele of Pol δ made by changing a leucine that is homologous to the methionine changed in Pol ε also has reduced fidelity but otherwise retains wild-type catalytic properties (98). Yeast strains with this allele retain relatively normal replicative capacity and growth rate (98, 99), but have an elevated mutation rate and a unique mutational signature (98). As in the Pol ε study (Pursell), this Pol δ mutational signature is both position- and orientation-dependent and is consistent with a primary role for Pol δ in replicating the lagging strand. Taken together, the results of both studies imply that under normal circumstances, Pol ε is primarily responsible for replicating the leading strand and Pol δ is primarily responsible for replicating the lagging strand (Fig. 5).

It remains to be seen if this pattern is observed throughout the yeast genome, or changes with chromatin status, genomic location, or other parameters. It also remains to be seen if the roles of Pol ε in yeast differ in any way from its roles in higher organisms.

VI. The Role of Pol ε in Checkpoint Control

A. Cell Cycle Progression and Replication

The cell cycle is an orderly progression of events that allows the genome to be completely duplicated prior to the onset of mitosis. Unrepaired DNA damage sustained at any point in the cell cycle can disrupt this progression. DNA repair and checkpoint activation are the two means by which cells can avoid the potentially deleterious uncoupling of normal cell cycle progression. Checkpoints are cellular pathways that involve slowing or blocking cell cycle progression to allow time for repair of the lesion. There are four main DNA damage checkpoints: G1/S, intra-S, S/M, and G2/M [reviewed in (100)]. DNA damage incurred prior to the cell fully committing to entering S phase and subsequent DNA replication activates the G1/S arrest primarily by inhibiting origin activation. DNA damage incurred once the cell has fully committed to replication activates the intra-S checkpoint, which acts by inhibiting late-firing origins, thus prolonging S phase, and stabilizing stalled replication forks to allow for productive restart once the replication block is removed. The S/M checkpoint is one of the least well understood, preventing catastrophic mitosis from occurring in the absence of complete replication. The G2/M checkpoint arrests the cell cycle prior to cell division in the presence of DNA damage. Replication in general, and Pol ε in particular, is implicated primarily in the intra-S and S/M checkpoints.

Replication forks arising at replication origins proceed until termination on meeting a fork traveling in the opposite direction from an adjacent origin. There are a number of reasons why a fork may stop prior to termination, and so mechanisms exist to ensure fork progression, to stabilize stalled forks and prevent collapse and strand breaks, and to signal for a delay in cell division. The two most well-studied causes of fork stalling are low dNTP pools resulting from treatment with hydroxyurea (HU) and blocking lesions in DNA. In both cases, stalled forks are sensed by a PI(3)K-like kinase, ATR in humans, and Mec1 in yeast, which phosphorylates downstream components of the checkpoint pathway. The ultimate responses to both types of damage are to inhibit the initiation of late-firing origins, to stabilize the stalled forks, and to allow the cell sufficient time to restart the stalled fork prior to proceeding with cell division. Failure to achieve any of these outcomes can result in fork collapse followed by double-strand breaks (DSBs), hyperrecombination, genome rearrangement, and catastrophic attempts at cell division prior to replicating the complete genome.

B. Pol ε Linked to Cell Cycle Progression and Replication

Pol ε was initially implicated as a component of the S/M replication checkpoint because mutants in the extreme C-terminus of the yeast catalytic subunit are sensitive to HU and are unable to activate the RNR3 transcriptional response and prevent catastrophic mitosis as a consequence of HU treatment (77). These defects in RNR3 transcriptional activation and in Rad53 phosphorylation were subsequently shown to be specific to S phase (101), similar to what is seen in Mec1 mutants. Rad53 is a protein kinase and checkpoint transducer that, when phosphorylated, in turn phosphorylates targets directly involved in initiating checkpoint responses. Additionally, the human Pol ε C-terminal half interacts with Mdm2 (39) and Mdm2 in turn stimulates Pol ε activity in vitro (40). As mentioned above, Mdm2 is an E3 ubiquitin ligase for p53 (38) and is involved in a complex regulatory loop resulting in activation or inhibition of p53 activity depending on Mdm2 interaction with a network of proteins (37). Thus, Pol ε is linked to the regulation of cell cycle progression in human cells, though its involvement is less well understood than in the model yeast system.

Yeast Pol ε involvement in the intra-S checkpoint was first proposed based on its genetic association with Sgs1. Sgs1 is the lone budding yeast RecQ family helicase and is nonessential. Deletion mutants are sensitive to MMS-induced replication blocks and allow replication fork progression in the presence of HU. Intriguingly, sgs1 mutants also show elevated levels of rDNA recombination (102). Humans have five RecQ family members, including BLM and WRN, and mutants in these helicases can cause genomic instability and cancer. C-terminal Pol ε mutants are epistatic with the HU-induced Δsgs1 defects, and mutants in both the helicase and polymerase lose viability when removed from HU, pointing to an intra-S checkpoint defect (103). Pol ε associates with stalled forks at early- and late-firing origins in the presence of HU and this stable association is dependent on both helicase-active Sgs1 (104) and Rad51 (105) and is also a component of a paused replisome, along with Pol α and the CMG replicative helicase (106). However, Pol ε no longer associates with early-firing origins in the absence of a functional Mec1, which likely acts directly at the fork to aid in stabilizing stalled forks (107). That Pol ε associates with late origins inappropriately firing in mec1 mutants in the presence of HU indicates that the assembly of Pol ε at replication forks remains intact, but it is the stabilization defect that causes loss of Pol ε (104).

C. Dpb2 and the Cell Cycle

Dpb2 phosphorylation is another means of regulating Pol ε function at the replication fork. It is phosphorylated in S phase by Cdc28 (17), the yeast CDK1 homologue that phosphorylates a number of replication proteins in order to help control replication initiation and prevent rereplication (108, 109). Dpb2 dephosphorylation is linked to proper S-phase progression and exit from mitosis. Cdc14 is a phosphatase whose targets generally include Cdc28 phosphorylation substrates, of which Dpb2 is an S-phase-specific substrate (17). A cdc14 mutant that mislocalizes away from its normal, nucleolar site deregulates Cdc14 activity, allowing Dpb2 and Sld2 to be prematurely dephosphorylated and causing defective S-phase progression (110). While Dpb2 phosphorylation is not essential for yeast growth, it was hypothesized that Dpb2 phosphorylation can regulate holoenzyme formation and DNA-binding, much like other B family accessory subunit phosphorylation events (111).

VII. Pol ε Involvement in Regulating Chromatin States

A. Pol ε Influence on Gene Silencing

Heterochromatic DNA is transcriptionally silent, and silenced regions are able to epigenetically influence neighboring regions through the use of Sir proteins, including the NAD^+-deacetylase Sir2. Silencing involves establishment, maintenance, and inheritance of the silent state (112). Budding yeast establish, maintain, and propagate three different silent regions through differential involvement of the Sir proteins: rDNA, silent mating-type loci, and telomeres. The rDNA exists in an array of up to 200 repeats, each of which contains an origin of replication along with the heavily transcribed rRNA genes. More than half of these repeats may be silenced in a SIR2-dependent manner. The silent mating-type loci are located 16 and 23 kb from the left and right telomere, respectively, of yeast chromosome III. Silencing at these loci requires the full complement of Sir proteins, Sir1–4. Silencing at telomeric regions of the yeast genome requires only Sir2–4. Pol ε is implicated in silencing at all three types of silenced loci in yeast, indicating a fundamental role in the propagation of transcriptional states during replication. The role of Pol ε in determining chromatin status may involve the stability of the Pol ε complex, rather than a biochemical activity per se, because this function is affected by truncations at the C-terminus of Pol2, which lacks catalytic activity and is involved in forming a complex with the smaller subunits (14, 113), and by mutations in Dpb3, which contains no identified catalytic activity and is not essential for viability (21). DPB3 was identified in a screen for deletion mutants that lost the ability to silence a marker located in the rDNA (114), indicating a role for Pol ε in positively regulating gene silencing. Mutants in Pol α and RFC were also identified; however, these mutants resulted in shortened telomeres, whereas telomere length was unaffected in the Δdpb3 mutant (114). Ehrenhofer-Murray et al. (115) established an allele of the HMRa locus that was deficient in silencing, and the strain could thus not mate with MATα strains. They then screened a large set of replication mutants for those able to restore silencing and allow mating. A Pol ε C-terminal truncation mutant, pol2–12, restored silencing along with mutants of RFC, PCNA, and CDC45. This allele of Pol ε is temperature-sensitive and disrupts the four-subunit complex. Alleles of Pol α and δ were both tested and observed to have no effect on silencing. It should be noted that in one case silencing is lost by disrupting the pol ε holoenzyme, whereas in the other case, silencing is restored by disrupting the pol ε holoenzyme. This may reflect different roles for pol ε involved in DNA synthesis at different chromosomal locations with different silencing mechanisms.

B. Pol ε and Telomeres

Telomeres are protein-bound repetitive sequence elements that cap the ends of eukaryotic linear chromosomes (116). The normal replication machinery has the capacity to gradually shorten chromosomes due to the "end-replication" problem [reviewed in (117)], ultimately resulting in senescence or apoptosis (118). This shortening is prevented by telomerase, a reverse-transcriptase-like enzyme that adds repetitive sequence elements to chromosomal ends to maintain telomere length (119). Subtelomeric regions in both yeast and mammals are subject to epigenetic silencing [reviewed in (120)], specifically through a heterochromatic spreading mechanism in yeast (121). These subtelomeric regions in budding yeast contain dormant or very late-firing origins of replication that are suppressed, in part, by this silencing. When telomere length is artificially reduced, either through inactivation of Terc, the RNA component of telomerase, in mouse (122) or by recombination in yeast (123), the heterochromatic state is lost and yeast ARS sequences promote origin firing early in S phase. In telomerase-proficient yeast, this early origin firing in turn promotes telomere lengthening (123). This coupling of replication timing and telomere length was demonstrated through the use of wild-type Pol ε in ChIP experiments. The telomeres in pol2-16 mutants after passaging are shorter than those in wild-type yeast, and these cells senesce much earlier than wild-type cells (22). The role of Pol ε in establishing and promoting the epigenetic silencing of regions of DNA during replication might possibly be linked to maintaining telomere length.

C. Telomere-Proximal Effect

Pol ε also has a role in telomere-proximal silencing, as shown in colorimetric and viability assays using reporter genes immediately adjacent to the telomere in which both pol2 C-terminal truncation and Δdpb3 mutants are defective for silencing telomere-proximal reporter genes. Iida et al. (124) used a clever single-cell assay to further determine that Pol ε normally functions to silence this region in opposition to the yeast CHRomatin Accessibility Complex (CHRAC), which operates by activating this region. This is striking given that CHRAC and Pol ε share a subunit in all organisms studied (124, 125) (see below). Disruption of the shared subunit shows equal amounts of silencing and activation in this assay, clearly demonstrating an independent role in two complexes for the shared subunit.

VIII. Pol ε Relationship with Chromatin Remodeling Complexes

A. Chromatin Remodeling

In order to package the entire genome into the nucleus and to control gene expression, DNA in eukaryotic cells is organized into chromatin (126). At the nucleic acid level, 146 bp of DNA is wrapped in two turns around a histone octamer composed of a core of two H3/H4 histone dimers flanked by two pairs of H2A/H2B histone dimers forming the nucleosome. These nucleosomes are separated by histone-free linker DNA and are arrayed along the DNA, which coils into 30 nm helices known as chromatin fibers which in turn are packed upon each other to form the higher order structure known as chromatin. The net result of this packaging is that the DNA is rendered less accessible to DNA-binding proteins, including DNA polymerases involved in replication and DNA repair. Cells have a large complement of multisubunit protein complexes called chromatin remodeling complexes that are able to reorganize the histone octamers to grant enzymes like transcription factors and components of the replication machinery access to the DNA (127). These are distinct from histone-modifying enzymes as they do not covalently alter the histone proteins. Nucleosomes are stable structures due in large part to a high number of DNA–protein contacts either directly via the histone proteins or mediated through water. Each of these chromatin remodeling complexes couples the energy released through ATP hydrolysis with a rearrangement of the nucleosomal array.

All chromatin remodeling complexes contain a catalytic subunit that is a member of the DEAD/H ATPase Swi2/Snf2 family [see (128) and references contained therein]. These members are organized into four groups based on their homology to canonical family members containing a unique complement of protein domains: the SWI/SNF, CHD, INO80, and ISWI groups. Along with the ATPase domain, the ISWI class contains a SANT domain that has homology to a class of transcription factor DNA-binding domains and, importantly, may be involved in coupling interaction with histone tails to histone modification.

Drosophila has one member of the ISWI family that forms three different complexes *in vitro*, called NURF, ACF, and CHRAC, based on their associated subunits and biochemical activities. CHRAC from *Drosophila* is composed of four subunits: ISWI, the ATPase-containing subunit; Acf1, which contains, among other characterized and uncharacterized domains, a bromodomain involved in binding histone acetyl-lysine residues and a WAC domain that targets proteins to heterochromatin in mouse (129); and two small HFM-containing proteins, p14 and p16 (130). CHRAC was initially purified from *Drosophila* extracts as a complex that allowed restriction enzymes to digest their target sequence in chromatin (131). An additional activity of CHRAC is its ability to organize ordered arrays of nucleosomes in the presence of ATP from disorganized nucleosomes formed in the absence of ATP (131).

B. Dpb4/p17: The Shared Subunit

The first evidence that DNA pol ε played some role in these chromatin remodeling complexes came with the purification of HuCHRAC, the human homologue of the *Drosophila* CHRAC (125). Two homologues of the DmCHRAC HFM-containing subunits, called p15 and p17 in humans, were identified and cloned. At approximately the same time, the two HFM-containing subunits of human Pol ε, p12 and p17, were cloned (15) and it was independently determined that p17 from HuCHRAC was identical to p17 from human Pol ε.

Two yeast homologues of ISWI, ISW1 and ISW2, were identified in a chromatin remodeling complex (132) before the genes were cloned for any of the HFM subunits from either complex. It was later determined that ISW2 exists in an ATPase chromatin remodeling complex containing two HFM subunits: DPB4, a small subunit of yeast Pol ε and DLS1 (Dpb3-like subunit), a novel yeast protein with homology to the unique HFM subunits in CHRAC (124, 133). The shared subunit is conserved throughout evolution. Why two entirely separate enzymatic activities should have such a shared subunit remains an open question. The shared subunit could bridge an interaction between the two complexes, but this appears unlikely given that p12, the Pol ε-specific HFM protein, is not found in purified CHRAC in multiple species (134, 135), and given that hSNF2H is not found in purified human Pol ε (Pursell and Linn, unpublished observations). Based on sequence homology to the CCAAT-binding factor subunits, CBF-A and -C, it was originally proposed that the human p12/p17 heterodimer might form an interaction surface, much like CBF-A/-C enables binding of CBF-B and subsequent transactivation, thus allowing Pol ε to interact with other factors, possibly involved in influencing chromatin structure (15).

One possible explanation for the shared subunit is to coordinately regulate Pol ε and CHRAC. Both the CHRAC and Pol ε HFM subunits from humans (125) and from yeast (52) are able to bind dsDNA. This enables the four-subunit Pol ε holoenzyme to tightly bind dsDNA (136), unique among replicative DNA polymerases. The shared subunit in yeast, Dpb4, can be cross-linked to extranucleosomal DNA, but not nucleosomal (137). The HFM subunits may bind dsDNA, either in complex with the catalytic subunits or in their absence in a manner similar to the histone H2A/H2B heterodimer [see (138) and Fig. 6], to mark regions of DNA to be operated on by either Pol ε, likely to propagate a silenced state, or CHRAC, likely to propagate a derepressed nucleosomal state (115). The decision to target Pol ε or CHRAC may be made at the epigenetic level through histone modifications. Histones were found to coimmunoprecipitate with both yCHRAC and Pol ε through Dpb4, with a Dpb4-associated histone H4 hypoacetylation pattern different from bulk genomic H4 (135). A genome-wide

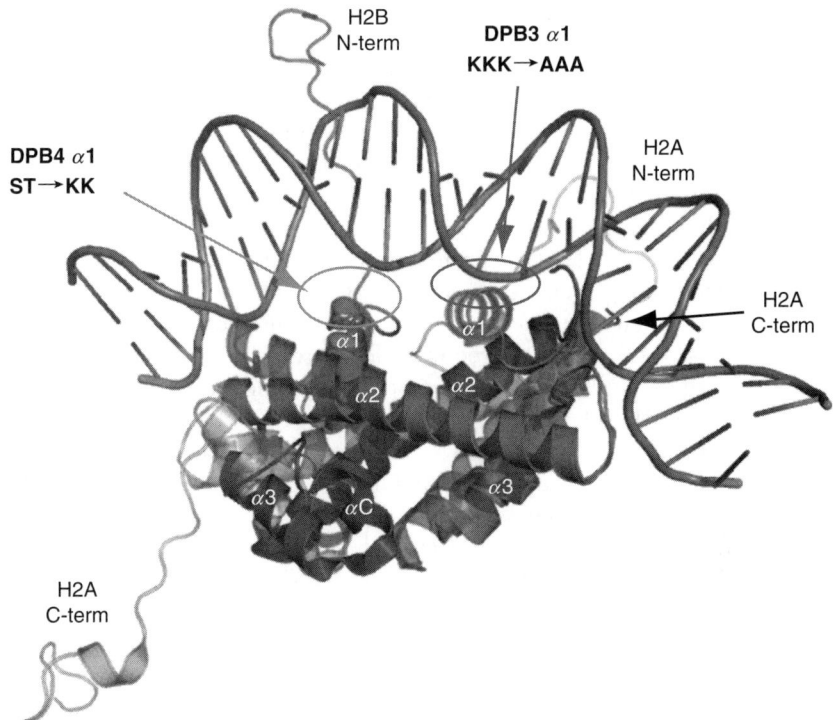

FIG. 6. Model of histone-fold subunits-DNA interaction. Shown is a structural alignment of the heterodimer DmCHRAC-14/DmCHRAC-16 from (138) with the DNA-bound heterodimer of histone H2A-H2B from (213). DmCHRAC-14 is the same as DmPol ε-p17. The dark gray circle indicates where the KKK→AAA triple mutant of Dpb3 that results in loss of DNA-binding and telomeric silencing (52) maps to the structure. The light gray circle represents where the S/T→KK mutant in Dpb4 that partially suppresses the KKK→AAA Dpb3 mutant maps to in the structure. The α1, α2, α3, and αC helices on each HFM subunit are indicated. N- and C-terminal ends of histones are indicated.

microarray of DNA bound to Dpb4–histone complexes showed enrichment near telomeres and boundary regions separating the silenced HM loci from the surrounding euchromatin, consistent with the silencing data observed by both Iida et al. and Ehrenhofer-Murray et al. (115). It may be that Pol ε and CHRAC are directed to silenced and nonsilenced regions of DNA, respectively, in order to ensure that the chromatin status is properly maintained during replication and then propagated to the progeny. Switching states, if necessary, would be carried out by other factors, possibly HATs or HDACs, whose actions would dictate the switching of the targeting signals for Pol ε and CHRAC.

C. Centromeres and Sister Chromatid Cohesion

Evidence from metazoan cells indicates that Pol ε and CHRAC may be targeted to regions of silenced DNA as well. Two components of CHRAC, Acf1 and Snf2h, colocalize with HP1β, which aids in the formation of pericentromeric heterochromatin by binding K9-methylated histone H3, and BrdU, during S phase (*139*). Additionally, reduction of both Acf1 and Snf2h causes a delay in S phase specific to elongation (*139*). Pericentromeric heterochromatin is a region of specialized heterochromatin surrounding the centromeres that is important for spindle attachment during mitosis. While budding yeast lack true pericentromeric heterochromatin, Pol ε physically interacts with and is stimulated by Trf4 (*140*), which is required for sister chromatid cohesion and completion of S phase (*141*). Trf4, and the redundant Trf5, was initially believed to be a novel DNA polymerase involved in this process, but was subsequently shown to have poly(A) polymerase activity (*142*). Snf2h has been found in a complex separate from CHRAC that contains hRad21 and is involved in loading cohesin onto DNA (*143*). Pol ε from nontransformed human fibroblasts was also found to colocalize with both PCNA and BrdU at sites of active DNA replication late in S phase, during which time heterochromatin is replicated in mammalian cells (*67*). Taken together, these data suggest that the shared p17 subunit may target both CHRAC and Pol ε to repetitive DNA, to facilitate replication of heterochromatin to ensure that the epigenetic state of heterochromatic DNA is maintained and propagated. The differences between yeast and human CHRAC-pol ε relationships may also suggest a difference in the balance of CHRAC and Pol ε activities between budding yeast, with its SIR2-dependent silencing and absence of true heterochromatin, and metazoan and fission yeast, with HP1-dependent heterochromatic DNA.

CHRAC may also contribute, directly or indirectly, to Pol ε during replication at the origin-unwinding stage. The initiation of SV40 origin-containing DNA into a nucleosomal template using cell-free extracts normally inhibits *in vitro* replication (*144*). Purified CHRAC was able to rearrange the nucleosomes on an SV40 template and relieve this inhibition of replication (*145*). While this system relies on Tag both for origin unwinding and for helicase activity, and while Pol ε is dispensable for this reaction (*88*), nevertheless it is intriguing that a nucleosome remodeling complex is able to facilitate replication origin firing and also shares a subunit with a DNA polymerase known to bind replication origins prior to origin firing (*25*).

IX. The Roles of Pol ε in Excision Repair of DNA Damage

Much of the early work implicating Pol ε in DNA repair synthesis was based in large part on studies making use of the DNA polymerase inhibitor, aphidicolin (*5, 146*). This was done primarily to distinguish aphidicolin-sensitive DNA synthesis from aphidicolin-insensitive DNA polymerases like Pol β. Unfortunately, aphidicolin inhibits both Pol δ and Pol ε, making distinctions between the two using this strategy difficult. Even today, it remains difficult to distinguish the involvement of the two polymerases in DNA repair synthesis, though there is much evidence to suggest that Pol ε does play a role in filling gaps generated during several types of excision repair.

A. Base Excision Repair

The cell is constantly exposed to a wide variety of agents that can chemically modify the DNA, posing a challenge to the normal progression of replication forks and transcription complexes. These insults include reactive oxygen species created within the cell and external sources like ultraviolet irradiation and chemicals that introduce DNA base adducts. In order to deal with these lesions and prevent mutagenesis and genome instability, multiple pathways exist to identify, excise, and correct these lesions (*147*). DNA base modifications that do not generally distort DNA helix geometry are primarily repaired by base excision repair (*148*). Multiple BER subpathways exist to repair the various nondistorting lesions [reviewed in (*149*)].

In mammals, Pol β has a major role in BER, being essential for the repair synthesis of a single nucleotide to replace the damaged nucleotide (*150*). However, Pols δ and ε have both been implicated in the PCNA-dependent BER, which involves synthesis of patch sizes greater than one nucleotide and likely serves as a backup to Pol β-dependent BER (*151, 152*). Pol ε was implicated in BER in yeast when extracts made from temperature-sensitive yeast Pol ε mutants were deficient in repairing three different lesions, a plasmid containing dUMP, an OsO_4-treated plasmid, and a UV-irradiated plasmid (*153*). Purified yeast Pol ε was able to restore DNA repair, thus implicating Pol ε in BER.

B. Nucleotide Excision Repair

Many lesions that strongly distort the DNA helix are repaired by nucleotide excision repair (NER). NER requires synthesis of about 30 nucleotides to fill the gap generated by excision of the lesion from the DNA (*147*). Early evidence that Pol ε plays a role in NER-dependent DNA synthesis came from studies in human cells that preceded the knowledge that pols δ and ε were distinct enzymes. Postconfluent, normal diploid human fibroblasts, which do not exhibit semiconservative (replicative) DNA synthesis, undergo conservative (repair) DNA synthesis after UV irradiation (*5*). When cells were permeabilized after UV treatment, a repair factor was lost along with the repair synthesis, which was complemented by the addition of HeLa extracts. Extensive fractionation and biochemical characterization revealed this soluble repair factor to be DNA pol ε (*6, 7*) (described above and below).

Complete reconstitution of mammalian NER was initially performed using calf thymus Pol ε as the gap-filling DNA polymerase (154). A subsequent study determined that Pol ε was better able than Pol δ to perform gap-filling synthesis that ultimately generated DNA ligase I-ligatable products (155). Unlike with pols α and δ, DNA synthesis efficiency by Pol ε increased as gap size decreased (156), another desirable characteristic for a gap-filling polymerase. Another study pointing to Pol ε and Pol δ redundancy in excision repair showed that only mutants in both polymerases showed an accumulation of single-strand DNA breaks in response to UV irradiation (157). Whereas extracts from temperature-sensitive Pol δ or Pol ε mutants show reduced NER activity, Pol α is not involved in yeast NER (158). At least the DNA polymerase component of Pol δ is able to completely compensate for the absence of Pol ε during budding yeast NER *in vivo*, as pol2–16 mutants lacking the catalytic domain are not sensitive to UV (94). However, extracts prepared from pol2–16 cells do show reduced NER synthesis *in vitro*, indicating that Pol ε does play a substantial role in NER (158).

A recent study combined whole-cell repair assays with ChIP and immunofluorescence microscopy in both proliferating and quiescent cells to identify XRCC1 and DNA ligase IIIα as novel core components of NER in humans (159). DNA ligase I is essential for chromosomal replication, plays an important role in NER (160), and has also been implicated in long-patch BER (113) along with the scaffold protein XRCC1 (161). For these reasons, DNA ligase I was used in the excision repair reconstitution assays that implicated both Pol δ and Pol ε, without providing a clear distinction between the two polymerases (154, 155). Interestingly, Pol δ was found to be present at sites of NER involving XRCC1-DNA ligase IIIα in both actively cycling and quiescent cells, whereas Pol ε was only found with DNA ligase I at sites of NER and only in cycling cells (159). This is curious given results with the original repair assays that identified Pol ε in confluent cells (5), which may indicate that the repair role of Pol ε in postmitotic cells relies on an as yet uncharacterized DNA ligase, because ligase I is not present in nonreplicating cells (159).

Interstrand cross-links (ICLs) are extremely toxic to cells and particularly relevant to human health as many anticancer drugs induce ICLs. In order to deal with this high toxicity, a number of repair pathways exist to accurately repair these lesions [reviewed in (162)]. The predominant ICL repair pathways involve the NER, recombinational repair, and translesion DNA synthesis systems; which pathway is used appears to vary considerably with both the cell cycle and type of lesion (163). While overall ICL repair is only poorly understood in eukaryotes, the use of NER and recombinational repair, both of which involve Pol ε to some degree, raises the possibility that Pol ε is involved in ICL repair.

C. Does Pol ε Function in Mismatch Repair?

Biochemical evidence has implicated Pol δ in MMR (*164–166*). Nonetheless, those studies noted that a possible role for Pol ε should not be excluded. In support of this possibility is one study that provides genetic data consistent with a role for the 3′ exonuclease activity of Pol ε in the excision step of MMR. In this study, exonuclease-deficient Pol ε mutants, which alone show very little effect on frameshift fidelity in long (≥ 8 nt) homonucleotide runs, showed a synergistic mutator effect in long homonucleotide runs with inactivation of exo1, which is known to be involved in mismatch repair (*51*). Based on the even more severe defect seen with exonuclease-deficient Pol δ mutants in combination with exo1 inactivation, the authors proposed that Exo1 competes with the 3′ exonuclease activities of both pols ε and δ in the excision step of mismatch repair. Biochemical studies with human cell extracts showed that the excision step of MMR is aphidicolin-insensitive, arguing against a direct role for either Pol ε or Pol δ. However, the authors note that the resistance varies somewhat with extract preparation and that their assay does not require the presence of a replication fork. If the 3′ excision step is dependent upon a replication fork, it would not be observed in their assay.

X. Pol ε in Recombination

A. Double-Strand Break Repair

Double-strand breaks (DSB) in DNA are among the most toxic of lesions; even one DSB can be lethal (*167*). DSBs arise through a number of different endogenous and exogenous sources including γ-irradiation, mating type switching, immunoglobulin rearrangement, crossing over during meiosis, and stalled replication forks [reviewed in (*147*)]. Two major types of repair abrogate the deleterious effects of DSBs, nonhomologous end-joining, and homologous recombination (*168, 169*). The latter can be subdivided into gene conversion and break-induced replication (BIR), both of which involve a number of replication fork proteins, likely including Pol ε.

B. Gene Conversion

DSBs repaired by gene conversion likely do so via a modified replication fork consisting of leading and lagging strand polymerases. Temperature-sensitive mutants of pols α, δ, and ε are each defective in gene conversion at the MAT locus in budding yeast upon generation of an HO-induced DSB (*170*). While an early study in asynchronous yeast cells reported that Pol α-primase mutants had the most serious defect, later examination of this mutant in synchronized cells revealed that Pol α is unnecessary, questioning the importance of Okazaki fragment processing in recombinational repair through gene conversion (*171*). A Pol ε mutant gave a more severe defect in gene conversion than did a Pol δ mutant, consistent with the idea that leading strand DNA synthesis might play a more prominent role in gene conversion, or that Pol ε may be able to more effectively substitute for Pol δ lagging strand synthesis than can Pol δ substitute for Pol ε. The gene conversion process occurs with fast kinetics and involves little, if any, checkpoint activation.

C. Break-Induced Replication

The break-induced replication (BIR) pathway may operate at replication forks that stall at a DNA lesion. Strand invasion mediated by Rad51 allows the reestablishment of a unidirectional replication fork that can proceed to the end of the chromosome (*172*). A Rad51-independent mode of BIR exists (*173*), which is dependent on Rad50 and Rad59, although no direct involvement of a replicative DNA polymerase has been reported for this pathway (*174*). However, in an assay designed to investigate the role of the replicative polymerases in Rad51-dependent BIR, yeast pols α and δ were found to be required for the initiation step (*175*). Pol ε was found to be dispensable for the initiation step but required for the elongation step and formation of long products of up to 30 kb. The Pol ε mutant used to probe this BIR was a C-terminal mutant that disrupts the holoenzyme. This region contains the portion of Pol ε implicated in damage sensing and checkpoint activation, indicating that perhaps these activities are the required activities of Pol ε in BIR.

D. Mammalian Recombination Complex

That Pol ε may participate in recombinational repair in metazoan cells is suggested by its association with recombination-like activities. Pol ε from bovine tissue copurifies with the multiprotein RC-1 complex that can transfer a homologous sequence from one closed, circular DNA template to another *in trans* (*176–179*). Also copurifying with Pol ε were DNA ligase III and a structure-specific endonuclease that may help to resolve Holliday junctions arising during recombination. During mouse testis development, Pol ε mRNA and protein levels persist through the late pachytene stage of meiosis, when Rad51 levels peak and the meiotic chromosomes are fully synapsed and recombination is completed (*180*). Human Pol ε may thus function as a DNA polymerase in a recombinational repair complex in a similar fashion to DNA synthesis during BIR in yeast, with a primary role in the elongation/completion phase of recombination repair.

E. rDNA Recombination

Pol ε is involved in silencing at the yeast rDNA loci (*114*), likely acting during replication to propagate the silenced state like it does at the boundary regions near the HM loci (*115*). The rDNA loci in yeast are recombinigenic and defects that reduce the replicative life span of yeast cause a hyperrecombination phenotype resulting in the accumulation of extrachromosomal rDNA minicircles, eventually causing the mother cell to die (*181*). This process of creating rDNA minicircles is promoted by Fob1, which stalls replication forks at Fob1-binding sites located within each rDNA repeat, and is repressed by Sir2 (and the Sir2-like Hst2), which silences rDNA through the formation of heterochromatin (*182*). rDNA in metazoans is heterochromatic and primarily replicates late in S phase (*183*), the same time that the replication-linked function of Pol ε is occurring (*67*). It is tempting to speculate that when replication forks are stalled to the point where a recombinational repair mechanism is invoked (e.g., a specific protein block, heavy transcriptional activity, a single-strand break at a replication fork), Pol ε is the primary elongation polymerase. In order to prevent premature mitosis, Pol ε is then able to signal through its C-terminal half the presence of the crossed-over replication intermediate, which is lethal if left unresolved. What makes this even more intriguing are observations with mutants lacking the Pol ε catalytic domain, which lose replicative capacity much earlier than do wild-type cells (*22*). A defect in Pol ε that causes reduced life span might thus be due to a defect in rDNA recombination. In addition, Pol ε physically interacts with Trf4, which is involved in sister chromatid cohesion (*141*). Trf4 is a component of a complex that binds the rDNA intergenic spacer region and helps regulate rDNA copy number (*184*), and Δtrf4 mutants are sensitive to MMS (*185*). This role is consistent with the major observations of Pol ε being essential for chromosomal replication, and that the essential portion in yeast is its C-terminal, subunit-interacting half. It is also consistent with Pol ε being important for Rad51-dependent BIR and the observation that Pol ε foci colocalize with replication foci only late in S phase. This may be a general mechanism that is observed primarily at the rDNA loci due to the naturally high degree of recombination occurring there and its regular assembly into nucleolar foci (*186*).

XI. *Schizosaccharomyces pombe* Pol ε

A. Catalytic Subunit

Schizosaccharomyces pombe Pol ε, or cdc20, was initially described as a mutant allele defective in the initiation of DNA replication (*187*) and with reduced sporulation efficiency (*188*). Thermosensitive mutants in cdc20, along with mutants of DNA polymerase δ and MCM4, undergo mitosis prior to completing replication (*189*), much like the *S. cerevisiae* C-terminal truncation mutants, linking Pol ε to the DNA damage checkpoint in fission yeast. When the cdc20 gene was cloned and found to encode Pol ε by virtue of sequence homology, disruption of the coding sequence was lethal, consistent with the idea that *S. pombe* Pol ε is essential for DNA replication (*190*). Haploids generated from two different mutants had reduced sporulation but showed no replication or other mitotic defects. Additionally, C-terminal mutants mimicking those C-terminal truncation mutants that are defective for the S-phase checkpoint arrest with the checkpoint intact, indicating that *S. pombe* Pol ε may differ from *S. cerevisiae* Pol ε in having little or no role in S-phase checkpoint integrity. However, when replication checkpoint-proficient cdc20 mutants are combined with a similarly replication checkpoint-proficient deletion of cid1, one of six *S. pombe* homologues of the budding yeast Trf4/Trf5 proteins involved in sister chromatid cohesion, the resulting cid1Δ cdc20 double mutant is defective for the replication checkpoint (*191*).

As seen in budding yeast, deletion of the catalytic portion of pombe Pol ε is not lethal but does prolong S phase (*93*). This supports the idea that Pol δ catalytic activity is able to partially compensate for Pol ε catalytic deficiency, and that the C-terminal half of the catalytic subunit contains the essential activity. This function likely involves the ability of the C-terminal half to interact with its own subunits and possibly other factors on DNA (*192*). The viability of this mutant depends on components of the DNA damage checkpoint, that is, the catalytic mutant is synthetically lethal when combined with deletion mutants of chk1, rad3 (ATR), and hus1 (9-1-1 clamp), as well as with cdc6 (pol δ). In contrast to observations in budding yeast (*94*), *S. pombe* catalytic deletion mutants are insensitive to HU and highly sensitive to MMS (*93*).

B. Accessory Subunits

As in budding yeast, the dpb2 subunit of *S. pombe* Pol ε is essential (*192*). In a system designed to downregulate dpb2 in the presence of thiamine, cells with reduced Dpb2 have delayed replication initiation and increased nuclear defects, including missegregated chromosomes and anucleate cells. Also as in budding yeast, dpb2 binds to ARS elements early in S phase and this binding depends on pre-RC formation, indicating a role in initiation or early elongation of replication. However, unlike budding yeast, *S. pombe* dpb3 is essential (*193*). Moreover, downregulation of dpb3 results in a cell cycle delay and an increase in multinucleate cells. This contrasts with *S. cerevisiae*, where Δdpb3 mutants are viable and the only observed phenotype is a moderate increase in mutation rate (*21*), a phenotype that has yet to be examined with *S. pombe* Pol ε mutants.

Dpb4 is not essential for viability, but unlike in *S. cerevisiae*, a dpb4 deletion mutant is synthetic lethal in combination with mutants in DNA replication initiation and DNA damage checkpoint signaling (*193*).

Rad4/Cut5 is a BRCT-repeat containing protein that is essential for replication initiation (*194*), likely through linking pre-RC formation to CDK signaling. It shares homology to the human TopBP1 and budding yeast DPB11, which has a role in checkpoint activation (*80*) as well as replication initiation (*79*) and physically associates with Pol ε (*25*). Although Rad4 binds chromatin in the absence of DNA damage, this association, as well as the stability of the Rad4 protein, depends on the presence of Pol ε (*194*), indicating that Rad4 and Pol ε may function together in replication initiation.

XII. Xenopus Pol ε

Understanding Pol ε function has been facilitated using the powerful cell-free replication system of Xenopus (*81*). Immunodepleting Pol ε from Xenopus extracts leads to a defect in bulk DNA synthesis (*81*) that is complemented by recombinant Xpol ε (*195*). This defect is more pronounced in extracts depleted of Pol δ, which also accumulated large amounts of ssDNA (*85*), consistent with a role for Pol δ in lagging strand synthesis. As in *S. cerevisiae* (*196*), Xenopus Pol ε binds to chromatin in a Cdc45-dependent manner (*197*). One major difference between the frog and yeast systems is in the division of labor involved in activating replication origins and stabilizing stalled replication forks. Whereas yeast separates these two processes, in part by maintaining the Sld2-Dpb11 origin activation pathway separate from the Sgs1 helicase, the Xenopus RecQ homologue, RecQ4, or xRTS, contains helicase and Sld2-like domains in a single polypeptide (*198*). This Sld2-like domain interacts with xCut5, the Xenopus Dpb11 homologue, though not with the replicative DNA polymerases. xWRN is another RecQ family member, in which mutations cause the premature aging disease Werner's syndrome (*199*). The WRN helicase associates with replication forks (*200*) but when mutated lacks the severe DNA synthesis defects observed in Bloom's syndrome-derived cells (*201*). A role was proposed for WRN in unwinding dsDNA byproducts generated during lagging strand synthesis (*202*). Consistent both with Pol ε carrying out leading strand synthesis and with Pol ε being involved in stalled replication fork signaling, xWRN/FFA-1 and Pol δ accumulate on chromatin after aphidicolin treatment, whereas the chromatin-bound level of Pol ε does not increase (*203*). Claspin, the Xenopus homologue of the mediator protein Mrc1, associates with replication forks during S phase (*204*) in a xCdc45-dependent manner. Claspin is also required for replication fork-stalled xChk1 activation by xAtr phosphorylation (205). Underscoring the specificity of Pol ε involvement in this checkpoint pathway, xClaspin physically interacts with xPol ε, but not with either xPol α or xPol δ (206).

XIII. Concluding Remarks

Pol ε plays a central role in replication fork establishment, progression, and maintenance of fork stability (Fig. 1). In addition, Pol ε plays important roles in the establishment and maintenance of a silenced chromatin state, the repair of DNA base damage, and the restart of stalled replication forks. Each of these processes must occur faithfully and in a regulated manner in order to allow duplication of the eukaryotic genome and stable transmission of this genetic information to the daughter cells. Disruptions to these processes can lead to mutagenesis, genome instability, aneuploidy, and cell death. Pol ε also plays a role in DNA transactions at the rDNA locus and telomeres, both of which are important in regulating the aging process. It will therefore be important to continue to characterize in detail the precise functions of Pol ε in order to determine how Pol ε contributes directly to genome stability and ultimately to human health.

Acknowledgments

The authors thank Drs. Kasia Bebenek and Stephanie Nick McElhinny for thoughtful discussion and comments on the manuscript. This work was supported by the Intramural Research Program of the National Institutes of Health, National Institute of Environmental Health Sciences.

References

1. Bessman, M. J., Kornberg, A., Lehman, I. R., and Simms, E. S. (1956). Enzymic synthesis of deoxyribonucleic acid. *Biochim. Biophys. Acta* **21**, 197–198.
2. Bollum, F. J., and Potter, V. R. (1958). Incorporation of thymidine into deoxyribonucleic acid by enzymes from rat tissues. *J. Biol. Chem.* **233**, 478–482.
3. Byrnes, J. J., Downey, K. M., Black, V. L., and So, A. G. (1976). A new mammalian DNA polymerase with 3′ to 5′ exonuclease activity: DNA polymerase delta. *Biochemistry* **15**, 2817–2823.
4. Crute, J. J., Wahl, A. F., and Bambara, R. A. (1986). Purification and characterization of two new high molecular weight forms of DNA polymerase delta. *Biochemistry* **25**, 26–36.
5. Nishida, C., Reinhard, P., and Linn, S. (1988). DNA repair synthesis in human fibroblasts requires DNA polymerase delta. *J. Biol. Chem.* **263**, 501–510.
6. Syvaoja, J., and Linn, S. (1989). Characterization of a large form of DNA polymerase delta from HeLa cells that is insensitive to proliferating cell nuclear antigen. *J. Biol. Chem.* **264**, 2489–2497.

7. Syvaoja, J., Suomensaari, S., Nishida, C., Goldsmith, J. S., Chui, G. S., Jain, S., and Linn, S. (1990). DNA polymerases alpha, delta, and epsilon: Three distinct enzymes from HeLa cells. *Proc. Natl. Acad. Sci. USA* **87,** 6664–6668.
8. Kesti, T., Frantti, H., and Syvaoja, J. E. (1993). Molecular cloning of the cDNA for the catalytic subunit of human DNA polymerase epsilon. *J. Biol. Chem.* **268,** 10238–10245.
9. Li, Y., Asahara, H., Patel, V. S., Zhou, S., and Linn, S. (1997). Purification, cDNA cloning, and gene mapping of the small subunit of human DNA polymerase epsilon. *J. Biol. Chem.* **272,** 32337–32344.
10. Bebenek, K., and Kunkel, T. A. (2004). Functions of DNA polymerases. *Adv. Protein Chem.* **69,** 137–165.
11. Franklin, M. C., Wang, J., and Steitz, T. A. (2001). Structure of the replicating complex of a pol alpha family DNA polymerase. *Cell* **105,** 657–667.
12. Kamtekar, S., Berman, A. J., Wang, J., Lazaro, J. M., de Vega, M., Blanco, L., Salas, M., and Steitz, T. A. (2004). Insights into strand displacement and processivity from the crystal structure of the protein-primed DNA polymerase of bacteriophage phi29. *Mol. Cell* **16,** 609–618.
13. Doublie, S., Sawaya, M. R., and Ellenberger, T. (1999). An open and closed case for all polymerases. *Structure* **7,** R31–R35.
14. Dua, R., Levy, D. L., and Campbell, J. L. (1998). Role of the putative zinc finger domain of *Saccharomyces cerevisiae* DNA polymerase epsilon in DNA replication and the S/M checkpoint pathway. *J. Biol. Chem.* **273,** 30046–30055.
15. Li, Y., Pursell, Z. F., and Linn, S. (2000). Identification and cloning of two histone fold motif-containing subunits of HeLa DNA polymerase epsilon. *J. Biol. Chem.* **275,** 23247–23252.
16. Dua, R., Levy, D. L., Li, C. M., Snow, P. M., and Campbell, J. L. (2002). In vivo reconstitution of *Saccharomyces cerevisiae* DNA polymerase epsilon in insect cells. Purification and characterization. *J. Biol. Chem.* **277,** 7889–7896.
17. Kesti, T., McDonald, W. H., Yates, J. R., 3rd, and Wittenberg, C. (2004). Cell cycle-dependent phosphorylation of the DNA polymerase epsilon subunit, Dpb2, by the Cdc28 cyclin-dependent protein kinase. *J. Biol. Chem.* **279,** 14245–14255.
18. Chilkova, O., Jonsson, B. H., and Johansson, E. (2003). The quaternary structure of DNA polymerase epsilon from *Saccharomyces cerevisiae*. *J. Biol. Chem.* **278,** 14082–14086.
19. Jaszczur, M., Flis, K., Rudzka, J., Kraszewska, J., Budd, M. E., Polaczek, P., Campbell, J. L., Jonczyk, P., and Fijalkowska, I. J. (2008). Dpb2p, a noncatalytic subunit of DNA polymerase {varepsilon}, contributes to the fidelity of DNA replication in *Saccharomyces cerevisiae*. *Genetics* **178,** 633–647.
20. Li, Y. (1998). Ph.D. Thesis. University of California, Berkeley, Berkeley, CA.
21. Araki, H., Hamatake, R. K., Morrison, A., Johnson, A. L., Johnston, L. H., and Sugino, A. (1991). Cloning DPB3, the gene encoding the third subunit of DNA polymerase II of *Saccharomyces cerevisiae*. *Nucleic Acids Res.* **19,** 4867–4872.
22. Ohya, T., Kawasaki, Y., Hiraga, S., Kanbara, S., Nakajo, K., Nakashima, N., Suzuki, A., and Sugino, A. (2002). The DNA polymerase domain of pol(epsilon) is required for rapid, efficient, and highly accurate chromosomal DNA replication, telomere length maintenance, and normal cell senescence in *Saccharomyces cerevisiae*. *J. Biol. Chem.* **277,** 28099–28108.
23. Asturias, F. J., Cheung, I. K., Sabouri, N., Chilkova, O., Wepplo, D., and Johansson, E. (2006). Structure of *Saccharomyces cerevisiae* DNA polymerase epsilon by cryo-electron microscopy. *Nat. Struct. Mol. Biol.* **13,** 35–43.
24. Shamoo, Y., and Steitz, T. A. (1999). Building a replisome from interacting pieces: Sliding clamp complexed to a peptide from DNA polymerase and a polymerase editing complex. *Cell* **99,** 155–166.

25. Masumoto, H., Sugino, A., and Araki, H. (2000). Dpb11 controls the association between DNA polymerases alpha and epsilon and the autonomously replicating sequence region of budding yeast. *Mol. Cell. Biol.* **20**, 2809–2817.
26. Araki, H., Leem, S. H., Phongdara, A., and Sugino, A. (1995). Dpb11, which interacts with DNA polymerase II(epsilon) in *Saccharomyces cerevisiae*, has a dual role in S-phase progression and at a cell cycle checkpoint. *Proc. Natl. Acad. Sci. USA* **92**, 11791–11795.
27. Tanaka, S., Umemori, T., Hirai, K., Muramatsu, S., Kamimura, Y., and Araki, H. (2007). CDK-dependent phosphorylation of Sld2 and Sld3 initiates DNA replication in budding yeast. *Nature* **445**, 328–332.
28. Zegerman, P., and Diffley, J. F. (2007). Phosphorylation of Sld2 and Sld3 by cyclin-dependent kinases promotes DNA replication in budding yeast. *Nature* **445**, 281–285.
29. Kubota, Y., Takase, Y., Komori, Y., Hashimoto, Y., Arata, T., Kamimura, Y., Araki, H., and Takisawa, H. (2003). A novel ring-like complex of Xenopus proteins essential for the initiation of DNA replication. *Genes Dev.* **17**, 1141–1152.
30. Takayama, Y., Kamimura, Y., Okawa, M., Muramatsu, S., Sugino, A., and Araki, H. (2003). GINS, a novel multiprotein complex required for chromosomal DNA replication in budding yeast. *Genes Dev.* **17**, 1153–1165.
31. Kanemaki, M., and Labib, K. (2006). Distinct roles for Sld3 and GINS during establishment and progression of eukaryotic DNA replication forks. *EMBO J.* **25**, 1753–1763.
32. Moyer, S. E., Lewis, P. W., and Botchan, M. R. (2006). Isolation of the Cdc45/Mcm2–7/GINS (CMG) complex, a candidate for the eukaryotic DNA replication fork helicase. *Proc. Natl. Acad. Sci. USA* **103**, 10236–10241.
33. Makiniemi, M., Hillukkala, T., Tuusa, J., Reini, K., Vaara, M., Huang, D., Pospiech, H., Majuri, I., Westerling, T., Makela, T. P., and Syvaoja, J. S. (2001). BRCT domain-containing protein TopBP1 functions in DNA replication and damage response. *J. Biol. Chem.* **276**, 30399–30406.
34. Kim, J. E., McAvoy, S. A., Smith, D. I., and Chen, J. (2005). Human TopBP1 ensures genome integrity during normal S phase. *Mol. Cell. Biol.* **25**, 10907–10915.
35. Maldonado, E., Shiekhattar, R., Sheldon, M., Cho, H., Drapkin, R., Rickert, P., Lees, E., Anderson, C. W., Linn, S., and Reinberg, D. (1996). A human RNA polymerase II complex associated with SRB and DNA-repair proteins. *Nature* **381**, 86–89.
36. Rytkonen, A. K., Vaara, M., Nethanel, T., Kaufmann, G., Sormunen, R., Laara, E., Nasheuer, H. P., Rahmeh, A., Lee, M. Y., Syvaoja, J. E. *et al.* (2006). Distinctive activities of DNA polymerases during human DNA replication. *FEBS J.* **273**, 2984–3001.
37. Iwakuma, T., and Lozano, G. (2003). MDM2, an introduction. *Mol. Cancer Res.* **1**, 993–1000.
38. Haupt, Y., Maya, R., Kazaz, A., and Oren, M. (1997). Mdm2 promotes the rapid degradation of p53. *Nature* **387**, 296–299.
39. Vlatkovic, N., Guerrera, S., Li, Y., Linn, S., Haines, D. S., and Boyd, M. T. (2000). MDM2 interacts with the C-terminus of the catalytic subunit of DNA polymerase epsilon. *Nucleic Acids Res.* **28**, 3581–3586.
40. Asahara, H., Li, Y., Fuss, J., Haines, D. S., Vlatkovic, N., Boyd, M. T., and Linn, S. (2003). Stimulation of human DNA polymerase epsilon by MDM2. *Nucleic Acids Res.* **31**, 2451–2459.
41. Steitz, T. A., Smerdon, S. J., Jager, J., and Joyce, C. M. (1994). A unified polymerase mechanism for nonhomologous DNA and RNA polymerases. *Science* **266**, 2022–2025.
42. Kunkel, T. A., and Bebenek, K. (2000). DNA replication fidelity. *Annu. Rev. Biochem.* **69**, 497–529.
43. Moon, A. F., Garcia-Diaz, M., Batra, V. K., Beard, W. A., Bebenek, K., Kunkel, T. A., Wilson, S. H., and Pedersen, L. C. (2007). The X family portrait: Structural insights into biological functions of X family polymerases. *DNA Repair* **6**, 1709–1725.

44. Chui, G., and Linn, S. (1995). Further characterization of HeLa DNA polymerase epsilon. *J. Biol. Chem.* **270**, 7799–7808.
45. Burgers, P. M., and Bauer, G. A. (1988). DNA polymerase III from *Saccharomyces cerevisiae*. II. Inhibitor studies and comparison with DNA polymerases I and II. *J. Biol. Chem.* **263**, 925–930.
46. Hamatake, R. K., Hasegawa, H., Clark, A. B., Bebenek, K., Kunkel, T. A., and Sugino, A. (1990). Purification and characterization of DNA polymerase II from the yeast *Saccharomyces cerevisiae*. Identification of the catalytic core and a possible holoenzyme form of the enzyme. *J. Biol. Chem.* **265**, 4072–4083.
47. Richey, B., Cayley, D. S., Mossing, M. C., Kolka, C., Anderson, C. F., Farrar, T. C., and Record, M. T., Jr. (1987). Variability of the intracellular ionic environment of *Escherichia coli*. Differences between in vitro and *in vivo* effects of ion concentrations on protein-DNA interactions and gene expression. *J. Biol. Chem.* **262**, 7157–7164.
48. Griep, M. A., and McHenry, C. S. (1989). Glutamate overcomes the salt inhibition of DNA polymerase III holoenzyme. *J. Biol. Chem.* **264**, 11294–11301.
49. Chilkova, O., Stenlund, P., Isoz, I., Stith, C. M., Grabowski, P., Lundstrom, E. B., Burgers, P. M., and Johansson, E. (2007). The eukaryotic leading and lagging strand DNA polymerases are loaded onto primer-ends via separate mechanisms but have comparable processivity in the presence of PCNA. *Nucleic Acids Res.* **35**, 6588–6597.
50. McCulloch, S. D., Kokoska, R. J., Chilkova, O., Welch, C. M., Johansson, E., Burgers, P. M., and Kunkel, T. A. (2004). Enzymatic switching for efficient and accurate translesion DNA replication. *Nucleic Acids Res.* **32**, 4665–4675.
51. Tran, H. T., Gordenin, D. A., and Resnick, M. A. (1999). The $3'\rightarrow 5'$ exonucleases of DNA polymerases delta and epsilon and the $5'\rightarrow 3'$ exonuclease Exo1 have major roles in post-replication mutation avoidance in *Saccharomyces cerevisiae*. *Mol. Cell. Biol.* **19**, 2000–2007.
52. Tsubota, T., Tajima, R., Ode, K., Kubota, H., Fukuhara, N., Kawabata, T., Maki, S., and Maki, H. (2006). Double-stranded DNA binding, an unusual property of DNA polymerase epsilon, promotes epigenetic silencing in *Saccharomyces cerevisiae*. *J. Biol. Chem.* **281**, 32898–32908.
53. Maki, S., Hashimoto, K., Ohara, T., and Sugino, A. (1998). DNA polymerase II (epsilon) of *Saccharomyces cerevisiae* dissociates from the DNA template by sensing single-stranded DNA. *J. Biol. Chem.* **273**, 21332–21341.
54. Kunkel, T. A., Sabatino, R. D., and Bambara, R. A. (1987). Exonucleolytic proofreading by calf thymus DNA polymerase delta. *Proc. Natl. Acad. Sci. USA* **84**, 4865–4869.
55. Thomas, D. C., Roberts, J. D., Sabatino, R. D., Myers, T. W., Tan, C. K., Downey, K. M., So, A. G., Bambara, R. A., and Kunkel, T. A. (1991). Fidelity of mammalian DNA replication and replicative DNA polymerases. *Biochemistry* **30**, 11751–11759.
56. Pursell, Z. F., Isoz, I., Lundstrom, E. B., Johansson, E., and Kunkel, T. A. (2007). Regulation of B family DNA polymerase fidelity by a conserved active site residue: Characterization of M644W, M644L and M644F mutants of yeast DNA polymerase epsilon. *Nucleic Acids Res.* **35**, 3076–3086.
57. Shcherbakova, P. V., Pavlov, Y. I., Chilkova, O., Rogozin, I. B., Johansson, E., and Kunkel, T. A. (2003). Unique error signature of the four-subunit yeast DNA polymerase epsilon. *J. Biol. Chem.* **278**, 43770–43780.
58. Beard, W. A., and Wilson, S. H. (2003). Structural insights into the origins of DNA polymerase fidelity. *Structure* **11**, 489–496.
59. Goodman, M. F., and Fygenson, K. D. (1998). DNA polymerase fidelity: From genetics toward a biochemical understanding. *Genetics* **148**, 1475–1482.
60. Johnson, K. A. (1993). Conformational coupling in DNA polymerase fidelity. *Annu. Rev. Biochem.* **62**, 685–713.

61. Joyce, C. M., and Benkovic, S. J. (2004). DNA polymerase fidelity: Kinetics, structure, and checkpoints. *Biochemistry* **43**, 14317–14324.
62. Kool, E. T. (2002). Active site tightness and substrate fit in DNA replication. *Annu. Rev. Biochem.* **71**, 191–219.
63. Kunkel, T. A. (2004). DNA replication fidelity. *J. Biol. Chem.* **279**, 16895–16898.
64. Morrison, A., Bell, J. B., Kunkel, T. A., and Sugino, A. (1991). Eukaryotic DNA polymerase amino acid sequence required for 3′→5′ exonuclease activity. *Proc. Natl. Acad. Sci. USA* **88**, 9473–9477.
65. Shcherbakova, P. V., and Pavlov, Y. I. (1996). 3′→5′ exonucleases of DNA polymerases epsilon and delta correct base analog induced DNA replication errors on opposite DNA strands in *Saccharomyces cerevisiae*. *Genetics* **142**, 717–726.
66. Garcia-Diaz, M., and Kunkel, T. A. (2006). Mechanism of a genetic glissando: Structural biology of indel mutations. *Trends Biochem. Sci.* **31**, 206–214.
67. Fuss, J., and Linn, S. (2002). Human DNA polymerase epsilon colocalizes with proliferating cell nuclear antigen and DNA replication late, but not early, in S phase. *J. Biol. Chem.* **277**, 8658–8666.
68. DePamphilis, M. L. (2003). The 'ORC cycle': A novel pathway for regulating eukaryotic DNA replication. *Gene* **310**, 1–15.
69. Diffley, J. F. (2004). Regulation of early events in chromosome replication. *Curr. Biol.* **14**, R778–R786.
70. Boskovic, J., Coloma, J., Aparicio, T., Zhou, M., Robinson, C. V., Mendez, J., and Montoya, G. (2007). Molecular architecture of the human GINS complex. *EMBO Rep.* **8**, 678–684.
71. Kornberg, A., and Baker, T. (1992). *DNA Replication*. 2. W.H. Freeman and Company, New York.
72. Morrison, A., Araki, H., Clark, A. B., Hamatake, R. K., and Sugino, A. (1990). A third essential DNA polymerase in *S. cerevisiae*. *Cell* **62**, 1143–1151.
73. Johnson, L. M., Snyder, M., Chang, L. M., Davis, R. W., and Campbell, J. L. (1985). Isolation of the gene encoding yeast DNA polymerase I. *Cell* **43**, 369–377.
74. Newlon, C. S. (1988). Yeast chromosome replication and segregation. *Microbiol. Rev.* **52**, 568–601.
75. Araki, H., Ropp, P. A., Johnson, A. L., Johnston, L. H., Morrison, A., and Sugino, A. (1992). DNA polymerase II, the probable homolog of mammalian DNA polymerase epsilon, replicates chromosomal DNA in the yeast *Saccharomyces cerevisiae*. *EMBO J.* **11**, 733–740.
76. Budd, M. E., and Campbell, J. L. (1993). DNA polymerases delta and epsilon are required for chromosomal replication in *Saccharomyces cerevisiae*. *Mol. Cell. Biol.* **13**, 496–505.
77. Navas, T. A., Zhou, Z., and Elledge, S. J. (1995). DNA polymerase epsilon links the DNA replication machinery to the S phase checkpoint. *Cell* **80**, 29–39.
78. Spellman, P. T., Sherlock, G., Zhang, M. Q., Iyer, V. R., Anders, K., Eisen, M. B., Brown, P. O., Botstein, D., and Futcher, B. (1998). Comprehensive identification of cell cycle-regulated genes of the yeast *Saccharomyces cerevisiae* by microarray hybridization. *Mol. Biol. Cell* **9**, 3273–3297.
79. Kamimura, Y., Masumoto, H., Sugino, A., and Araki, H. (1998). Sld2, which interacts with Dpb11 in *Saccharomyces cerevisiae*, is required for chromosomal DNA replication. *Mol. Cell. Biol.* **18**, 6102–6109.
80. Wang, H., and Elledge, S. J. (1999). DRC1, DNA replication and checkpoint protein 1, functions with DPB11 to control DNA replication and the S-phase checkpoint in *Saccharomyces cerevisiae*. *Proc. Natl. Acad. Sci. USA* **96**, 3824–3829.
81. Waga, S., Masuda, T., Takisawa, H., and Sugino, A. (2001). DNA polymerase epsilon is required for coordinated and efficient chromosomal DNA replication in Xenopus egg extracts. *Proc. Natl. Acad. Sci. USA* **98**, 4978–4983.

82. Johnson, A., and O'Donnell, M. (2005). Cellular DNA replicases: Components and dynamics at the replication fork. *Annu. Rev. Biochem.* **74**, 283–315.
83. McHenry, C. S. (1991). DNA polymerase III holoenzyme. Components, structure, and mechanism of a true replicative complex. *J. Biol. Chem.* **266**, 19127–19130.
84. Podust, V. N., and Hubscher, U. (1993). Lagging strand DNA synthesis by calf thymus DNA polymerases alpha, beta, delta and epsilon in the presence of auxiliary proteins. *Nucleic Acids Res.* **21**, 841–846.
85. Fukui, T., Yamauchi, K., Muroya, T., Akiyama, M., Maki, H., Sugino, A., and Waga, S. (2004). Distinct roles of DNA polymerases delta and epsilon at the replication fork in Xenopus egg extracts. *Genes Cells* **9**, 179–191.
86. Garg, P., Stith, C. M., Sabouri, N., Johansson, E., and Burgers, P. M. (2004). Idling by DNA polymerase delta maintains a ligatable nick during lagging-strand DNA replication. *Genes Dev.* **18**, 2764–2773.
87. Jin, Y. H., Obert, R., Burgers, P. M., Kunkel, T. A., Resnick, M. A., and Gordenin, D. A. (2001). The 3′→5′ exonuclease of DNA polymerase delta can substitute for the 5′ flap endonuclease Rad27/Fen1 in processing Okazaki fragments and preventing genome instability. *Proc. Natl. Acad. Sci. USA* **98**, 5122–5127.
88. Waga, S., and Stillman, B. (1994). Anatomy of a DNA replication fork revealed by reconstitution of SV40 DNA replication *in vitro*. *Nature* **369**, 207–212.
89. Zlotkin, T., Kaufmann, G., Jiang, Y., Lee, M. Y., Uitto, L., Syvaoja, J., Dornreiter, I., Fanning, E., and Nethanel, T. (1996). DNA polymerase epsilon may be dispensable for SV40- but not cellular-DNA replication. *EMBO. J.* **15**, 2298–2305.
90. Dodson, M., Dean, F. B., Bullock, P., Echols, H., and Hurwitz, J. (1987). Unwinding of duplex DNA from the SV40 origin of replication by T antigen. *Science* **238**, 964–967.
91. Huberman, J. A. (1991). Cell cycle control of initiation of eukaryotic DNA replication. *Chromosoma* **100**, 419–423.
92. Dua, R., Levy, D. L., and Campbell, J. L. (1999). Analysis of the essential functions of the C-terminal protein/protein interaction domain of *Saccharomyces cerevisiae* pol epsilon and its unexpected ability to support growth in the absence of the DNA polymerase domain. *J. Biol. Chem.* **274**, 22283–22288.
93. Feng, W., and D'Urso, G. (2001). *Schizosaccharomyces pombe* cells lacking the amino-terminal catalytic domains of DNA polymerase epsilon are viable but require the DNA damage checkpoint control. *Mol. Cell. Biol.* **21**, 4495–4504.
94. Kesti, T., Flick, K., Keranen, S., Syvaoja, J. E., and Wittenberg, C. (1999). DNA polymerase epsilon catalytic domains are dispensable for DNA replication, DNA repair, and cell viability. *Mol. Cell* **3**, 679–685.
95. Diede, S. J., and Gottschling, D. E. (1999). Telomerase-mediated telomere addition *in vivo* requires DNA primase and DNA polymerases alpha and delta. *Cell* **99**, 723–733.
96. Karthikeyan, R., Vonarx, E. J., Straffon, A. F., Simon, M., Faye, G., and Kunz, B. A. (2000). Evidence from mutational specificity studies that yeast DNA polymerases delta and epsilon replicate different DNA strands at an intracellular replication fork. *J. Mol. Biol.* **299**, 405–419.
97. Pursell, Z. F., Isoz, I., Lundstrom, E. B., Johansson, E., and Kunkel, T. A. (2007). Yeast DNA polymerase epsilon participates in leading-strand DNA replication. *Science* **317**, 127–130.
98. Nick McElhinny, S. A., Gordenin, D. A., Stith, C. M., Burgers, P. M., and Kunkel, T. A. (2008). Division of labor at the eukaryotic replication fork. *Mol. Cell.* **30**, 259–260.
99. Venkatesan, R. N., Hsu, J. J., Lawrence, N. A., Preston, B. D., and Loeb, L. A. (2006). Mutator phenotypes caused by substitution at a conserved motif A residue in eukaryotic DNA polymerase delta. *J. Biol. Chem.* **281**, 4486–4494.
100. Sancar, A., Lindsey-Boltz, L. A., Unsal-Kacmaz, K., and Linn, S. (2004). Molecular mechanisms of mammalian DNA repair and the DNA damage checkpoints. *Annu. Rev. Biochem.* **73**, 39–85.

101. Navas, T. A., Sanchez, Y., and Elledge, S. J. (1996). RAD9 and DNA polymerase epsilon form parallel sensory branches for transducing the DNA damage checkpoint signal in *Saccharomyces cerevisiae*. *Genes Dev.* **10,** 2632–2643.
102. Gangloff, S., McDonald, J. P., Bendixen, C., Arthur, L., and Rothstein, R. (1994). The yeast type I topoisomerase Top3 interacts with Sgs1, a DNA helicase homolog: A potential eukaryotic reverse gyrase. *Mol. Cell. Biol.* **14,** 8391–8398.
103. Frei, C., and Gasser, S. M. (2000). The yeast Sgs1p helicase acts upstream of Rad53p in the DNA replication checkpoint and colocalizes with Rad53p in S-phase-specific foci. *Genes Dev.* **14,** 81–96.
104. Cobb, J. A., Bjergbaek, L., Shimada, K., Frei, C., and Gasser, S. M. (2003). DNA polymerase stabilization at stalled replication forks requires Mec1 and the RecQ helicase Sgs1. *EMBO J.* **22,** 4325–4336.
105. Bjergbaek, L., Cobb, J. A., Tsai-Pflugfelder, M., and Gasser, S. M. (2005). Mechanistically distinct roles for Sgs1p in checkpoint activation and replication fork maintenance. *EMBO J.* **24,** 405–417.
106. Calzada, A., Hodgson, B., Kanemaki, M., Bueno, A., and Labib, K. (2005). Molecular anatomy and regulation of a stable replisome at a paused eukaryotic DNA replication fork. *Genes Dev.* **19,** 1905–1919.
107. Lopes, M., Cotta-Ramusino, C., Pellicioli, A., Liberi, G., Plevani, P., Muzi-Falconi, M., Newlon, C. S., and Foiani, M. (2001). The DNA replication checkpoint response stabilizes stalled replication forks. *Nature* **412,** 557–561.
108. Henneke, G., Koundrioukoff, S., and Hubscher, U. (2003). Phosphorylation of human Fen1 by cyclin-dependent kinase modulates its role in replication fork regulation. *Oncogene* **22,** 4301–4313.
109. Ubersax, J. A., Woodbury, E. L., Quang, P. N., Paraz, M., Blethrow, J. D., Shah, K., Shokat, K. M., and Morgan, D. O. (2003). Targets of the cyclin-dependent kinase Cdk1. *Nature* **425,** 859–864.
110. Bloom, J., and Cross, F. R. (2007). Novel role for Cdc14 sequestration: Cdc14 dephosphorylates factors that promote DNA replication. *Mol. Cell. Biol.* **27,** 842–853.
111. Voitenleitner, C., Fanning, E., and Nasheuer, H. P. (1997). Phosphorylation of DNA polymerase alpha-primase by cyclin A-dependent kinases regulates initiation of DNA replication in vitro. *Oncogene* **14,** 1611–1615.
112. Lau, A., Blitzblau, H., and Bell, S. P. (2002). Cell-cycle control of the establishment of mating-type silencing in S. cerevisiae. *Genes Dev.* **16,** 2935–2945.
113. Levin, D. S., McKenna, A. E., Motycka, T. A., Matsumoto, Y., and Tomkinson, A. E. (2000). Interaction between PCNA and DNA ligase I is critical for joining of Okazaki fragments and long-patch base-excision repair. *Curr. Biol.* **10,** 919–922.
114. Smith, J. S., Caputo, E., and Boeke, J. D. (1999). A genetic screen for ribosomal DNA silencing defects identifies multiple DNA replication and chromatin-modulating factors. *Mol. Cell. Biol.* **19,** 3184–3197.
115. Ehrenhofer-Murray, A. E., Kamakaka, R. T., and Rine, J. (1999). A role for the replication proteins PCNA, RF-C, polymerase epsilon and Cdc45 in transcriptional silencing in *Saccharomyces cerevisiae*. *Genetics* **153,** 1171–1182.
116. Chan, S. R., and Blackburn, E. H. (2004). Telomeres and telomerase. *Philos. Trans. R. Soc. Lond.* **359,** 109–121.
117. Gilson, E., and Geli, V. (2007). How telomeres are replicated. *Nat. Rev.* **8,** 825–838.
118. Zhang, X., Mar, V., Zhou, W., Harrington, L., and Robinson, M. O. (1999). Telomere shortening and apoptosis in telomerase-inhibited human tumor cells. *Genes Dev.* **13,** 2388–2399.
119. Greider, C. W., and Blackburn, E. H. (1987). The telomere terminal transferase of Tetrahymena is a ribonucleoprotein enzyme with two kinds of primer specificity. *Cell* **51,** 887–898.

120. Blasco, M. A. (2007). The epigenetic regulation of mammalian telomeres. *Nat. Rev. Genet.* **8**, 299–309.
121. Hecht, A., Strahl-Bolsinger, S., and Grunstein, M. (1996). Spreading of transcriptional repressor SIR3 from telomeric heterochromatin. *Nature* **383**, 92–96.
122. Benetti, R., Garcia-Cao, M., and Blasco, M. A. (2007). Telomere length regulates the epigenetic status of mammalian telomeres and subtelomeres. *Nat. Genet.* **39**, 243–250.
123. Bianchi, A., and Shore, D. (2007). Early replication of short telomeres in budding yeast. *Cell* **128**, 1051–1062.
124. Iida, T., and Araki, H. (2004). Noncompetitive counteractions of DNA polymerase epsilon and ISW2/yCHRAC for epigenetic inheritance of telomere position effect in *Saccharomyces cerevisiae*. *Mol. Cell. Biol.* **24**, 217–227.
125. Poot, R. A., Dellaire, G., Hulsmann, B. B., Grimaldi, M. A., Corona, D. F., Becker, P. B., Bickmore, W. A., and Varga-Weisz, P. D. (2000). HuCHRAC, a human ISWI chromatin remodelling complex contains hACF1 and two novel histone-fold proteins. *EMBO J.* **19**, 3377–3387.
126. Henikoff, S. (2008). Nucleosome destabilization in the epigenetic regulation of gene expression. *Nat. Rev. Genet.* **9**, 15–26.
127. Groth, A., Rocha, W., Verreault, A., and Almouzni, G. (2007). Chromatin challenges during DNA replication and repair. *Cell* **128**, 721–733.
128. Wang, G. G., Allis, C. D., and Chi, P. (2007). Chromatin remodeling and cancer, Part II: ATP-dependent chromatin remodeling. *Trends Mol. Med.* **13**, 373–380.
129. Tate, P., Lee, M., Tweedie, S., Skarnes, W. C., and Bickmore, W. A. (1998). Capturing novel mouse genes encoding chromosomal and other nuclear proteins. *J. Cell Sci.* **111**(Pt. 17), 2575–2585.
130. Corona, D. F., Eberharter, A., Budde, A., Deuring, R., Ferrari, S., Varga-Weisz, P., Wilm, M., Tamkun, J., and Becker, P. B. (2000). Two histone fold proteins, CHRAC-14 and CHRAC-16, are developmentally regulated subunits of chromatin accessibility complex (CHRAC). *EMBO J.* **19**, 3049–3059.
131. Varga-Weisz, P. D., Wilm, M., Bonte, E., Dumas, K., Mann, M., and Becker, P. B. (1997). Chromatin-remodelling factor CHRAC contains the ATPases ISWI and topoisomerase II. *Nature* **388**, 598–602.
132. Tsukiyama, T., Palmer, J., Landel, C. C., Shiloach, J., and Wu, C. (1999). Characterization of the imitation switch subfamily of ATP-dependent chromatin-remodeling factors in *Saccharomyces cerevisiae*. *Genes Dev.* **13**, 686–697.
133. McConnell, A. D., Gelbart, M. E., and Tsukiyama, T. (2004). Histone fold protein Dls1p is required for Isw2-dependent chromatin remodeling *in vivo*. *Mol. Cell. Biol.* **24**, 2605–2613.
134. Kukimoto, I., Elderkin, S., Grimaldi, M., Oelgeschlager, T., and Varga-Weisz, P. D. (2004). The histone-fold protein complex CHRAC-15/17 enhances nucleosome sliding and assembly mediated by ACF. *Mol. Cell* **13**, 265–277.
135. Tackett, A. J., Dilworth, D. J., Davey, M. J., O'Donnell, M., Aitchison, J. D., Rout, M. P., and Chait, B. T. (2005). Proteomic and genomic characterization of chromatin complexes at a boundary. *J. Cell Biol.* **169**, 35–47.
136. Tsubota, T., Maki, S., Kubota, H., Sugino, A., and Maki, H. (2003). Double-stranded DNA binding properties of *Saccharomyces cerevisiae* DNA polymerase epsilon and of the Dpb3p-Dpb4p subassembly. *Genes Cells* **8**, 873–888.
137. Dang, W., Kagalwala, M. N., and Bartholomew, B. (2007). The Dpb4 subunit of ISW2 is anchored to extranucleosomal DNA. *J. Biol. Chem.* **282**, 19418–19425.
138. Hartlepp, K. F., Fernandez-Tornero, C., Eberharter, A., Grune, T., Muller, C. W., and Becker, P. B. (2005). The histone fold subunits of Drosophila CHRAC facilitate nucleosome sliding through dynamic DNA interactions. *Mol. Cell. Biol.* **25**, 9886–9896.

139. Collins, N., Poot, R. A., Kukimoto, I., Garcia-Jimenez, C., Dellaire, G., and Varga-Weisz, P. D. (2002). An ACF1-ISWI chromatin-remodeling complex is required for DNA replication through heterochromatin. *Nat. Genet.* **32,** 627–632.
140. Carson, D. R., and Christman, M. F. (2001). Evidence that replication fork components catalyze establishment of cohesion between sister chromatids. *Proc. Natl. Acad. Sci. USA* **98,** 8270–8275.
141. Edwards, S., Li, C. M., Levy, D. L., Brown, J., Snow, P. M., and Campbell, J. L. (2003). *Saccharomyces cerevisiae* DNA polymerase epsilon and polymerase sigma interact physically and functionally, suggesting a role for polymerase epsilon in sister chromatid cohesion. *Mol. Cell. Biol.* **23,** 2733–2748.
142. Haracska, L., Johnson, R. E., Prakash, L., and Prakash, S. (2005). Trf4 and Trf5 proteins of *Saccharomyces cerevisiae* exhibit poly(A) RNA polymerase activity but no DNA polymerase activity. *Mol. Cell. Biol.* **25,** 10183–10189.
143. Hakimi, M. A., Bochar, D. A., Schmiesing, J. A., Dong, Y., Barak, O. G., Speicher, D. W., Yokomori, K., and Shiekhattar, R. (2002). A chromatin remodelling complex that loads cohesin onto human chromosomes. *Nature* **418,** 994–998.
144. Cheng, L., and Kelly, T. J. (1989). Transcriptional activator nuclear factor I stimulates the replication of SV40 minichromosomes *in vivo* and *in vitro*. *Cell* **59,** 541–551.
145. Alexiadis, V., Varga-Weisz, P. D., Bonte, E., Becker, P. B., and Gruss, C. (1998). *In vitro* chromatin remodelling by chromatin accessibility complex (CHRAC) at the SV40 origin of DNA replication. *EMBO J.* **17,** 3428–3438.
146. Coverley, D., Kenny, M. K., Lane, D. P., and Wood, R. D. (1992). A role for the human single-stranded DNA binding protein HSSB/RPA in an early stage of nucleotide excision repair. *Nucleic Acids Res.* **20,** 3873–3880.
147. Friedberg, E. C., Walker, G. C., Siede, W., Wood, R. D., Schultz, R. A., and Ellenberger, T. (2006). *DNA Repair and Mutagenesis*. 2. ASM Press, Washington, DC.
148. Beard, W. A., and Wilson, S. H. (2006). Structure and mechanism of DNA polymerase Beta. *Chem. Rev.* **106,** 361–382.
149. Horton, J. K., Watson, M., Stefanick, D. F., Shaughnessy, D. T., Taylor, J. A., and Wilson, S. H. (2008). XRCC1 and DNA polymerase beta in cellular protection against cytotoxic DNA single-strand breaks. *Cell Res.* **18,** 48–63.
150. Sobol, R. W., Horton, J. K., Kuhn, R., Gu, H., Singhal, R. K., Prasad, R., Rajewsky, K., and Wilson, S. H. (1996). Requirement of mammalian DNA polymerase-beta in base-excision repair. *Nature* **379,** 183–186.
151. Parlanti, E., Locatelli, G., Maga, G., and Dogliotti, E. (2007). Human base excision repair complex is physically associated to DNA replication and cell cycle regulatory proteins. *Nucleic Acids Res.* **35,** 1569–1577.
152. Stucki, M., Pascucci, B., Parlanti, E., Fortini, P., Wilson, S. H., Hubscher, U., and Dogliotti, E. (1998). Mammalian base excision repair by DNA polymerases delta and epsilon. *Oncogene* **17,** 835–843.
153. Wang, Z., Wu, X., and Friedberg, E. C. (1993). DNA repair synthesis during base excision repair *in vitro* is catalyzed by DNA polymerase epsilon and is influenced by DNA polymerases alpha and delta in *Saccharomyces cerevisiae*. *Mol. Cell. Biol.* **13,** 1051–1058.
154. Aboussekhra, A., Biggerstaff, M., Shivji, M. K., Vilpo, J. A., Moncollin, V., Podust, V. N., Protic, M., Hubscher, U., Egly, J. M., and Wood, R. D. (1995). Mammalian DNA nucleotide excision repair reconstituted with purified protein components. *Cell* **80,** 859–868.
155. Shivji, M. K., Podust, V. N., Hubscher, U., and Wood, R. D. (1995). Nucleotide excision repair DNA synthesis by DNA polymerase epsilon in the presence of PCNA, RFC, and RPA. *Biochemistry* **34,** 5011–5017.

156. Mozzherin, D. J., and Fisher, P. A. (1996). Human DNA polymerase epsilon: Enzymologic mechanism and gap-filling synthesis. *Biochemistry* **35**, 3572–3577.
157. Budd, M. E., and Campbell, J. L. (1995). DNA polymerases required for repair of UV-induced damage in *Saccharomyces cerevisiae*. *Mol. Cell. Biol.* **15**, 2173–2179.
158. Wu, X., Guo, D., Yuan, F., and Wang, Z. (2001). Accessibility of DNA polymerases to repair synthesis during nucleotide excision repair in yeast cell-free extracts. *Nucleic Acids Res.* **29**, 3123–3130.
159. Moser, J., Kool, H., Giakzidis, I., Caldecott, K., Mullenders, L. H., and Fousteri, M. I. (2007). Sealing of chromosomal DNA nicks during nucleotide excision repair requires XRCC1 and DNA ligase III alpha in a cell-cycle-specific manner. *Mol. Cell* **27**, 311–323.
160. Wu, X., Braithwaite, E., and Wang, Z. (1999). DNA ligation during excision repair in yeast cell-free extracts is specifically catalyzed by the CDC9 gene product. *Biochemistry* **38**, 2628–2635.
161. Cappelli, E., Taylor, R., Cevasco, M., Abbondandolo, A., Caldecott, K., and Frosina, G. (1997). Involvement of XRCC1 and DNA ligase III gene products in DNA base excision repair. *J. Biol. Chem.* **272**, 23970–23975.
162. Dronkert, M. L., and Kanaar, R. (2001). Repair of DNA interstrand cross-links. *Mutat. Res.* **486**, 217–247.
163. McHugh, P. J., and Sarkar, S. (2006). DNA interstrand cross-link repair in the cell cycle: A critical role for polymerase zeta in G1 phase. *Cell Cycle (Georgetown, Tex.)* **5**, 1044–1047.
164. Constantin, N., Dzantiev, L., Kadyrov, F. A., and Modrich, P. (2005). Human mismatch repair: Reconstitution of a nick-directed bidirectional reaction. *J. Biol. Chem.* **280**, 39752–39761.
165. Dzantiev, L., Constantin, N., Genschel, J., Iyer, R. R., Burgers, P. M., and Modrich, P. (2004). A defined human system that supports bidirectional mismatch-provoked excision. *Mol. Cell* **15**, 31–41.
166. Longley, M. J., Pierce, A. J., and Modrich, P. (1997). DNA polymerase delta is required for human mismatch repair *in vitro*. *J. Biol. Chem.* **272**, 10917–10921.
167. Valerie, K., and Povirk, L. F. (2003). Regulation and mechanisms of mammalian double-strand break repair. *Oncogene* **22**, 5792–5812.
168. Haber, J. E. (2006). Transpositions and translocations induced by site-specific double-strand breaks in budding yeast. *DNA Repair* **5**, 998–1009.
169. Weterings, E., and Chen, D. J. (2008). The endless tale of non-homologous end-joining. *Cell Res.* **18**, 114–124.
170. Holmes, A. M., and Haber, J. E. (1999). Double-strand break repair in yeast requires both leading and lagging strand DNA polymerases. *Cell* **96**, 415–424.
171. Wang, X., Ira, G., Tercero, J. A., Holmes, A. M., Diffley, J. F., and Haber, J. E. (2004). Role of DNA replication proteins in double-strand break-induced recombination in *Saccharomyces cerevisiae*. *Mol. Cell. Biol.* **24**, 6891–6899.
172. Bianco, P. R., Tracy, R. B., and Kowalczykowski, S. C. (1998). DNA strand exchange proteins: A biochemical and physical comparison. *Front. Biosci.* **3**, D570–D603.
173. Malkova, A., Ivanov, E. L., and Haber, J. E. (1996). Double-strand break repair in the absence of RAD51 in yeast: A possible role for break-induced DNA replication. *Proc. Natl. Acad. Sci. USA* **93**, 7131–7136.
174. Chen, Q., Ijpma, A., and Greider, C. W. (2001). Two survivor pathways that allow growth in the absence of telomerase are generated by distinct telomere recombination events. *Mol. Cell. Biol.* **21**, 1819–1827.
175. Lydeard, J. R., Jain, S., Yamaguchi, M., and Haber, J. E. (2007). Break-induced replication and telomerase-independent telomere maintenance require Pol32. *Nature* **448**, 820–823.
176. Jessberger, R., and Berg, P. (1991). Repair of deletions and double-strand gaps by homologous recombination in a mammalian in vitro system. *Mol. Cell. Biol.* **11**, 445–457.

177. Jessberger, R., Chui, G., Linn, S., and Kemper, B. (1996). Analysis of the mammalian recombination protein complex RC-1. *Mutat. Res.* **350**, 217–227.
178. Jessberger, R., Podust, V., Hubscher, U., and Berg, P. (1993). A mammalian protein complex that repairs double-strand breaks and deletions by recombination. *J. Biol. Chem.* **268**, 15070–15079.
179. Jessberger, R., Riwar, B., Rolink, A., and Rodewald, H. R. (1995). Stimulation of defective DNA transfer activity in recombination deficient SCID cell extracts by a 72-kDa protein from wild-type thymocytes. *J. Biol. Chem.* **270**, 6788–6797.
180. Kamel, D., Mackey, Z. B., Sjoblom, T., Walter, C. A., McCarrey, J. R., Uitto, L., Palosaari, H., Lahdetie, J., Tomkinson, A. E., and Syvaoja, J. E. (1997). Role of deoxyribonucleic acid polymerase epsilon in spermatogenesis in mice. *Biol. Reprod.* **57**, 1367–1374.
181. Kaeberlein, M., McVey, M., and Guarente, L. (1999). The SIR2/3/4 complex and SIR2 alone promote longevity in *Saccharomyces cerevisiae* by two different mechanisms. *Genes Dev.* **13**, 2570–2580.
182. Rusche, L. N., Kirchmaier, A. L., and Rine, J. (2003). The establishment, inheritance, and function of silenced chromatin in *Saccharomyces cerevisiae*. *Annu. Rev. Biochem.* **72**, 481–516.
183. Coffman, F. D., He, M., Diaz, M. L., and Cohen, S. (2005). DNA replication initiates at different sites in early and late S phase within human ribosomal RNA genes. *Cell Cycle (Georgetown, Tex.)* **4**, 1223–1226.
184. Houseley, J., Kotovic, K., El Hage, A., and Tollervey, D. (2007). Trf4 targets ncRNAs from telomeric and rDNA spacer regions and functions in rDNA copy number control. *EMBO J.* **26**, 4996–5006.
185. Gellon, L., Carson, D. R., Carson, J. P., and Demple, B. (2008). Intrinsic 5'-deoxyribose-5-phosphate lyase activity in *Saccharomyces cerevisiae* Trf4 protein with a possible role in base excision DNA repair. *DNA Repair* **7**, 187–198.
186. Guarente, L. (1999). Diverse and dynamic functions of the Sir silencing complex. *Nat. Genet.* **23**, 281–285.
187. Nasmyth, K., and Nurse, P. (1981). Cell division cycle mutants altered in DNA replication and mitosis in the fission yeast *Schizosaccharomyces pombe*. *Mol. Gen. Genet.* **182**, 119–124.
188. Grallert, B., and Sipiczki, M. (1991). Common genes and pathways in the regulation of the mitotic and meiotic cell cycles of *Schizosaccharomyces pombe*. *Curr. Genet.* **20**, 199–204.
189. Francesconi, S., De Recondo, A. M., and Baldacci, G. (1995). DNA polymerase delta is required for the replication feedback control of cell cycle progression in *Schizosaccharomyces pombe*. *Mol. Gen. Genet.* **246**, 561–569.
190. D'Urso, G., and Nurse, P. (1997). *Schizosaccharomyces pombe* cdc20+ encodes DNA polymerase epsilon and is required for chromosomal replication but not for the S phase checkpoint. *Proc. Natl. Acad. Sci. USA* **94**, 12491–12496.
191. Wang, S. W., Toda, T., MacCallum, R., Harris, A. L., and Norbury, C. (2000). Cid1, a fission yeast protein required for S-M checkpoint control when DNA polymerase delta or epsilon is inactivated. *Mol. Cell. Biol.* **20**, 3234–3244.
192. Feng, W., Rodriguez-Menocal, L., Tolun, G., and D'Urso, G. (2003). *Schizosacchromyces pombe* Dpb2 binds to origin DNA early in S phase and is required for chromosomal DNA replication. *Mol. Biol. Cell* **14**, 3427–3436.
193. Spiga, M. G., and D'Urso, G. (2004). Identification and cloning of two putative subunits of DNA polymerase epsilon in fission yeast. *Nucleic Acids Res.* **32**, 4945–4953.
194. Siam, R., Gomez, E. B., and Forsburg, S. L. (2007). *Schizosaccharomyces pombe* Rad4/Cut5 protein modification and chromatin binding changes in DNA damage. *DNA Cell Biol.* **26**, 565–575.
195. Shikata, K., Sasa-Masuda, T., Okuno, Y., Waga, S., and Sugino, A. (2006). The DNA polymerase activity of Pol epsilon holoenzyme is required for rapid and efficient chromosomal DNA replication in Xenopus egg extracts. *BMC Biochem.* **7**, 21.

DNA POLYMERASE ε   145

196. Aparicio, O. M., Stout, A. M., and Bell, S. P. (1999). Differential assembly of Cdc45p and DNA polymerases at early and late origins of DNA replication. *Proc. Natl. Acad. Sci. USA* **96**, 9130–9135.
197. Mimura, S., Masuda, T., Matsui, T., and Takisawa, H. (2000). Central role for cdc45 in establishing an initiation complex of DNA replication in Xenopus egg extracts. *Genes Cells* **5**, 439–452.
198. Matsuno, K., Kumano, M., Kubota, Y., Hashimoto, Y., and Takisawa, H. (2006). The N-terminal noncatalytic region of Xenopus RecQ4 is required for chromatin binding of DNA polymerase alpha in the initiation of DNA replication. *Mol. Cell. Biol.* **26**, 4843–4852.
199. Bohr, V. A. (2005). Deficient DNA repair in the human progeroid disorder, Werner syndrome. *Mutat. Res.* **577**, 252–259.
200. Lebel, M., Spillare, E. A., Harris, C. C., and Leder, P. (1999). The Werner syndrome gene product co-purifies with the DNA replication complex and interacts with PCNA and topoisomerase I. *J. Biol. Chem.* **274**, 37795–37799.
201. Spanos, A., Holliday, R., and German, J. (1986). Bloom's syndrome. XIII. DNA-polymerase activity of cultured lymphoblastoid cells. *Hum. Genet.* **73**, 119–122.
202. Brosh, R. M., Jr., Waheed, J., and Sommers, J. A. (2002). Biochemical characterization of the DNA substrate specificity of Werner syndrome helicase. *J. Biol. Chem.* **277**, 23236–23245.
203. Sasakawa, N., Fukui, T., and Waga, S. (2006). Accumulation of FFA-1, the Xenopus homolog of Werner helicase, and DNA polymerase delta on chromatin in response to replication fork arrest. *J. Biochem.* **140**, 95–103.
204. Lee, J., Kumagai, A., and Dunphy, W. G. (2003). Claspin, a Chk1-regulatory protein, monitors DNA replication on chromatin independently of RPA, ATR, and Rad17. *Mol. Cell* **11**, 329–340.
205. Kumagai, A., and Dunphy, W. G. (2000). Claspin, a novel protein required for the activation of Chk1 during a DNA replication checkpoint response in Xenopus egg extracts. *Mol. Cell* **6**, 839–849.
206. Lee, J., Gold, D. A., Shevchenko, A., Shevchenko, A., and Dunphy, W. G. (2005). Roles of replication fork-interacting and Chk1-activating domains from Claspin in a DNA replication checkpoint response. *Mol. Biol. Cell* **16**, 5269–5282.
207. Fortune, J. M., Pavlov, Y. I., Welch, C. M., Johansson, E., Burgers, P. M., and Kunkel, T. A. (2005). *Saccharomyces cerevisiae* DNA polymerase delta: High fidelity for base substitutions but lower fidelity for single- and multi-base deletions. *J. Biol. Chem.* **280**, 29980–29987.
208. Fortune, J. M., Stith, C. M., Kissling, G. E., Burgers, P. M., and Kunkel, T. A. (2006). RPA and PCNA suppress formation of large deletion errors by yeast DNA polymerase delta. *Nucleic Acids Res.* **34**, 4335–4341.
209. Kunkel, T. A., Hamatake, R. K., Motto-Fox, J., Fitzgerald, M. P., and Sugino, A. (1989). Fidelity of DNA polymerase I and the DNA polymerase I-DNA primase complex from *Saccharomyces cerevisiae*. *Mol. Cell. Biol.* **9**, 4447–4458.
210. Bebenek, K., Garcia-Diaz, M., Patishall, S. R., and Kunkel, T. A. (2005). Biochemical properties of *Saccharomyces cerevisiae* DNA polymerase IV. *J. Biol. Chem.* **280**, 20051–20058.
211. McCulloch, S. D., Wood, A., Garg, P., Burgers, P. M., and Kunkel, T. A. (2007). Effects of accessory proteins on the bypass of a cis-syn thymine-thymine dimer by *Saccharomyces cerevisiae* DNA polymerase eta. *Biochemistry* **46**, 8888–8896.
212. Nick McElhinny, S. A., Stith, C. M., Burgers, P. M., and Kunkel, T. A. (2007). Inefficient proofreading and biased error rates during inaccurate DNA synthesis by a mutant derivative of *Saccharomyces cerevisiae* DNA polymerase delta. *J. Biol. Chem.* **282**, 2324–2332.
213. Harp, J. M., Hanson, B. L., Timm, D. E., and Bunick, G. J. (2000). Asymmetries in the nucleosome core particle at 2.5 A resolution. *Acta Crystallogr.* **56**, 1513–1534.

Site-directed Spin Labeling Studies on Nucleic Acid Structure and Dynamics

Glenna Z. Sowa and Peter Z. Qin

Department of Chemistry, University of Southern California, Los Angeles, California 90089

I.	Overview	148
II.	Basic Physics Underlying SDSL	150
	A. The Zeeman and Hyperfine Interaction	150
	B. The Dipolar Interaction Between a Pair of Electron Spins	152
III.	Site-Specific Attachment of Nitroxides to Nucleic Acids	153
	A. Nitroxide Labeling Methods Relying on Solid-Phase Chemical Synthesis	154
	B. Nitroxide Labeling Using Enzymatic Methods	161
	C. Future Developments	162
IV.	Distance Measurements Using SDSL	162
	A. Short-Range Distances Measured Using cw-EPR	163
	B. Nanometer Distances Measured Using Pulsed EPR	164
	C. Interpretation of the Measured Internitroxide Distances, Computational Approaches	168
	D. Distance Measurement Summary	172
V.	Site-Specific Structural and Dynamic Information from a Single-Labeled Nitroxide	172
	A. Nitroxide Dynamics and EPR Spectral Lineshape	173
	B. Characterizing Nitroxide Dynamics at the Slow-Motion Regime	174
	C. Correlating Nitroxide Motion to Local Structural and Dynamic Features	177
	D. Examples of Application	179
VI.	Beyond Distance Measurements and Nitroxide Dynamics Analysis	187
	A. Nitroxide-Solvent Accessibility	187
	B. Polarity Effects	188
	C. Paramagnetic Relaxation Enhancement	188
	D. Probing Paramagnetic Metal Ions in RNA Folding and Function	189
VII.	Future Directions	189
	References	190

Site-directed spin labeling (SDSL) uses electron paramagnetic resonance (EPR) spectroscopy to monitor the behavior of a stable nitroxide radical attached at specific locations within a macromolecule such as protein, DNA, or RNA. Parameters obtained from EPR measurements, such as internitroxide distances and descriptions of the rotational motion of a nitroxide, provide unique information on features near the labeling site. With recent advances in solid-phase synthesis of nucleic acids and developments in EPR methodologies, particularly pulsed EPR technologies, SDSL has been increasingly used to study the structure and dynamics of DNA and RNA at the level of the individual nucleotides. This chapter summarizes the current SDSL studies on nucleic acids, with discussions focusing on literature from the last decade.

I. Overview

In site-directed spin labeling (SDSL) (*1*), a nitroxide molecule (spin label) is covalently attached at a specific site within a macromolecule such as protein, DNA, or RNA (Fig. 1). Electron paramagnetic resonance (EPR) spectroscopy is used to characterize the behavior of a chemically inert, unpaired electron that is localized at the nitroxyl group of the nitroxide. Structural and dynamic features at the attachment site are inferred from EPR measurements (Fig. 1). In protein studies, SDSL has matured as a tool for studying the structure–function relationship, and has been used to provide information at the level of the protein backbone (*2–8*). SDSL is useful in studying high molecular weight systems under physiological conditions. It has been particularly successful in studying systems (e.g., membrane proteins) that are difficult to investigate using other methods, such as X-ray crystallography and NMR spectroscopy (*2–8*).

SDSL has been used to study nucleic acids, and data suggest that one can obtain unique structural and dynamic information about DNA and RNA at the level of individual nucleotides (*9*). Use of nitroxides to study nucleic acids was first reported in the 1970s (*10*). In the 1970s and 1980s, studies were done using DNA and RNA that were uniformly labeled with nitroxides (*11*), as well as using tRNAs labeled at naturally occurring modified nucleotides (*12, 13*). In the 1980s and 1990s, with developments in solid-phase chemical synthesis, nitroxides were incorporated site-specifically into DNA strands, and EPR was used to study DNA duplexes (*11, 14*). In the last decade, a number of methods were reported for attaching nitroxides at specific sites within RNA strands, and SDSL was used to study folded RNA molecules (*9*).

The majority of nucleic acid SDSL studies used one of two types of EPR measurements (Fig. 1) (*9*). Distance measurements between pairs of nitroxides provide direct structural constraints in nucleic acid systems. In addition, the mobility of a single-labeled nitroxide can be measured to yield structural and dynamic information at the labeling site. Because nucleic acids are different from proteins in the nature of the basic chemical constituents (4 nucleotides vs 20 amino acids) and their secondary structural units (B-form/A-form

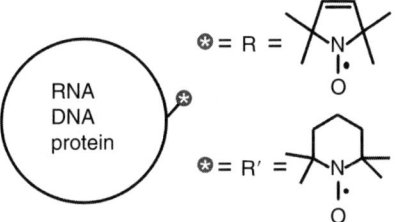

FIG. 1. The general strategy of site-directed spin labeling (SDSL). The first step in SDSL is to covalently attach nitroxides (yellow star) to specific sites of a macromolecule. The chemical structures of the two most commonly used nitroxides (R and R') are shown. In step 2, EPR spectroscopy is used to monitor the behavior of the nitroxide. In step 3, the nitroxide behavior is used to obtain information on the local environment of the labeling site(s). The table shows the two types of EPR measurements that serve as the primary source of information in nucleic acid SDSL studies.

doubled-stranded helix vs α-helix/β-sheet), SDSL of nucleic acids requires unique methodologies, particularly in the areas of nitroxide attachment and the correlation of the nitroxide behavior to that of the parent molecule.

In this chapter, we summarize the current state of the use of SDSL to study nucleic acids, with the discussions focusing primarily on literature from the last decade. The physical principles underlying EPR measurements are first described, followed by an overview of available methods for site-specific attachment of nitroxides to nucleic acids. Then we discuss the use of distance measurements and nitroxide spectral lineshape analysis to study nucleic acids. We conclude with a brief overview of possible future directions in nucleic acid SDSL.

II. Basic Physics Underlying SDSL

A spin label is a molecule that has at least one unpaired electron. Nitroxides, such as those containing a moiety of either R = 1-oxyl-2,2,5,5-tetramethyl-3-pyrroline (with a five-membered unsaturated ring) or R' = 1-oxyl-2,2,6,6-tetramethylpiperidine (with a six-membered saturated ring) (Fig. 1), are commonly used as spin labels in studies of proteins and nucleic acids. These probes are stable radicals and contain one unpaired electron that localizes primarily within the nitroxyl group (15). This unpaired electron has a spin quantum number S = 1/2. The physics describing the behavior of this unpaired electron in the presence of a magnetic field is analogous to that of proton NMR describing the behavior of a proton (nuclear spin I = 1/2) in the presence of a magnetic field. In this section, we outline the basic physics necessary to understand SDSL. The readers are referred to several excellent books and review articles for more detailed descriptions of the physics relevant to SDSL (4, 8, 16–18).

A. The Zeeman and Hyperfine Interaction

The Zeeman interaction is the interaction between an external magnetic field and the magnetic moment of an electron. The magnetic dipole moment of an electron (μ_e) is related to its spin (S)

$$\mu_e = g\beta_e S \tag{1}$$

where β_e is the Bohr magneton (the intrinsic unit of an electron magnetic moment, β_e = 9.274 × 10^{-24} J T^{-1}), and g is the electron spin g-factor. The g-factor of a nitroxide is very close to that of a free electron ($g_e \sim 2.0023$).

In the presence of a static magnetic field B_0, an electron occupies one of two states, which can be thought of as the electron aligning with or against the magnetic field. These two states are described by the quantum number S_z which is the projection of the electron spin angular momentum (S) on an arbitrary axis, z (usually chosen as the direction of the external magnetic field B_0). The nitroxide has S_z = +1/2 or S_z = −1/2, and the energy of each state can be written as

$$E = \mu B_0 = g\beta_e S_z B_0 \tag{2}$$

The energy difference, ΔE between these two states depends on the strength of the magnetic field, and can be expressed as

$$\Delta E = g\beta_e B_0 \tag{3}$$

At resonance, the electron absorbs energy from the external electromagnetic radiation. The resonance condition is defined by

$$\Delta E = g\beta_e B_0 = h\nu \qquad (4)$$

where h is the Planck constant (h = 6.626 × 10^{-34} J s) and ν is the frequency of the electromagnetic radiation. Notice that the electron magnetic moment (μ_e = 9.285 × 10^{-24} J T^{-1}) is 658 times larger than the proton magnetic moment (μ_p = 1.411 × 10^{-26} J T^{-1}). This gives rise to a high sensitivity in EPR detection compared to NMR.

When the resonance condition is met, electrons can be promoted from a low energy state (S_z = −1/2) to a higher energy state (S_z = +1/2), and absorption of the microwave radiation occurs (Fig. 2). This is the basis of the continuous-wave (cw) EPR measurement. Conventionally, the cw-EPR spectrum is measured by sweeping the external magnetic field (B_0) at a fixed microwave frequency. Most studies are done in the X-band, with the microwave frequency at ~9.5 GHz and resonance occurring at B_0 ~ 3400 Gauss (0.34 T). Spectrum is generally obtained using source modulation and the lock-in amplification technique, which increases the signal-to-noise ratio (4, 16). The resulting spectrum is the first derivative of the absorption spectrum (Fig. 2B).

In addition to the Zeeman effect, the unpaired electron also experiences a small local magnetic field produced by nearby nuclei. This is known as the hyperfine interaction, and may counter or enhance the externally applied field. In a nitroxide, the dominant hyperfine interaction is between the unpaired electron and the ^{14}N nucleus of the nitroxyl group. The I = 1 ^{14}N nucleus splits the nitroxide EPR signal into $2I + 1$ = 3 lines, each corresponding to a ^{14}N nuclear spin state (m_I = −1, 0, +1) (Fig. 2A). The resonance condition is modified to include the hyperfine interaction, A

$$h\nu = g\beta_e B_0 + m_I A \qquad (5)$$

The hyperfine interaction is due to the dipolar interaction between the electron spin and the nuclear spin, and includes both isotropic and anisotropic components. The isotropic component is proportional to the electron spin density at the nitrogen and the oxygen nucleus of the nitroxyl group. The anisotropic components arise from the orientation dependence of the electron–nuclei interactions due to the asymmetric orbital of an unpaired electron. For example, the dipolar interaction is stronger when the orbital z-axis is aligned with the external field, and weaker when the alignment is perpendicular. In fact, the Zeeman interaction is also anisotropic, as the asymmetric orbital results in orientation dependence in the g-factor.

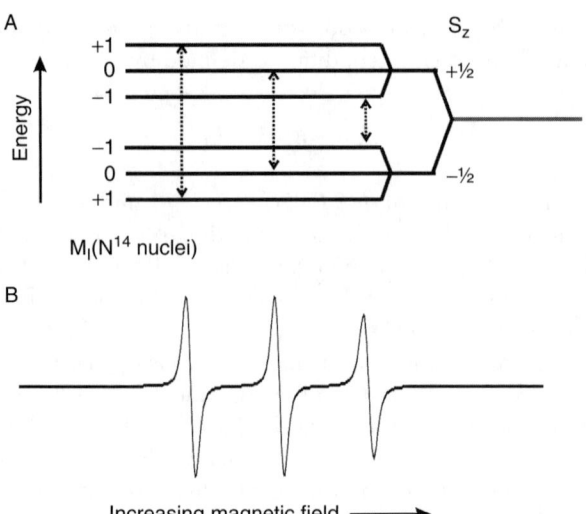

FIG. 2. Electron spin energy level diagram and the continuous-wave EPR spectrum. (A) A schematic of the energy levels of an electron spin. In the absence of an applied magnetic field, the energy levels corresponding to the two electron spin states ($S_z = \pm 1/2$) are degenerate (purple line). In the presence of an applied magnetic field, the electron energy levels separate according to the spin quantum number S_z (the Zeeman effect) and the nuclear spin quantum number M_I (the hyperfine interaction). (B) A representative EPR spectrum. The spectrum is obtained by varying the applied magnetic field strength at a fixed microwave frequency, and is shown as the first derivative of the absorption spectrum. The three peaks correspond to the three transitions indicated by the dotted arrows in the energy diagram shown in (A).

Rotational motions of the nitroxide (dynamics) lead to averaging between the elements of the g tensor and the A tensor, and give rise to variations in the observed cw-EPR spectrum. When the nitroxide dynamics are coupled to the macromolecule, the cw-EPR spectrum can be used to infer information about the local environment around the labeling site (see Section V).

B. The Dipolar Interaction Between a Pair of Electron Spins

When the dipolar interaction between electron spins A and B is considered, an E_{dip} term is added to each electron energy level (32, 80):

$$E_{dip} = D_{dip} \cdot S_z^A \cdot S_z^B \cdot \frac{(1 - 3\cos^2\theta)}{r_{AB}^3} \qquad (6)$$

and

$$D_{dip} = \frac{\mu_0 g_A g_B \beta_e^2}{4\pi} = h\nu_{dip} \qquad (6a)$$

where μ_0 is the vacuum permeability, r_{AB} is the distance between spin A and spin B, and θ is the angle between the distance vector $\mathbf{r}_{AB}$ and the external magnetic field $\mathbf{B}_0$ (Fig. 3A). D_{dip} represents the dipolar splitting of an EPR transition. When $g_A = g_B = 2$, the dipolar frequency $v_{dip} = 51921$ (MHz)·(Å^3) (32, 80).

Eq. 6 indicates that the dipolar interaction depends on θ. When $\mathbf{r}_{AB}$ adopts a static and isotropic distribution with respect to $\mathbf{B}_0$, all θ values are equally populated. This yields a Pake Pattern (Fig. 3B) (19). On the other hand, if the frequency of the spin pair rotation is rapid relative to v_{dip}, no net dipolar interaction can be observed.

III. Site-Specific Attachment of Nitroxides to Nucleic Acids

The first step in SDSL is to attach a nitroxide at a specific site within the macromolecule (Fig. 1). In protein studies, the residue at the intended labeling site is generally mutated to a cystein, and a thiol-reactive nitroxide derivative is then reacted with the cystein –SH group (1, 2, 8). This strategy cannot be applied directly to nucleic acids. There are 4 common nucleotides compared to 20 amino acids, and these nucleotides do not contain functional groups that are reactive under conditions compatible with nucleic acid stability (i.e., aqueous solution and physiological pH).

Currently, the most efficient spin labeling strategies rely on solid-phase chemical synthesis to achieve site-specific attachment of the nitroxide (Table I). Solid-phase chemical synthesis of nucleic acids, primarily based on phosphoramidite chemistry (20, 21), assembles either DNA or RNA strands from the 3′ to the 5′ direction one nucleotide at a time. It is possible to modify the synthesis cycle to introduce a modification at a specific nucleotide. The modified nucleotide may contain a reactive functional group (such as –SH or –NH$_2$), which can be further derivatized with nitroxide. Alternatively, the modified nucleotide may include a nitroxide moiety. Current state of the art solid-phase synthesis is capable of routinely synthesizing DNA oligonucleotides that are approximately 100-nucleotide (nt) long and RNAs that are approximately 50-nt long. Methods have been reported to attached nitroxides to specific phosphate (22), sugar (23, 24), and base positions (25–34) within a chemically synthesized DNA or RNA oligonucleotide (Table I). These methods will be discussed in Section III.A.

Currently, there is no general enzymatic approach to substitute one nitroxide-modified nucleotide at a given site within a DNA or RNA. Enzymatic incorporations of nitroxides at the terminus of nucleic acids as well as at a limited number of internal positions of particular systems have been reported. These methods will be discussed in Section III.B.

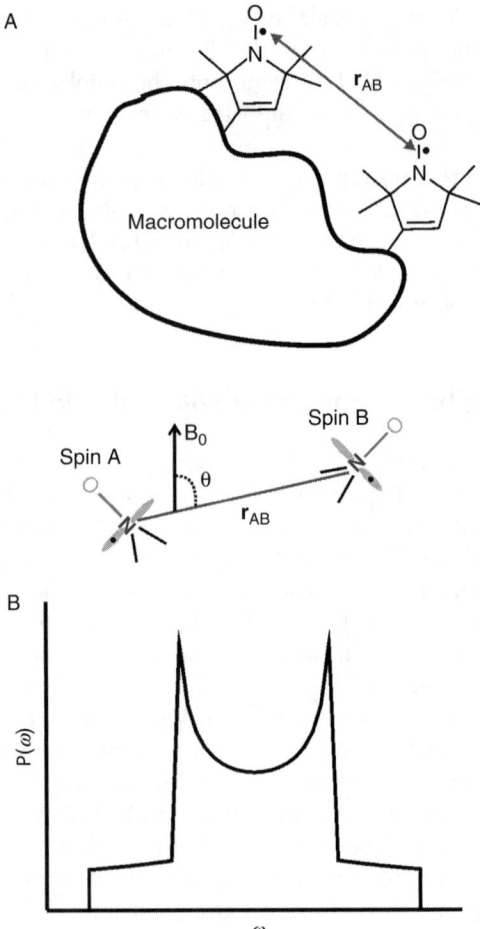

Fig. 3. Dipolar coupling between a pair of electron spins. (A) Schematic diagrams illustrating the attachment of two nitroxides to a macromolecule and the relationship between the distance vector (r_{AB}) and the magnetic field vector (B_0). (B) A Pake pattern illustrating the probability distribution of the dipolar interaction energy expressed in terms of the angular frequency ω. This Pake pattern represents the case where the distance vector adopts a static and isotropic distribution with respect to the magnetic field vector.

A. Nitroxide Labeling Methods Relying on Solid-Phase Chemical Synthesis

1. Nitroxide Labeling at a Specific Internal Phosphate Site

A phosphate group is present in every nucleotide, and is an ideal site for the attachment of a nitroxide at any desired nucleotide within a nucleic acid sequence. Qin et al. have reported a phosphorothioate scheme for attaching a

SUMMARY OF NITROXIDE LABELING SCHEMES THAT USE SOLID-PHASE CHEMICAL SYNTHESIS

Nitroxide attachment site	Examples of chemical structure	References
1 Phosphate		[22]
2 Sugar: 2′-position		[23, 24]

(*Continues*)

Table I (continued)

Nitroxide attachment site	Examples of chemical structure	References
3 Base: 4-position of uridine	3a, 3b	[31, 46]
4 Base: acetylene linkage	4a (T*), 4b, 4c	[27, 30, 32, 34]

5 Base: 5-position of pyrimidines with variable tethers (2-11 atom) [11]

5a

5b

6 Fused base [28, 33]

6a (dQ)

6b (Ç)

nitroxide to a specific modified phosphate position in DNA or RNA (Table I, entry 1) (22, 35–37). There is also a report where an H-phosphonate modification was used to attach a nitroxide to a nucleotide near the 5' terminus of DNA (38).

In the phosphorothioate scheme, the phosphate group at the desired labeling site is modified to a phosphorothioate (substituting a sulfur for one of the nonbridging oxygens) during solid-phase chemical synthesis (22). The phosphorothioate containing oligonucleotide is reacted with an iodomethyl derivative of a nitroxide. The product is a nitroxide linked to the phosphorous of the oligonucleotide by a thioether bond. This nitroxide is designated as R5, and a detailed labeling protocol has been recently published (37).

The phosphorothioate labeling scheme targets a functional group that is present in all nucleotides. Therefore, the same chemistry can be used to attach a nitroxide to any nucleotide position within an arbitrary DNA or RNA sequence. This enables one to efficiently conduct nitroxide scanning, where the probe is systematically attached to different positions in a given DNA or RNA sequence. Studies have demonstrated that R5 labeling is 90–100% efficient (22, 35, 36), and that the presence of R5 does not severely perturb the conformation of DNA and RNA duplexes (35, 36). The cost of introducing a phosphorothioate is less than 1/20 of the cost of adding a modified nucleotide, and R5 labeling is a one-step reaction in the aqueous solution. These features facilitate spin labeling of large molecules. The phosphorothioate scheme has been demonstrated on a 68-nt-long DNA (MW ~ 21 kDa) (35).

There are several additional features of R5 that should be taken into account when designing an experiment (37). First, the pyrroline ring is linked to the nucleic acid (represented by the phosphorous atom at the labeling site) through three rotatable single bonds (Table I, entry 1). While such flexibilibty may minimize the effect of the nitroxide on the local structure of the parent molecule, there is not a uniform, direct correlation between the the location of the nitroxide and the attachment point at the parent molecule. Second, solid-phase synthesis introduces two phosphorothioate diastereomers (R_p and S_p) in an approximately 50/50 ratio, and nitroxides attached to different disastereomeric centers at the same nucleotide may experience different local environments. In short oligonucleotides (up to ~15 nt), the diastereomers can be separated using HPLC (37, 82), and one can carry out SDSL studies on either mixed diastereomers or on pure diastereomers (82, Popova and Qin, manuscript in preparation). Third, at the R5-labeled nucleotide, one negative charge at the phosphate group is lost, which may pose problems at certain labeling sites or in some studies (e.g., electrostatic measurements). Finally, when using the phosphorothioate scheme to label RNA, one has to remove the 2'-hydroxyl group adjacent to the phosphorothioate to prevent strand scission upon labeling (37). This may lead to additional perturbation to the structure, dynamics, and function of the parent molecule.

The R5 nitroxide has been used to measure distances in both DNA and RNA (35, 36) (Section IV.B), and an efficient computation program has been established to interpret the measured distances based on the structure of the parent molecule (37, 39) (Section IV.C.1). Together these provide the necessary elements for using R5 to map the global structure (as defined by the relative positioning of helical axes) of nucleic acids. In addition, studies have shown that cw-EPR spectra of nitroxides labeled using the phosphorothioate scheme can provide information on the local environment in DNA and RNA (Popova and Qin, manuscripts in preparation).

2. Attaching Nitroxide at the 2′ Position of the Sugar

Schemes for attaching a nitroxide to the 2′ position of the sugar of RNA have been reported by the Sigurddsson (23) (Table I, entry 2a) and DeRose groups (24) (Table I, entry 2b). In these schemes, a modified phosphoroamidite containing a 2′-NH_2 group is substituted at the desired labeling site. The 2′-NH_2 group is then reacted with an amine-reactive nitroxide derivative. This connects the nitroxide to the sugar of the nucleotide by a semirigid linker (Table I, entry 2).

Attachment of a nitroxide at the 2′-sugar position does not significantly alter the structure and function of the RNA (23, 24). Distance measurements (24, 40) (Section IV.A) and EPR lineshape analysis (Sections V.D.4 and V.D.5) have been used to study RNA in this fashion (23, 41–45). In principle, this labeling scheme can be applied regardless of the base identity. However, incorporating 2′-NH_2-modified purines (A and G) is difficult, and currently this labeling scheme has only been used to attach nitroxides to specific uridine (23, 24) and cytosine nucleotides (44) within RNAs.

3. Nitroxide Labeling at base Positions

The reported methods for attaching nitroxides to nucleic acid bases can be grouped into three classes. One set of methods attaches nitroxides postsynthetically to an oligonucleotide containing a modification at one specific base. The second group of methods involves on-column nitroxide derivatization during chemical synthesis. The third group of methods directly incorporates nitroxide-containing phosphoramidites into nucleic acid strands.

The postsynthetic derivatization approach uses a similar strategy to that used in backbone (Section III.A.1) and sugar labeling (Section III.A.2). Two groups have reported introducing a 4-thio uridine (4-thioU) at specific sites within RNA oligonucleotides, and then labeling the RNA with various thio-reactive nitroxides (Table I, entry 3) (31, 46). The 4-thioU method has been used to probe structural and dynamic features of folded RNA (31) (Section V.D.3) and to obtain long-range distance constraints in protein/RNA complexes by measuring the paramagnetic relaxation enhancement (PRE) effects (46–49) (Section VI.C). The 4-thioU scheme has the advantage that the coupling between the nitroxide and the oligonucleotide can be carried out at near physiological pH in aqueous solution. However, upon nitroxide labeling at the 4-position of the uridine base, the N3 proton is lost (Table I, entry 3), which affects the Watson–Crick base-pairing between A and U (*31*).

Another class of base labeling methods uses an on-column derivatization strategy. The nitroxide is reacted with the modified base during the solid-phase chemical synthesis, and the labeled oligonucleotide is separated from the reagents and by-products in the wash step during chemical synthesis. Schiemann *et al.* have used this approach to attach nitroxides to 5-iodouridine, 5-iodocytidine, and 2-iodoadenine (Table I, entry 4) (*32, 34, 50, 51*). The resulting nitroxide moiety does not interfere with Watson–Crick base pairing, and is rigidly attached to the base with an acetylene bond. Long-range distance measurements using this class of nitroxides have been demonstrated in both DNA (*32*) and RNA (*34, 50*) (Section IV.B). However, during chemical synthesis, the nitroxide is exposed to various chemicals as it goes through multiple synthesis cycles and deprotection steps. This may lead to undesired products such as reduced nitroxides that have no EPR signal (*32, 34, 50*).

The third class of methods directly inserts a nitroxide-containing phosphoroamidite into the oligonucleotide during chemical synthesis. In the 1980s and 1990s, Bobst *et al.* synthesized a series of nitroxide-modified 2′-deoxy pyrimidine nucleosides, where the nitroxide moiety was tethered to the C-5 positions by linkers of various lengths and flexibilities (see two examples in Table I, entry 5) (*11*). These modified nucleosides are subsequently converted to building blocks used in solid-phase chemical synthesis, and are incorporated into DNA strands (Section V.D.2) (*11*). Hopkins *et al.* synthesized oligonucleotides using a phosphoroamidite called T°, in which the nitroxide is linked to the base by an acetylene bond (Table I, entry 4a) (*27*). Recently, Gannett *et al.* reported a simplified scheme for the synthesis of a T° variant that contains a six-membered ring nitroxide (*30*).

Hopkins *et al.* developed a heavily modified nucleoside called dQ (Table I, entry 6a) after their work on T° (*28*). dQ has a nitroxide fused to an oxoquinoline and can base pair with 2-aminopurine. It has been used extensively in studying DNA duplexes (Section V.D.2). Very recently, Sigurdsson *et al.* reported a new probe, Ç (C-spin), in which the nitroxide is fused to a phenoxazine derivative (Table I, entry 6b) (*33*). This modified nucleoside can form hydrogen bond with a guanine in the same fashion as that of a Watson–Crick G/C pair. Upon reduction of the nitroxide, Ç is found to be highly fluorescent (*33*), allowing multiple studies to be done on the same sample. However, synthesis of both dQ and Ç is very challenging, and currently has only been reported for DNA.

B. Nitroxide Labeling Using Enzymatic Methods

1. Nitroxide Labeling at the Nucleic Acid Terminus

Grant and Qin recently reported a 5ps labeling scheme for nitroxide labeling at the 5′ terminus of arbitrary sized nucleic acids (52). In this method, T4-polynucleotide kinase was used to enzymatically substitute a phosphorothioate group at the 5′ terminus of either DNA or RNA. The resulting phosphorothioate was reacted with an iodomethyl derivative of a nitroxide. The method was successfully demonstrated on both chemically synthesized and naturally occurring nucleic acids. Duplex formation and tertiary folding of nucleic acids were observed using these nitroxide labels.

Shin et al. reported a method for attaching nitroxides to the 5′ terminus of RNAs synthesized by the T7 RNA polymerase (53). In this method, a modified nucleotide, guanosine 5′-O-(3-thiomonophosphate) (GMPS), is added to a mixture of unmodified nucleoside triphosphates. T7 RNA polymerase places GTP or GMPS at the 5′ terminus of the RNA transcript, but incorporates only GTP at the subsequent internal positions. With a large excess of GMPS to GTP (3:1), a majority of the RNA transcripts possesses a 5′ phosphorothioate instead of a 5′ triphosphate, and the phosphorothioate group is subsequently reacted with a thiol-reactive nitroxide derivative.

Labeling at the 3′ terminus was reported on tRNA (54). In this method, the ribose at the 3′ terminus was oxidized with periodate, and then reacted with a hydrazine derivative of nitroxide. However, this method cannot be applied directly to DNA.

2. Enzymatic Incorporation of Nitroxide-Modified Nucleotides

A number of nitroxide-labeled nucleotide triphosphates (NTPs) have been reported to serve as acceptable substrates for nucleic acid-synthesizing enzymes. For example, Bobst and co-workers synthesized a variety of triphosphate derivatives of modified uridine, deoxyuridine, and deoxycytidine, in which the nitroxide is attached to the 5 or 4 position of the base by flexible tethers (e.g., Table I, entry 5) (11). Most of these nitroxide-modified NTPs were successfully incorporated into nucleic acid strands by enzymes, including template-independent enzymes, such as terminal deoxynucleotide transferase (55), or template-dependent enzymes, such as DNA polymerase I from Escherichia coli. (56) and Taq polymerase (57). Studies found that nitroxide-modified NTPs with 1–2 atom tethers do not incorporate well, indicating that some flexibility in the tether is required (11).

Perhaps the biggest challenge in enzymatic incorporation of nitroxides is to achieve site-specificity. For example, one can attach nitroxides to all Ts within a given DNA sequence by replacing dTTP with a nitroxide-labeled deoxyuridine triphosphate, or one can randomly replace a certain percentage of Ts by mixing dTTP and the nitroxide-labeled variant. However, to replace only one specific T, one generally has to pick a sequence that has only one T present (58). This limits experimental design.

C. Future Developments

In general, the chemical synthesis approach allows one to direct the nitroxide to a specific location, but is limited by the size of the DNA and RNA that can be synthesized. The enzymatic method is more suitable for large DNA and RNA molecules, but it is difficult to place the nitroxide at one specific internal site. Combining the two approaches may allow site-specific nitroxide labeling of large DNA and RNA molecules, and should be explored in the future.

IV. Distance Measurements Using SDSL

In SDSL, internitroxide distance measurements provide structural constraints, which can be used to determine the global structure of a macromolecule and to monitor conformational changes. SDSL distance measurements do not require crystalline samples and are not limited by the molecular weight of the system. This makes SDSL useful for systems that are difficult to study using X-ray crystallography or NMR spectroscopy.

While both SDSL and flouroscence resonance energy transfer (FRET) use a pair of molecular probes to measure distances in macromolecules, there are a number of features that distinguish SDSL from FRET. In SDSL, distances are measured using a pair of chemicially identical nitroxide probes, which simplifies the labeling procedure as compared to attaching two chemically distinct fluorophores. Nitroxides are smaller than most fluorophores, causing less structural perturbation as compared with larger probes. As shown in Eq. (6), EPR measurements explicitly account for the angle (θ) between the internitroxide distance vector and the external magnetic field, whereas in FRET, a complete average of the angular dependence is generally assumed. Interpretation of distances measured using SDSL is also facilitated by the fact that the unpaired electron is localized to the nitroxyl group in a nitroxide (15), whereas relationships between the dipole moment and the chemical structure of the label vary with the choice of fluorophophore. On the other hand, single-molecule FRET measurements in biological systems are now routine, while single-molecule detection of electron spins is still being developed (59).

The fact that a nitroxide is an extrinsic molecular probe has important implications in distance measurement and data analysis and interpretation. Hustrdt and Beth have categorized the majority of SDSL distance measurements into three cases (60, 61). In the first case, called statically arranged spins, each nitroxide adopts a unique conformation with respect to the parent macromolecule, while rotational reorientation of the macromolecule can be treated as static on the EPR timescale (rotational correlation time is 1 μs or longer for X-band EPR). This means that there is a unique internitroxide distance vector with respect to the macromolecule. The magnitude and orientation of this vector can be determined analytically (62). However, most studies of proteins and nucleic acids do not fall into this category due to nitroxide motions.

In the second case, the macromolecule can be treated as static on the EPR timescale, and a distribution is observed in both the magnitude and the orientation of the internitroxide distance vector. This case, called statically disordered spins, describes the majority of SDSL distance measurements in proteins and all reported studies in nucleic acids. The goal is then to describe a distance and orientation distribution $P(r, \theta)$ that best models the true distribution and provides the best fit to experimental data. The determination of $P(r, \theta)$ may be limited by both theoretical and experimental factors such as available information on nitroxide conformer distributions, noise in the experimental data, and the number of parameters that can be uniquely determined.

The third case is when the nitroxides or macromolecules or both are undergoing significant rotational reorientation on the EPR timescale. These motions partially or completely average the dipolar interaction, resulting in a reduction or elimination of observable dipolar interaction effects. An example in this case has been reported in SDSL distance measurements in the protein T4 lysozyme (63), where global tumbling of the small protein leads to complete averaging of the dipolar interaction. In this study, it was shown that perturbations due to dipolar interactions give rise to line broadening in the cw-EPR spectrum. The amount of broadening has a $1/r^6$ dependence on the internitroxide distance (63).

In the following sections, we will describe distance measurements in nucleic acids using cw-EPR (Section IV.A) and pulsed EPR (Section IV.B). Pulsed EPR techniques are used to measure distances between approximately 20 and 80 Å, whereas cw-EPR is used to measure distances below 25 Å. Reported methods for correlating measured internitroxide distances to the stucture of the parent molecule will be discussed in Section IV.C.

A. Short-Range Distances Measured Using cw-EPR

Continuous wave EPR (cw-EPR) measures the absorption of external radiation by an unpaired electron, which is determined by the electron's spin energy levels (section II.A). The electron-electron dipolar interaction can be thought of as a perturbation to the energy levels of the non-interacting spin, and therefore, this interaction affects the cw-EPR spectrum. At a distance of 25 Å, dipolar interaction splits EPR absorption lines by ~ 2 Guass (64, 65). This is small compared to the cw-EPR spectrum linewidth of a frozen ^{14}N nitroxide, and is difficult to measure. Therefore, cw-EPR measures distances < 25 Å.

All reported cw-EPR distance measurements in nucleic acids are examples of statically disordered spins, with the dipolar interaction manifesting itself as an overall spectral broadening (24, 53). To use cw-EPR to measure the dipolar interaction between sites A and B of a macromolecule, three spectra should ideally be measured. These three spectra are of nitroxides attached to site A only, to site B only, and simultaneously to A and B. By comparing the lineshape of the double-labeled spectrum (labels at both A and B) to the sum of the two single-labeled spectra (A only + B only), the broadening function can be determined. Distances can be determined from these measurements using deconvolution (64–66) or convolution (67) methods.

There are two reports in the nucleic acid literature on cw-EPR distance measurements. Shin *et al.* measured distances between the HIV Rev peptide and the Rev response element (RRE) RNA (53). One nitroxide was attached to a fixed residue within the Rev peptide, and the other nitroxide was attached to 5′ terminus of the RNA using the GMPS scheme (Section III.B.1). A combination of circular permutation and mutation was used to move the 5′ terminus of the RNA to three different locations within the tertiary structure. The three measured distances are in agreement with the available structures of the RRE/Rev complex (53). DeRose *et al.* used nitroxides attached to 2′-sugar positions of specific uridines (Table I, entry 2b) to measure distances in RNA (24). Measurements in a 10-mer of RNA were found to be in qualitative agreement with values expected based on an idealized A-form duplex. Furthermore, cw-EPR distance measurements were used to study ion-dependent conformational changes in the TAR RNA (24).

Distances measured using cw-EPR are technologically simple, but are limited to distances <25 Å. cw-EPR may be most suitable for structural probing in a local region, as well as for monitoring conformational changes in which some or all distances involved are less than 25 Å. To ensure that the condition of statically disordered spins was met, the reported nucleic acid studies measured samples in a frozen glassy state (24, 53). However, it is also possible to measure distances in samples at room temperature (66).

B. Nanometer Distances Measured Using Pulsed EPR

Pulsed EPR uses short (ns) but intense (kW) pulses of microwave radiation to probe the behavior of the unpaired electron(s), providing much better spectral and time resolution than cw-EPR (68). While the theory and technology of pulsed EPR is closely related to that of modern (pulsed) NMR spectroscopy, EPR specific modifications are required due to unique features of electron spins, such as their large magnetic moment and short relaxation time (approximately three orders of magnitude shorter in EPR than in NMR) (68). Currently, a commercially available pulsed EPR spectrometer (e.g., ELEXSYS

E580 from Bruker Biospin, Inc.) is capable of repeated detection (every 1 s) of tiny signals (<1 nW) that occur tens of nanoseconds, after a powerful (>1 kW) applied microwave pulse. More powerful instrumentation is also being developed (69). This will provide many opportunities to use pulsed EPR in studying biological systems (68).

A number of EPR pulse schemes have been developed for measuring distances between two paramagnetic centers (70–79) [see also a recent review by Schiemann and Prisner (80)]. These schemes aim to achieve precise distance measurements by completely separating the electron dipolar interaction from all other interactions. Two general classes of pulse schemes have been used to measure distances in nucleic acids. One class of methods, developed by the Freed group, measures dipolar interactions by monitoring the time-dependence in the generation of double-quantum coherence (DQC) involving both spins (5, 69, 75, 76, 78, 81). The measurements use strong, nonselective pulses to excite all radical populations in the system. The DQC signal shows an oscillating pattern with the oscillation frequency giving a direct measurement of the dipolar interaction. DQC schemes with 4, 5, and 6 pulses have been developed (69). A variant of the six-pulse scheme, called double-quantum filtered refocused electron spin echo, has been used to measure a distance in a 26-base-pair RNA duplex (81). In this study, the nitroxides were attached to two 4-thioU bases located at the termini of the duplex, and a distance of 72 Å was measured. This represents the longest measured distance reported in nucleic acids to date.

Another pulsed EPR method, based on electron–electron double resonance (ELDOR) techniques, has been applied to measure distance in a number of DNA and RNA systems (32, 34–36, 40, 83). In ELDOR measurements, the sample is irradiated with microwave pulses at two frequencies, called the observer and pump frequencies, corresponding to two different regions of the EPR spectrum (see Fig. 4 for an example). The spectral region corresponding to the observer frequency is monitored, and any pump pulse-dependent change is called an ELDOR response.

The dipolar interaction between electron spins is one of the mechanisms that give rise to an ELDOR signal. One ELDOR scheme that allows selective measurement of the dipolar interaction is pulsed double electron–electron resonance (DEER, also referred to as pulsed ELDOR). The first published DEER scheme used a three-pulse sequence that was based on a method for detecting nuclear spin coupling (the Hahn sequence) (70, 72). An extended four-pulse DEER scheme was subsequently developed and has the advantage of being able to provide dead-time free time domain data (79, 84). The pulse scheme for the four-pulse DEER is shown in Fig. 4. At the observer frequency (v_1), a refocused echo with a fixed position in time is detected for spins in resonance (defined as spin A). Between the second and third v_1 pulses, an

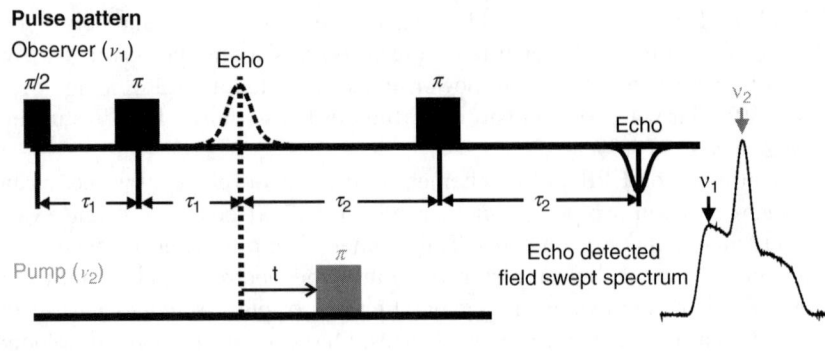

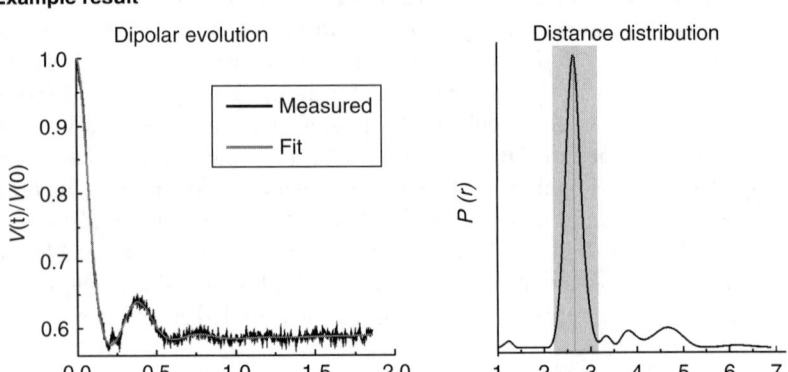

FIG. 4. Double electron-electron resonance (DEER) spectroscopy. Shown on top is the pulse sequence for the four-pulse DEER. Shown at the bottom is a set of DEER data measured in a dodecameric DNA duplex (35). The DEER measurement gives an average distance $\langle r_{DEER} \rangle = 26.4$ Å. The standard deviation (σ_{DEER}), which characterizes the width of the distance distribution, is 1.7 Å. The predicted distance, computed using the NASNOX, is $\langle r_{NASNOX} \rangle = 26.8$ Å and $\langle \sigma_{NASNOX} \rangle = 2.8$ Å (35). Reproduced from references (35, 37) with permission.

inversion pulse is applied at the pump frequency (ν_2), which flips spins that are in resonance with ν_2 (defined as spin B). Dipolar coupling between spin A and B results in the modulation of the amplitude of the refocused echo (Fig. 4). The modulation frequency is a function of the interspin distance.

DEER has been used to measure distances from 20 to 70 Å in DNA and RNA. Distances in DNA duplexes have been measured using the four-pulse scheme. In 2004, Schiemann et al. measured distances between nitroxides rigidly attached to specific base positions (Table I, entry 4) in a DNA duplex (32). Five distances, ranging from 20 to 55 Å, were measured and were shown to agree very well with values predicted by MD simulations of a standard B-DNA. In 2006, the Qin group used nitroxides attached via the phosphorothioate scheme (Table I, entry 1) to measure distances in a DNA dodecamer and a 68-base-pair duplex (35). Eight distances, ranging from 20 to 40 Å, were measured on the dodecamer (see an example in Fig. 4). The measured distances were found to correlate strongly (R^2 = 0.98) with values calculated using the NMR structure and a conformer search program (Section IV.C). In 2007, Ward *et al.* used nitroxides attached to the 2′-sugar position (Table I, entry 2a) to measure distances in DNA duplexes. Five distances were reported, with the largest distance being 68 Å (83).

Distance measurements in RNA have also been reported. In 2003, Schiemann *et al.* used a three-pulse scheme to measure a distance between nitroxides attached to 2′-sugar positions (Table I, entry 2a) in a RNA duplex (40). This was the first report of SDSL measurements of distances greater than 30 Å in an RNA system. In 2007, distance measurements in RNA duplexes were reported by Schiemann *et al.* using nitroxides rigidly attached to base positions (Table I, entry 4) (34) and by the Qin group using flexible nitroxides attached to the phosphate backbone (Table I, entry 1) (36). These studies found that the measured distances agree very well with those predicted based on either the crystal structure (36) or a standard A-form conformation (34). It was also shown that SDSL measured distances clearly distinguish between A-form duplexes and B-form duplexes (34).

While distance measurements using DEER have been demonstrated in many nucleic acid systems, a number of practical issues are worth noting. The concentrations of the double-labeled sample are generally between 100 and 150 μM (32, 34–36, 40), although concentrations as low as 25 μM have been used (83). Additionally, in DEER, the lower limit of the measured distance is approximately 15 Å, which is set primarily by the excitation bandwidth requirement (84). The upper limit is approximately 80 Å, and is determined by the phase memory time (T_m) characterizing the dephasing of the electron spin transverse magnetization (84). To achieve a sufficiently long T_m for nanometer distance measurements, samples are measured in a frozen glassy state. In all reported nucleic acid studies, samples were rapidly frozen in liquid nitrogen with glycerol or ethylene glycol added to samples, and EPR measurements were carried out at temperature between 10 and 80 K. Using deuterated solvents can lead to a longer T_m and facilitates long distance measurements (83). Under these experimental conditions, one needs to guard against formation of clusters (e.g., ice crystals), which would negate the assumption of an isotropic distribution of macromolecular orientations (32, 35, 40, 83). Possible interference due to intermolecular spin coupling also needs to be considered (32, 35, 40, 83).

It is also important to consider the procedure for obtaining interspin distances from measured DEER signals, as well as the interpretation of these distances. In studies with rigidly attached nitroxides (Table I, entry 4), the interspin distance distribution is presumably rather narrow, and the interspin distances have been calculated directly from measured Pake patterns (32). However, in most cases, fitting of the DEER signal is required to obtain a distribution, $P(r)$, of the interspin distances. The Tikhonov regularization approach, which does not assume a specific form of $P(r)$, has been the method of choice (85, 86), though options are available for fitting data using model functions such as one or two Gaussians (85). The resulting $P(r)$ may show one population with a narrow width, one population with a broad width, or multiple populations. The interpretation of $P(r)$ is still debated. In a recent study, Ward et al. showed that under carefully controlled conditions, a $P(r)$ with features of multiple populations can be analyzed to recover five distance populations that match the five DNA species in the sample (83). In a number of studies, the average distance (35, 36) or the maximum probable distance (34, 83) is used to represent the interspin distance.

C. Interpretation of the Measured Internitroxide Distances, Computational Approaches

Although the distance between the unpaired electrons in a pair of nitroxides can be accurately measured, a challenge in SDSL studies is to correlate the measured internitroxide distances to the three-dimensional structure of the target macromolecule. The unpaired electron in a nitroxide is predominantly localized to the nitroxyl group (15), which is spatially distinct from the site of interest (the nitroxide attachment point). The measured internitroxide distances (r_{NN}, between the two unpaired electrons) are generally different from the actual distances at the parent molecule (r_{target}). The r_{NN}/r_{target} relationship depends on the bonds that link the nitroxide to the target molecule, which varies significantly depending on the nitroxide labeling scheme used (Section III). Furthermore, even if the same nitroxide labeling scheme is considered, the nitroxide conformation may vary significantly between different sites of a folded macromolecule, and the r_{NN}/r_{target} relationship may vary from site to site within the same target molecule. The key issue in correlating r_{NN} and r_{target} is to identify the nitroxide conformers that are allowed at a given labeling site. In SDSL distance measurements in nucleic acid systems, this issue has been adressed using a number of computational approaches, including stepwise conformer searches (35, 36, 39), molecular dynamics (MD) simulations (24, 32, 34, 39), and simple geometry modeling (83).

1. NASNOX, A Discrete Conformer Search

Haworth et al. have developed a program, called NASNOX (35, 36, 39), that predicts the distances between two R5 nitroxide probes (nitroxides that are attached to DNA or RNA molecules using the phosphorothioate scheme, Table I, entry 1) (Fig. 5). In the NASNOX program, the target macromolecule is input as a pdb file. The target may have an atomic resolution structure that has been experimentally determined by X-ray crystallography or NMR spectroscopy, or it may be a theoretical model generated using computational approaches.

Once a pair of nitroxide labeling sites is specifed at the target molecule, NASNOX models R5 at these sites using experimentally determined bond lengths and bond angles (Fig. 5A) (39). As directed by the user, the R5 probe

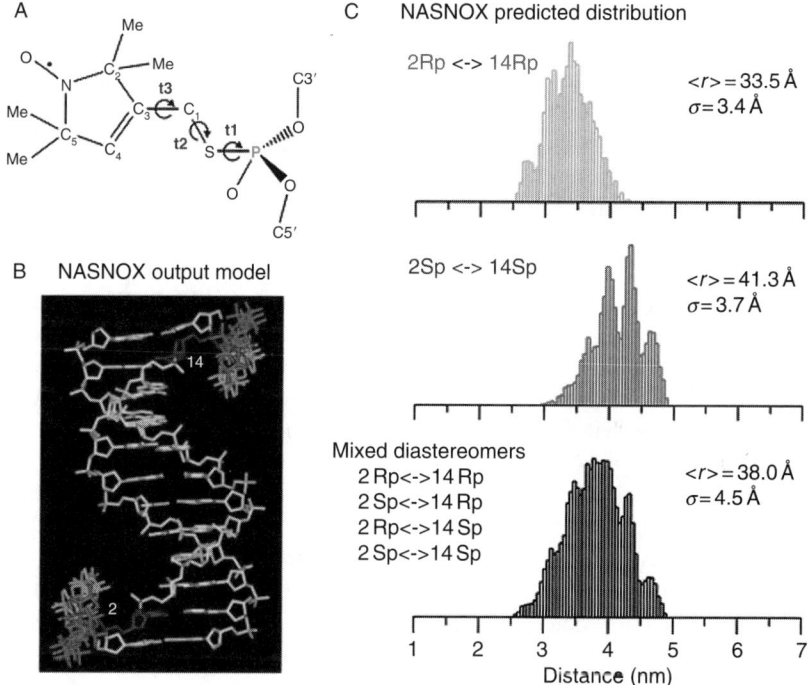

FIG. 5. The NASNOX program for predicting inter-R5 distances. (A) A schematic of the R5 nitroxide probe. The configuration shown is the R_p diastereomer. The three torsion angles (t1, t2, and t3) that are varied in the conformer search are marked. (B) An example of structural output from NASNOX. Shown as colored sticks are the predicted allowable R5 conformers at two sites near the termini of a dodecameric DNA duplex. At each site, nitroxides attached to the R_p diastereomer are shown in green, and nitroxides attached to the S_p diastereomer are shown in red. (C) The predicted internitroxide distance distribution obtained from the output structure shown in (B). The DEER measurement using mixed diastereomers gave a $\langle r_{DEER} \rangle$ of 38.8 Å (35), which is consistent with the NASNOX prediction of $\langle r \rangle = 38.0$ Å. Data reproduced from (35) with permission.

can be attached to either the R_p, the S_p, or both phosphorothioate diastereomers at a given site. To identify the allowable nitroxide conformers at each labeling site, the program systematically varies the three torsion angles that characterize the rotations of the single bonds linking the pyrroline ring to the nucleic acid (t1, t2, and t3, Fig. 5A). For each nitroxide conformer, defined by a set of t1, t2, and t3, with the nucleic acid coordinates fixed, the program checks for steric collisions between any atom of the nitroxide and any atom of the parent molecule. The conformer without steric collision is recorded as an allowable conformer (Fig. 5B). Once the allowable conformers are identified, the ensemble of inter-R5 distances is computed, and the corresponding mean and standard deviation of the distances are calculated (Fig. 5C).

The NASNOX program has been successfully tested on model DNA and RNA molecules that have atomic resolution structures available (35, 36). For inter-R5 distances >20 Å, the mean distances computed by NASNOX agree very well with those measured experimentally (35, 36). The biggest advantage of NASNOX is its speed—each R5 conformer distribution and the corresponding inter-R5 distances can be computed in seconds to minutes on a desktop PC. An internet-accessible version of NASNOX, called NASNOX-W, is available at http://pzqin.usc.edu/NASNOX/ (37).

To ensure the efficiency of the conformer search, NASNOX makes two key assumptions. First, the allowable conformers are determined solely by steric exclusion from DNA, while other forces, such as electrostatic interactions, hydrophobic effects, and hydrogen bonding, are not considered. This assumption is supported by all atom MD simulations of R5-labeled DNA (39) (Section IV.C.2). The second assumption in NASNOX is that the parent molecule coordinates are fixed during the nitroxide conformer search. In the model DNA and RNA studies, the dynamics of the parent molecule seem to have only minor effects on the average inter-R5 distances, as the values determined by DEER, computed by NASNOX, and predicted by MD agree well with each other (35, 36, 39). Less-satisfying correlations were found between the NASNOX predicted widths of the distance distributions (the standard deviation of the distribution) and those measured from DEER. This may be due, in part, to dynamics of the parent molecule (35, 36, 39). Currently, the NASNOX program works only with the R5 label, but the underlying principles should be applicable to other probes.

2. MD Simulations

MD simulations have been used in interpreting measured distances in SDSL studies of nucleic acids (24, 32, 34, 39) and proteins (87). MD simulations account for the physical forces at an atomistic level, and should, in theory, provide a complete description of the behavior of the nitroxide. In practice, in order for a MD simulation to serve as an accurate representation of the behavior of the experimental system, a number of issues need to be addressed. These include the force field (for both the nucleic acid and the nitroxide), the solvent model, and ergodicity (sufficient sampling of the conformational space).

Price *et al.* used MD simulations to examine internitroxide distances between a pair of R5s attached to a DNA dodecamer (39). The simulations used amber98 force field and a full atomistic solvent model. The R5 nitroxide was parameterized based on an empirical approach, and the force field was reported (39). Simulations with different nitroxide starting conformations give convergent results in both the average internitroxide distances and the nitroxide conformer distributions, indicating sufficient sampling of the conformational space. The average distances obtained from the MD simulations agreed with both those determined experimentally by DEER and those predicted using NASNOX. The MD simulations revealed discrete transitions between chemically preferred values (*gauche*$^+$, trans, *gauche*$^-$) of the torsion angles t1, t2, and t3 (Fig. 5A), as well as steric exclusion between the nitroxide and the DNA. Both of these features are adequately represented in the NASNOX search algorithm, accounting for its success. Simulations also revealed a possible sequence-specific nitroxide–DNA interaction, which was recently confirmed by experiments (Popova and Qin, manuscript in preparation).

Schiemann *et al.* have used MD simulations to interpret distances measured using a class of nitroxides that are linked to a nucleic acid base by an acetylene bond (Table I, entry 4) (32, 34). Their studies used the amber98 force field and an atomistic solvent model. Quantum mechanical calculations based on density functional theory were used to parameterize the nitroxide. The results show a good correlation between the predicted internitroxide distances compared to those measured using DEER (32, 34). For DNA duplexes, it was reported that the B-form structure is maintained in nitroxide-labeled DNA, and that fluctuations in internitroxide distances are largely due to DNA dynamics (32). In the RNA studies, nitroxide-labeled duplexes were found to adopt a nonstandard conformation that is in between the A-form and the B-form, and transient base pair opening in nitroxide-labeled RNA duplexes was detected in the 50-ns trajectories (34).

DeRose *et al.* have carried out short MD simulations on a pair of nitroxides attached to the 2′-positions of RNAs (Table I, entry 2b) (24). Simulations were carried out on idealized A-form RNA duplexes and NMR-derived structures of the TAR RNA. Results were used to interpret cw-EPR measured distances.

Darian and Garnet also reported MD simulations on a single nitroxide that is rigidly attached to base positions of DNA duplexes (88). The reported work includes studies on both five- and six-membered ring nitroxides. The nitroxide was parameterized based on quantum mechanic calculations and crystallographic data. Using data from the MD simulations, the authors examined the influences of the nitroxide on the DNA structural parameters, such as the helical groove width, helical twist, and displacement of bases.

3. GEOMETRY-BASED MODELING

To interpret distances measured in DNA duplexes, Ward *et al.* modeled the nitroxide as a pseudoatom attached to a simple cylinder that is parameterized based on standard B-form DNA (*83*). The measured internitroxide distances were fit to the modeled distances by adjusting a set of geometric parameters that describe the positioning of the nitroxide pseudoatom with respect to the DNA and the other nitroxide. A good agreement between the measured and modeled distances was obtained with one set of parameters, suggesting this geometrical model provides a sufficiently accurate representation of nitroxide-labeled DNAs. The geometrical model also agreed with studies that examine the nitroxide pseudoatom in the context of an atomistic model of standard B-form DNA.

D. Distance Measurement Summary

Methods for measuring and interpreting internitroxide distances have now been established. This makes it possible to use SDSL to probe the conformation of a nucleic acid molecule with no known atomic structure. Using a molecular probe, such as a nitroxide, it may not be possible to define an atomic-resolution structure. Instead, internitroxide distances may reveal the global structure of nucleic acids, as defined by the relative spatial orientations between various duplexes within a folded molecule. Work toward this goal is currently underway.

V. Site-Specific Structural and Dynamic Information from a Single-Labeled Nitroxide

CW-EPR spectral analysis is the method most frequently used to obtain information on the local environment of a spin label (*2–4, 6–9*). Both the g and A tensors of the unpaired electron within the nitroxide are orientation dependent. The cw-EPR spectrum is influenced by the reorientation dynamics of the nitroxide, which average the g and A tensors. It is possible to analyze the EPR spectrum to obtain information on nitroxide dynamics, and to use this information to better understand features of the local environment.

For a nitroxide covalently attached to a macromolecule, such as a DNA or RNA, the nitroxide dynamics can be thought of as being due to a combination of three types of motions (Fig. 6A). These are the overall rotational motion of the macromolecule (characterized by a rotational correlation time τ_R), the

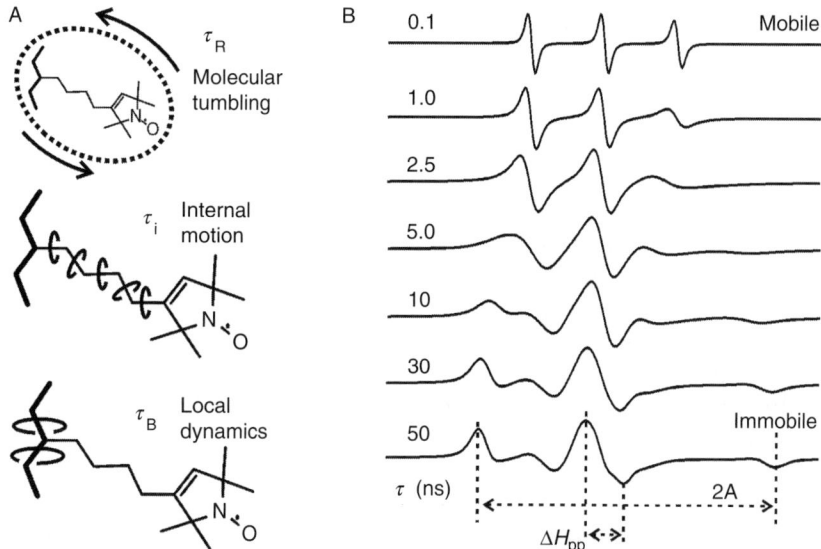

FIG. 6. The relationship between nitroxide dynamics and cw-EPR spectral lineshapes. (A) The three modes of motion that contribute to nitroxide dynamics. (B) Simulated X-band EPR spectra of nitroxides undergoing isotropic rotation at different rotational correlation time τ.

torsional rotation about bonds that connect the nitroxide ring to the macromolecule (τ_i), and the motion of the segment of the macromolecule at or near the labeling point (τ_B) (2). The τ_R motion is uniform at all labeling sites, whereas the τ_i and τ_B motions are specific to the labeling site. The combined effects of τ_R, τ_i, and τ_B determine the observed EPR spectrum. Experimental conditions can be manipulated to enhance spectral effects that are specific to particular mode(s) of motion. For example, site-specific features (τ_i and/or τ_B) can be enhanced by reducing the overall tumbling of the macromolecule, which can be accomplished by increasing solvent viscosity (2, 8, 9).

In the following sections, we will first discuss methods for characterizing nitroxide dynamics from cw-EPR spectra. Then we will discuss approaches for extracting information about the local environment from the nitroxide dynamics. Specific examples of using cw-EPR to study nucleic acids will then be presented. Our discussions will focus on X-band cw-EPR (~9.5 GHz), as most SDSL studies were carried out at this frequency.

A. Nitroxide Dynamics and EPR Spectral Lineshape

Figure 6B shows simulated X-band EPR spectra of nitroxides undergoing isotropic tumbling. Three regimes can be defined by the sensitivity of the EPR spectral lineshape to the nitroxide dynamics, which are characterized by an overall rotational correlation time τ. The fast motion limit corresponds to $\tau \sim 10^{-11}$–10^{-9} s (4, 11, 17). In this regime, only the isotropic average of the magnetic interactions will be observed. The spectrum shows three sharp lines because of nearly complete averaging of the g and A tensors (Fig. 6B, $\tau = 0.1$ ns), and the peak heights and peak widths of the spectrum are sufficient to characterize the nitroxide motion (4, 8, 11, 17, 89).

In the rigid limit ($\tau > 3 \times 10^{-8}$ s at X-band), a static distribution of all possible nitroxide orientations is present, and the cw-EPR spectrum is the sum of lineshapes from all possible values of the g and A tensors (a powder spectrum). The splitting between the outermost peaks (2A) is readily measured (Fig. 6B) and can be used as an empirical parameter to characterize the nitroxide dynamics (4, 8, 11, 17, 89).

In the slow-motion regime (1×10^{-9} s $< \tau < 3 \times 10^{-8}$ s), g and A tensor averaging is incomplete. The cw-EPR spectrum is most sensitive to nitroxide rotation, and drastic lineshape changes can be observed depending on nitroxide motions. As the nitroxide motion decreases (τ increases), the central line becomes broader, new features become apparent at the low-field and the high-field regions, and splitting between the outer peaks increases (Fig. 6B). It should be noted that as the EPR frequency changes, the time domain in which spectral variations are observed also changes.

In the slow-motion regime, quantitative spectral analysis is much more challenging. Given a particular physical model of nitroxide dynamics, one can compute the EPR spectrum. However, given an experimentally observed EPR spectrum, it is difficult to obtain a quantitative and unambiguous description of the nitroxide dynamics. A number of approaches, with varying levels of precision, have been used to characterize nitroxide dynamics from a given EPR spectrum. These will be discussed in the following section.

B. Characterizing Nitroxide Dynamics at the Slow-Motion Regime

1. SEMIEMPIRICAL APPROACHES

Many SDSL studies have used a semiquantitative approach to spectral analysis, in which simple parameters measured directly from the EPR spectrum are used to characterize the nitroxide dynamics (3). These parameters include the width of the central line (ΔH_{pp}, Fig. 6B), the splitting of the resolved hyperfine extrema (2A, Fig. 6B), and the second moment of the spectrum ($\langle H_2 \rangle$, characterizing how broad the spectrum is) (90). These parameters report on the nitroxide mobility describing a combined effect of the rate and the amplitude of motion. For example, a broad center line gives a small $(\Delta H_{pp})^{-1}$ value and indicates low mobility, which can result from low frequency but large amplitude motions, or small amplitude motions with fast rates.

Although the lineshape parameters do not directly give the rate and amplitude of motion, they provide a means to quickly access relative nitroxide mobility at different sites. The semiempirical nitroxide mobility characterization approach has been widely used in protein studies (2, 3, 6, 91). It has also been shown to be capable of qualitatively capturing structural and dynamic features of nucleic acids at the level of individual nucleotides (9), and has been used to probe nucleic acid structures and to monitor conformational changes (Sections V.D.3 to V.D.5).

2. SPECTRAL SIMULATIONS IN THE SLOW-MOTION REGIME

Two approaches have been used to simulate EPR spectra in the slow-motion regime. One approach uses MD to simulate the time-dependent trajectories of axes that are fixed with respect to the nitroxide (the nitroxide frame), and uses these trajectories to calculate the EPR spectrum directly (92–94, 134–137). However, this approach is very demanding in terms of computational resources, and its success depends on generating trajectories that sufficiently sample the free energy landscape (95). The trajectory-based approach has not been widely used in conjunction with nucleic acid SDSL, although this may change with advances in computer speed and simulations techniques.

A more widely used approach in EPR spectral simulation is based on the stochastic Liouville equation (SLE) (96). In the SLE, the electronic and nuclear spins are treated quantum mechanically, whereas the nitroxide reorientation motion is treated classically and parameterized in terms of rotational diffusion constants. The SLE approach is extremely efficient and capable of computing a spectrum in a fraction of a second. This enables iterative fitting of experimental spectra.

Two SLE-based simulation approaches have been developed in studies of nitroxides attached to bases within DNA duplexes (25, 97). The Bobst group developed a base disk model, in which the spin-labeled base is considered an axially symmetric diffusing system (25, 29). The principle motion is characterized by two correlation times: $\tau_{\parallel}$ describes rotational diffusion about the principle axis as defined by the average position of the bonds that connect the base to the nitroxide, and $\tau_{\perp}$ describes the motion about the axis perpendicular to the tether bond. $\tau_{\perp}$ is used to characterize motions of the labeled base, which includes contributions from DNA global tumbling, the collective bending and twisting of the DNA segment, and twisting/tilting of the individual base.

Robinson et al. used a SLE-based approach to simulate spectra of nitroxides that are rigidly tethered to bases within a DNA duplex (97). In this approach, the DNA helix is treated as a cylinder with internal motion. The overall tumbling of the DNA duplex is treated explicitly according to hydrodynamic models. All the internal motions are treated in the fast motion limit, and their collective effects are accounted for using motionally averaged effective g and A tensors. Spectra are simulated using a single parameter, the mean-square oscillation amplitude of the nitroxide attached to the ith base pair. The mean-square oscillation amplitude is directly related to the order parameter, but the simulation does not produce any estimation of the correlation time for the internal dynamics. For nitroxides that are more flexibly attached to bases within a DNA duplex (e.g., Table I, entry 5), the Bobst group has taken a similar approach and developed a dynamic cylinder model (29), which yields an order parameter that describes the amplitude of the nitroxide motion. It was suggested that the order parameter reports on both the base dynamics and the motion of the nitroxide with respect to the base.

Simulation models used in DNA duplex studies described above can be regarded as special cases in a family of SLE-based EPR simulation programs developed in the Freed group (98–100). These programs describe the diffusive motion of the nitroxide under the constraint of an orienting potential. The orienting potential is expressed as a function of the polar angles of a director axis in a rotational diffusion frame, and yields an order parameter (S) for the motion being described. In each system, the rotational diffusion frame is fixed with respect to the structure of the nitroxide, although its exact orientation may vary from system to system. Similarly, the director is fixed at the target macromolecule, such as the helical axis of a DNA duplex (101), in a system-dependent fashion. Simulations provide both a rate (τ) and an order parameter (S) to describe the partially ordered (restrictive) diffusive motion of the nitroxide with respect to the target macromolecule.

One of the programs developed by the Freed group is called the *microscopic ordered macroscopic disordered* (MOMD) model (99). In MOMD simulations, the spin label undergoes microscopic molecular ordering with respect to a local director, which is a frame fixed at the target macromolecule. The local directors (the macromolecule themselves), however, are oriented randomly and rigidly fixed with respect to the laboratory frame (generally defined by the applied magnetic field). The resulting EPR spectrum is obtained by integrating over the distribution of the local director orientations. The MOMD program is most useful in crystalline and liquid crystalline systems, but may also be used to describe systems where the global tumbling of the macromolecule is slow with respect to the EPR timescale. The MOMD approach has been used to simulate SDSL spectra in RNA (31, 102) and DNA (101).

Another program developed by the Freed group is called the *slowly relaxing local structure* (SRLS) model (100). In SRLS simulations, the orienting potential itself is allowed to undergo rotational diffusive motion on the EPR timescale. The model describes the situation where the spin label is reorienting with respect to the parent macromolecule, whereas the macromolecule is slowly tumbling. This model has been applied to DNA duplexes (101).

In MOMD and SRLS simulations, some of the parameters such as the g and A tensors are independently determined. However, it is well known that certain parameters in the simulation are correlated and not uniquely determined. For example, within a certain range, S and τ may covary (99, 103). When such correlations are identified, allowable ranges of each parameter are determined by exploration of the parameter space to estimate probable uncertainties. In protein studies, it has been demonstrated that a global fitting of EPR spectra at multiple frequencies may provide better constraints for the parameters (100).

One may also constrain the simulation parameters by simultaneously analyzing a group of spectra measured using chemically modified nitroxides. For example, in studies of protein (103) and RNA (31), simultaneous analyses have been carried out on spectra obtained from nitroxides that are modified at the 4-position of the pyrroline ring. These modifications systematically introduce larger functional groups (i.e., 4-H, 4-CH$_3$, 4-Br, and 4-phenol), and are expected to increasingly restrict rotations about the bonds that connect the pyrroline ring to the parent macromolecule. In the spectral simulations, such effects are accounted for by increasing the order parameter while keeping the rate of nitroxide motion fixed (31, 103).

C. Correlating Nitroxide Motion to Local Structural and Dynamic Features

Methods described in Section V.B allow one to characterize nitroxide motion with varying levels of precision. The challenge for SDSL lies in the determination of local environmental feature(s) that give rise to the observed nitroxide motion (or EPR spectrum).

1. Probing the Local Environment Using Semiempirical Nitroxide Mobility

Many studies have demonstrated qualitative correlations between features at the labeling site and the nitroxide mobility as characterized by the semiquantitative approach (Section V.B.1) (2, 3, 6–9). In nucleic acid SDSL, a high-mobility nitroxide generally reports a highly flexible labeling site, such as unpaired and unstacked nucleotides in single strand or loop regions (23, 31). A low-mobility nitroxide generally reveals structural and/or dynamic constraints at the labeling site, such as stacking and/or hydrogen-bonding (23, 31, 102), as well as tertiary contacts within a folded RNA (44).

In RNA SDSL studies using a nitroxide attached at a modified uridine base (Table I, entry 3a), Qin *et al.* have reported that different secondary structures (e.g., single-strand, stacked A/U pair, and U/U mismatch) give clearly distinct lineshapes, whereas the stacked A/U pair spectrum gives a very similar lineshape in two different RNA molecules (*102*). This has led to the proposal of building a library in which observed spectral lineshapes are categorized based on known RNA structures (*102*). The library will reveal both spectral *divergence* (distinctive lineshapes for different elements) and spectral *convergence* (similar lineshapes for the same element in different contexts). If successful, the library will enable lineshape-based structure identification at any RNA site.

In protein SDSL studies, semiquantitative mobility analysis is frequently coupled with nitroxide scanning (*2*). In this approach, nitroxide mobility is measured, one residue at a time, along a stretch of amino acid sequence, and the periodicity in the nitroxide mobility variation experimentally identifies the secondary structural element (e.g., 3.6 residue/period for the α-helix). However, similar nitroxide scanning experiments have not been reported in studies of nucleic acids. This may be due to the difficulty in scanning the nitroxide along a given nucleic acid sequence using most of the available nitroxide labeling methods. Recent data suggest that nitroxides attached using the phosphorothioate labeling scheme (Table I, entry 1) can be used to monitor the local environment at the level of individual nucleotides (Popova and Qin, manuscript in preparation). This nitroxide label may be a good candidate for scanning nucleic acids. Additionally, secondary structural elements in nucleic acids generally involve a much more extended segment of the primary sequence (e.g., ~10 base pairs per turn in A or B-form duplexes). Further studies are needed to determine what kind of information can be obtained from nitroxide scanning in nucleic acids.

2. Probing the Local Environment with Quantitative Spectral Simulations

Spectral simulations yield a set of quantitative parameters based on a given model. The simulated parameters define the correlation between the nitroxide motion and the features of the parent molecule. In each simulation, assumptions have to be made regarding what type of motions (parameters) are tractable and how these motions are related to the structural and dynamic features of the parent macromolecule. These assumptions need to be tested experimentally.

One approach for correlating parameters obtained from spectral simulations to local structural features is to carry out systematic molecular level mutations, either at the macromolecule or at the nitroxide probe, and to evaluate their effects on the simulation parameters (*11, 14*). For example, in studies of nitroxides attached to bases within DNA duplexes, the Robinson group explicitly accounts for the DNA duplex global tumbling, and extracts a parameter equivalent to the mean-square oscillation amplitude of the spin-labeled base pair (Section V.B.2) (*97, 104*). Studies were then carried out with varying DNA lengths, sequences, and nitroxide labeling positions. Effects of these changes on the mean-square oscillation amplitude were used to test assumptions made in the spectral simulation (*104*). Once validated, the mean-square oscillation amplitude was used to extract information on length-independent segmental deformation in DNA duplexes (*104–106*).

As noted in work from both the Bobst and Robinson groups (*11, 14*), nitroxide motions can be influenced by a wide variety of factors that may change between systems as well as between different chemical labeling schemes. Quantitative correlation of the observed EPR lineshape to local structural and dynamic features remains a challenging task. Advances on this front will likely require a combination of theoretical and experimental efforts.

D. Examples of Application

Selected cases from the literature are presented in this section to illustrate methods discussed in Sections V.A–V.C. In Section V.D.1, we will discuss the use of measuring the global tumbling of the nitroxide-labeled macromolecules to monitor molecular interactions. Then we will discuss examples of EPR lineshape analysis and their applications in probing site-specific features in DNA duplexes (Section V.D.2) and folded RNA molecules (Sections V.D.3–V.D.5). Currently, we are aware of few reported examples on using nitroxide-labeled nucleic acids to investigate protein–nucleic acid interactions (*11*). With the advances in SDSL, protein–nucleic acids interactions present an interesting area for future exploration.

1. Monitoring Molecular Interactions via the τ_R Effect

The τ_R effect refers to a change in the EPR lineshape due to a change in the overall tumbling of the macromolecule. A simple estimation based on the molecular weight of a 25-nt oligonucleotide (~7500 Da) gives an overall tumbling rate τ_R of approximately 4 ns in aqueous solution at room temperature (*31, 107*). One expects that tumbling of this short oligonucleotide affects the X-band EPR spectrum. Interactions between such an oligonucleotide and other macromolecules, such as a complimentary strand (*11*), a partner nucleic acid (*22*), or a protein (*11*), increase the molecular weight of the system. This leads to changes in the overall tumbling of the complex, which are reflected in the nitroxide lineshape.

The τ_R effect has been used to detect nucleic acid hybridizations (*11*). In the 1970s, Bobst *et al.* used spin-labeled polyU to detect the polyA tract in mRNAs (*10*). More recently, Gannet *et al.* reported detection of DNA triplex formation using spin-labeled oligonucleotides (*30*). In these studies, hybridization of a nitroxide-labeled oligonucleotide to its complimentary strand leads to a drastic reduction of the overall tumbling rate, which is reflected as line broadening in the EPR spectrum. The change of EPR lineshape is due to a combined τ_R and τ_i/τ_B effects, as hybridization reduces the overall tumbling of the oligonucleotide as well as motions of the nitroxide-labeled nucleotide(s) due to hydrogen bonding and stacking. Recently, Bobst *et al.* reported a reverse approach to detect DNA hybridization, in which a single genome is identified by monitoring the appearance of fast-tumbling fragments (*108*).

Qin *et al.* used the τ_R effect to study a frequently used RNA tertiary interaction motif—binding between a GAAA tetraloop and its RNA receptor (*22*) (Fig. 7). Upon binding of an RNA hairpin containing the GAAA tetraloop to its RNA receptor, increases in the molecular weight of the complex gave rise to EPR spectral broadening (Fig. 7). The dissociation constant (K_d) was measured to be 0.4 mM between the GAAA tetraloop and its receptor, which is a weak interaction and is difficult to measure using other methods (*22*). This study illustrates several facets of using the τ_R effect to monitor interactions in nucleic acids. First, the nitroxide is attached far away from the interface between the tetraloop and the receptor (Fig. 7A), allowing one to monitor the interaction with minimal perturbation. Second, as the spin label concentration in an EPR experiment is generally >10 μM, the method is suitable for measuring interactions where the K_d is approximately 10–1000 μM. Third, the EPR measurements can be carried out with a wide range of buffer conditions.

2. Dynamic Behavior of DNA Duplexes

The Bobst group studied a large number of double-stranded DNAs and RNAs using thymidylates and deoxycytidylates that are substituted at position 5 with nitroxides containing tethers ranging from 2 to 11 atoms (Table I, entry 5) (*11*). Data on B-form DNA duplexes were simulated using a model that characterizes nitroxide motions by two correlation times, $\tau_\parallel$ and $\tau_\perp$ (Section V.B.2) (*25, 29, 101, 109*). $\tau_\parallel$ was found to be tether dependent, was less than 1 ns, and was attributable to nitroxide motion that was independent of base motion (*11*). $\tau_\perp$ was thought to represent motions including twisting and tilting of the individual base, collective bending and twisting of the DNA segment, and global tumbling of DNA (*25, 29, 101, 109*). Values of $\tau_\perp$ from all double-stranded systems were reported to be in the 1–10 ns range (*11, 25, 29, 101, 109*), leading to the conclusion that there are significant base motions and these motions contribute to the dynamics of the nitroxide. However, questions have been raised as to the degree to which $\tau_\perp$ is influenced by the motion of the tether versus that of the base (see below) (*14*).

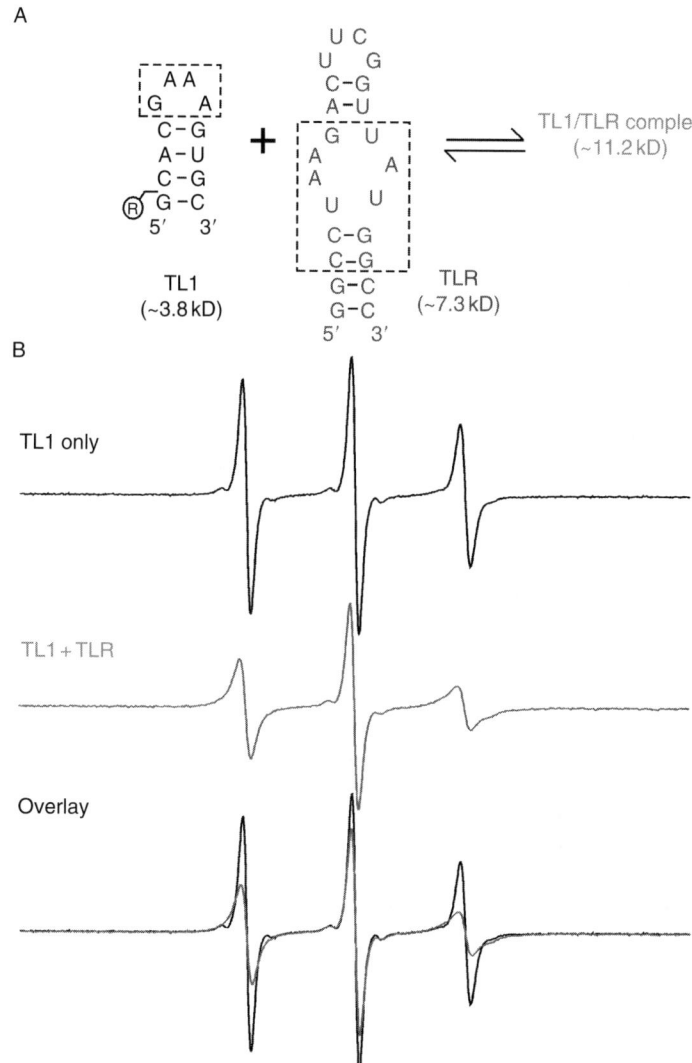

FIG. 7. SDSL studies of GAAA tetraloop binding to its RNA receptor. (A) A schematic diagram illustrating the experimental design. Shown in black is a 12-nt RNA hairpin (TL1) containing the GAAA tetraloop (boxed). The TL1 RNA was labeled with the phosphorothioate scheme (Table I, entry 1), with the labeling site represented by R. Shown in purple is a 23-nt RNA (TLR), which contains the 11-nt RNA receptor (boxed) for the GAAA tetraloop. (B) EPR spectra of the free and bound TL1. The spectra shown were normalized to the same number of spins. For TL1 by itself, the spectrum in aqueous solution (black) shows three sharp lines, consistent with that of a molecule undergoing fast tumbling. On formation of the TL1/TLR complex, the spectrum (red) shows a clear reduction in magnitude of the high-field line as well as broadening of all three lines, indicating a

The Robinson group used nitroxides that are rigidly coupled to a DNA base (T° and dQ, Table I, entry 4a and entry 6a) to examine dynamics in B-form DNA duplexes (*14, 97, 104–106, 110*). A mean-square oscillation amplitude, obtained by simulating the observed EPR spectra, was used to obtain information on the DNA (Section V.B.2) (*97*). The authors concluded that the global tumbling of the DNA duplex is accurately described by hydrodynamic theory (*97*), and the internal collective DNA deformation fits well to a modified weakly bending rod model (*104*). The observed spectra also depend on sequence variations that are distant from the label, and efforts have been made to account for these observations using a sequence-dependent force bending constant (*105, 106*). Furthermore, the Robinson group reported that DNA duplex flexibility is linked to phosphate backbone neutralization (*110*).

From studies with rigidly tethered nitroxide probes, Robinson *et al.* concluded that there are no large amplitude fluctuations of the labeled base (*14*), which contradicts conclusions drawn from studies using flexible nitroxides (*11*). There has been debate over whether the rigidly coupled nitroxides alter the behavior of the base, and if so, to what degree (*11, 14, 29, 101, 109*). On the other hand, for the flexible nitroxides, questions have been raised regarding whether motions of the flexible linker lead to an overestimation of base dynamics (*14*). Furthermore, Bobst *et al.* have reported that the short-tether nitroxide may give rise to a multiple component spectrum (*101*), which may further complicate quantitative analysis. These studies illustrate an important issue in SDSL. The information one can gain from the nitroxide depends on how strongly the nitroxide is coupled to the target molecule, and therefore the tether plays a very significant role. An ideal tether would completely couple the nitroxide to the base, yet would not at all alter its behavior. In practice, such a tether may not ever be developed.

Other studies have examined base mismatches and bulges in DNA duplexes (*11, 111*). The results generally show distinct nitroxide lineshape changes indicating variation in nitroxide dynamics. However, quantitative interpretations have not been reported.

3. THE TLR RNA

Qin *et al.* have reported SDSL studies on a 23-nt RNA, designated as TLR, that contains two structural elements—the UUCG loop and the 11-nt motif of the GAAA tetraloop receptor (*31*). Three nitroxide variants were attached, one at a time, to six uridine positions within the TLR molecule using the 4-thioU

---

reduction in nitroxide dynamics. The change in nitroxide dynamics reflects a reduction in the overall tumbling rate due to the formation of the TL1/TLR complex with a higher molecular weight. Data reproduced from reference (*22*) with permission.

scheme (Table I, entry 3a). The observed lineshapes vary significantly among the different uridine sites, which is consistent with the expected versatility in RNA structures. For example, at the $U_{11}U_{12}CG$ loop, mobility parameters obtained by semiquantitative analysis showed that the nitroxide at U11 has low mobility, whereas that at U12 is highly mobile (Fig. 8) (*31*). This is consistent with NMR studies showing that U11 is hydrogen-bonded and stacked, and U12 is structurally unrestrained and highly dynamic (*112, 113*). This suggests that the nitroxide is reporting RNA base motion.

MOMD simulations were carried out for spectra obtained at U16, which participates in a Watson-Crick base pairing interaction (Fig. 9) (*31*). The analyses led to a proposed motion model for nitroxides attached via the 4-thioU scheme, where in the absence of RNA–nitroxide interaction, rotations about the distal two bonds (the $X_3^r\,X_4^r$ torsion angles, Fig. 9) are the major contributors to the EPR spectrum. The characteristic lineshape at U16 was observed at a stacked Watson-Crick A/U base pair in another RNA molecule, a dodecameric RNA duplex (*102*), suggesting that this motion model may

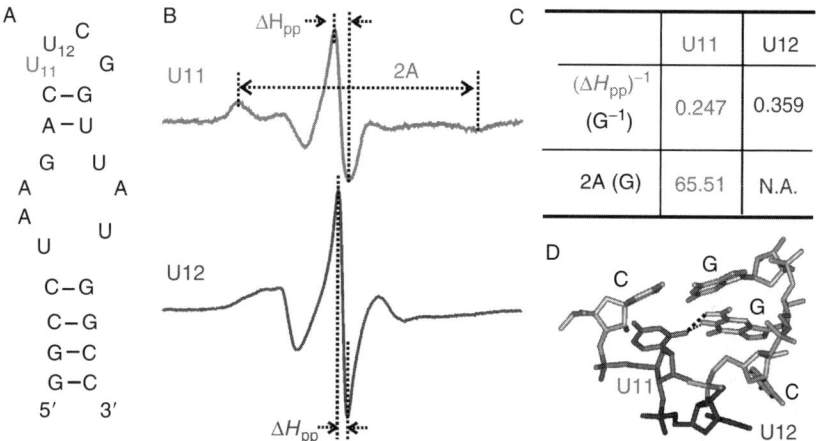

FIG. 8. SDSL studies of a UUCG loop. (A) The sequence and secondary structure of the TLR RNA. The two uridines within the UUCG loop are colored blue and red, respectively. (B) EPR spectra of nitroxides attached to U11 (red) and U12 (blue). The nitroxides were attached to the respective uridine base using the 4-thioU scheme (Table I, entry 3a). (C) Semiquantitative analysis of the spectra shown in (B). The EPR spectrum of U11 (red) shows broad centerline [large ΔH_{pp} and small $(\Delta H_{pp})^{-1}$] and large hyperfine splitting (2A), indicating a nitroxide with limited mobility. The spectrum for U12 shows narrow centerline [small ΔH_{pp} and large $(\Delta H_{pp})^{-1}$] and no clear hyperfine splitting, indicating a mobile nitroxide. (D) The NMR structure of the UUCG loop (*112*). The structure shows that U11 is hydrogen-bonded (dotted lines) and stacked, while U12 is structurally unrestrained. Data reproduced with permission from (*31*).

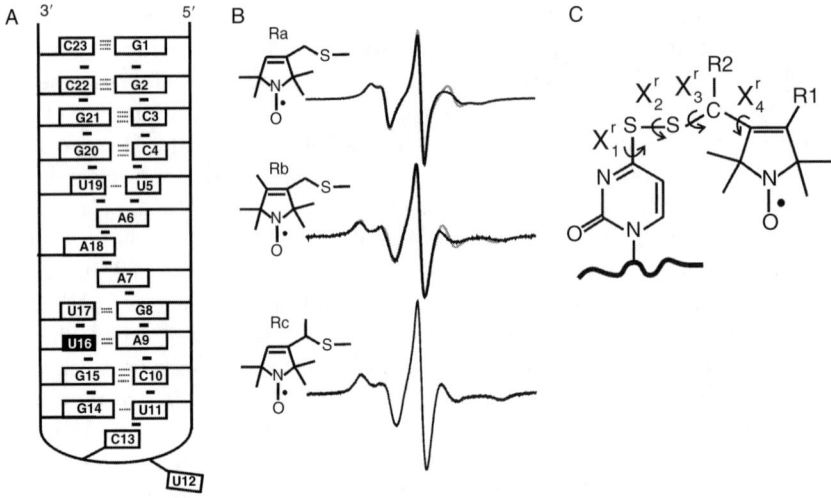

FIG. 9. Nitroxide motion at U16 of the TLR RNA. (A) A schematic of the sequence and secondary structure of the TLR RNA. The U16 base is shown in black. (B) EPR spectra of three variants of nitroxide (Ra, Rb, Rc) attached to the U16 base using the 4-thioU scheme. The experimental spectra are shown in black, and spectra simulated using MOMD are shown in red. (C) The chemical structure of the Rx series of spin labels. The designations of various atoms and dihedral angles are shown. Data reproduced with permission from (*31*).

be general. The U16 lineshape is also the first experimental evidence of spectral convergence, where the same secondary structure elements embedded in different molecules give the same characteristic lineshape (*102*).

The nitroxide-labeled TLR molecules allowed the detection of conformational changes in the GAAA tetraloop receptor upon tetraloop docking (*114*). Increases in nitroxide mobility were observed at U19 upon formation of the tetraloop/receptor complex (Fig. 10), and experiments linked the nitroxide mobility increase to increases in the dynamics of the U19 base (*114*). This suggests that U19 becomes unstacked in the complex, which is consistent with the crystal structure showing that U19 flipped out of the helix as the tetraloop docks (Fig. 10) (*115*).

4. HIV TRANS-ACTIVATION RESPONSIVE RNA

The *trans*-activation responsive (TAR) RNA is the 5′-leader sequence of the HIV-1 mRNA genome. Interaction between the TAR RNA and the Tat protein is a key determinant in viral transcription, and the Tat/Tat complex has long been a target for therapeutic intervention of HIV replication (*116*). Sigurdsson *et al.* have attached nitroxides to the 2′-sugar position (Table I, entry 2a) of four specific uridine nucleotides within a 27-nt RNA construct that

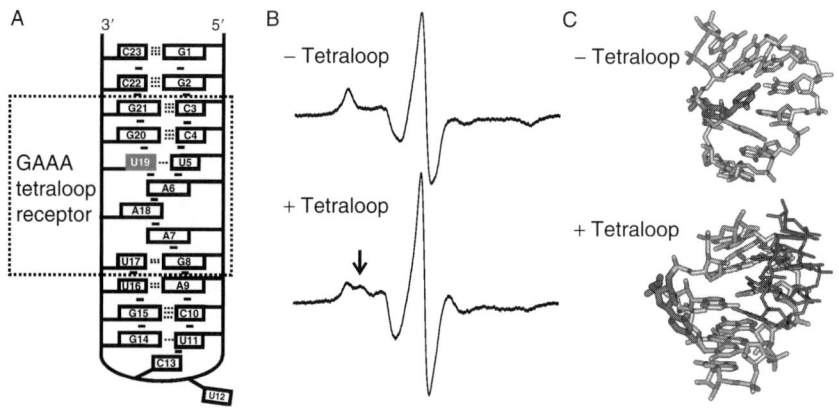

FIG. 10. Conformational changes in the GAAA tetraloop receptor upon tetraloop docking. (A) A schematic diagram showing the secondary structure of the TLR RNA. The GAAA tetraloop receptor is indicated by the box. The U19 base is shown in red. (B) EPR spectra at the U19 site. The nitroxide is attached to the U19 base using the 4-thioU scheme. The arrow indicates a mobile component that appears upon tetraloop-receptor interaction. Due to the weak affinity between the isolated tetraloop and receptor in solution, the "+ Tetraloop" spectrum is a composite of the bound and unbound receptors. (C) Structures of the unbound and bound receptor. Shown on top is the NMR structure of the unbound receptor (133), and shown at the bottom is the crystal structure of the tetraloop/receptor complex as observed in the P456 domain of group I intron *Tetrahymena* (115). In the unbound receptor (top), the U19 base (thick red line) is hydrogen-bonded and stacked within the helix. In the bound receptor (bottom), the U19 base (thick red line) is flipped out of the helix. Data reproduced from (114) with permission.

contains the TAR element (23) (Fig. 11). In the TAR RNA alone, nitroxides attached at the two base-paired sites were found to report the overall tumbling of the TAR RNA, while those at the bulge region showed significantly higher mobility, similar to that of a single strand (23). These studies established that the nitroxide attached at the 2′-sugar position is capable of reporting site-specific features of the RNA local environment at the level of a nucleotide.

Using the spin-labeled TAR RNAs, Sigurdsson *et al.* carried out extensive studies on local structural and dynamic changes in HIV TAR that are induced by interactions with ligands, including small molecules (42), metal ions (41, 43), and peptides (41, 117). In these studies, the semiquantitative mobility parameters are used to characterize the nitroxide spectra in the presence and absence of the ligand. A dynamic signature for each ligand is determined based on the pattern of nitroxide mobility changes as a function of nucleotide position. This is used as a qualitative but effective means to characterize the RNA response to various ligands. Figure 11 shows an example of dynamic signatures observed from complexes between TAR

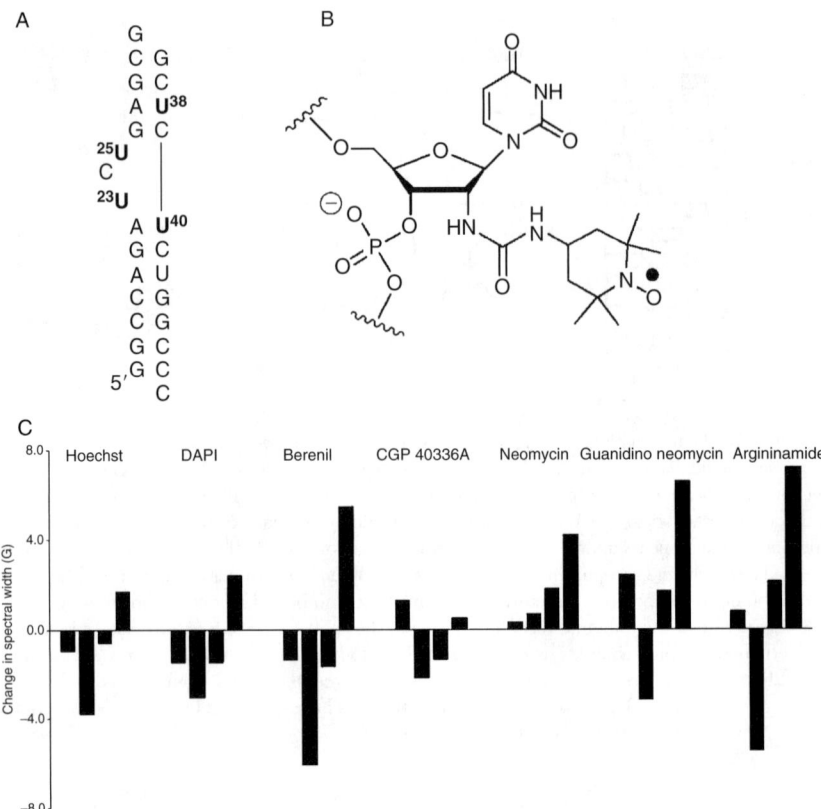

FIG. 11. EPR dynamic signatures in the TAR RNA. (A) The TAR RNA construct used in SDSL studies. The nitroxide-labeled uridine nucleotides are shown in bold. (B) A schematic of a nitroxide covalently linked to the 2′-position of a uridine nucleotide in RNA (23). (C) The dynamic signatures of TAR RNA complexed with small molecule inhibitors (42). Changes in spectral width (2A) are plotted as a function of nucleotide position (U23, U25, U38, and U40) for each compound. An increase in spectral width corresponds to a decrease in nitroxide mobility and a decrease in spectral width corresponds to an increase in nitroxide mobility. Molecules that bind in a similar fashion, such as Hoechst, DAPI, and berenil, show similar dynamic signatures, whereas different patterns of the dynamic signature are observed for molecules that bind differently. Reproduced from (9) with permission.

RNA and small molecules (42). The data indicated that ligands that are known to bind similarly to the TAR RNA give nearly identical patterns, whereas ligands that bind at different sites give clearly different patterns (Fig. 11). The dynamic signature was used in conjunction with other information to understand TAR–small molecule interactions.

5. Hammerhead Ribozyme

The hammerhead ribozyme is a small self-cleaving RNA motif that catalyzes a transesterification reaction at a specific site. It has been a model system for understanding RNA catalysis, structure and folding, conformational dynamics, and metal ion interaction (118). The DeRose group used SDSL to probe metal ion interactions in an extended hammerhead construct, in which an extended loop–loop interaction enhanced catalytic activity (44). Nitroxides were attached to 2′-sugar positions (Table I, entry 2b) of specific uridine and cytosine sites. Changes in the observed nitroxide lineshape at a specific site (U1.6) were assigned to tertiary contacts arising from interactions between the extended loops upon RNA folding. By analyzing the observed U1.6 spectra, it was found that this RNA folding event occurs at a much lower Mg^{2+} concentration compared to what is required for enhanced catalytic activity.

The Sigurdsson group used SDSL to study RNA dynamics in a truncated hammerhead (45). The nitroxide was attached to a uridine near the catalytic core. EPR was used to monitor variations in RNA conformations in response to metal ions, ribozyme inhibitors, and pH changes. The data indicate a two-step metal ion-dependent folding pathway, which is consistent with previous studies using other biophysical and biochemical techniques.

VI. Beyond Distance Measurements and Nitroxide Dynamics Analysis

In addition to nitroxide dynamics analysis and distance measurements, a number of additional EPR measurements have been used in SDSL studies of protein and membrane systems (2, 4, 8). We briefly describe some of these measurements in this section. Although these measurements have not been widely used in SDSL studies of nucleic acids, they should provide exciting opportunities for future investigations.

A. Nitroxide-Solvent Accessibility

The solvent accessibility of nitroxides has been used extensively to map protein secondary structures, to reveal the global topology of proteins, and to orient proteins with respect to their environments (e.g., lipid bilayers) (2, 4, 8). Collisions between a nitroxide and paramagnetic agents (e.g., oxygen, metal ion complexes, or other nitroxides) give rise to additional nitroxide relaxation pathways. By measuring these relaxation effects using either cw-EPR or pulsed EPR techniques, one can obtain information on the collision frequency between the nitroxide and the paramagnetic agent in solution, which is directly correlated to the solvent accessibility at the nitroxide.

There are few reports of measuring nitroxide-solvent accessibility in nucleic acids systems. Shin and Hubbell used nitroxide accessibility to study electrostatic potential distributions around a DNA duplex (119). In the study, a DNA duplex was randomly intercalated by a nitroxide-modified acridine. Collision rates were measured between the nitroxide and neutral and charged paramagnetic agents. The ratio between the collision rates allows one to calculate the average electrostatic potential sensed by the nitroxide. The study found that the electrostatic potential distribution around a DNA duplex agrees with that predicted by the nonlinear Poisson–Boltzmann model. In addition, Robinson et al. used pulsed EPR to measure the T_1 relaxation rate of nitroxides attached to DNA duplexes, and examined how the measured T_1 is affected by oxygen in the solution (120).

B. Polarity Effects

The polarity of the environment at the nitroxide tunes the localization of the unpaired electron within the nitroxyl group of the nitroxide, giving rise to changes in the A and g tensor. This enables one to use the nitroxide as a probe for local polarity, pH, and the formation of hydrogen bonds with the nitroxyl group (17). This effect has been used extensively in studies of membranes and proteins (17, 121). Currently, we are not aware of reports on polarity measurements in nucleic acid SDSL.

C. Paramagnetic Relaxation Enhancement

PRE refers to an increase in the relaxation rate of nuclear magnetization due to magnetic dipolar interaction between a nucleus (e.g., ^1H) and an unpaired electron. Because the magnetic moment of the unpaired electron is large, PRE can be effective over long distances, extending in the case of Mn^{2+}, for example, up to ~35 Å [see a recent review by Clore et al. (122)].

Nitroxides have been shown to be suitable for PRE measurements in studies of proteins and protein/nucleic acid complexes. For example, Varani et al. have reported PRE measurements in the structural determination of a protein/RNA complex. The study places a nitroxide at a modified uridine position (Table I, entry 3b) in an RNA hairpin. PRE was measured between the labeled RNA and a double-stranded RNA-binding domain (dsRBD) of the Staufen protein from Drasophila.(46–48). The measurements provided long-range distance constraints (up to ~15 Å) between the RNA and the protein, which aided in structural determination of the complex. More recently, PRE using nitroxide-labeled RNA was also reported in studies of the dsRBD of Rnt1p RNase III (123). Furthermore, nitroxides attached to proteins have been used to investigate the U1A protein/PIE RNA complex (49).

Recent work from the Clore group has used EDTA-Mn^{2+} complex as the paramagnetic probe to study low population encounter complexes in protein–DNA recognition, as well as large-scale dynamics in nonspecific protein–DNA interactions (122). In studies of RNA and protein/RNA complexes, divalent metal ions such as Mn^{2+} may be potentially involved in RNA folding and function (see Section VI.D), and the nitroxide spin label may provide an alternative probe for PRE.

D. Probing Paramagnetic Metal Ions in RNA Folding and Function

Because DNA and RNA are highly negatively charged polymers, metal ions, such as Na^+, K^+, and Mg^{2+}, play important roles in the structure and function of nucleic acids. For example, it is well known that Mg^{2+} plays critical roles in the folding and catalysis of ribozymes (124). The paramagnetic Mn^{2+} ion, similar to Mg^{2+} in ionic radius and enthalpy of hydration, is capable of supporting functions of many ribozymes and has been utilized as an EPR probe for studying Mg^{2+}–RNA interactions (124). Studies have been carried out on nucleotides (125) as well as large-structured RNAs, including the hammerhead ribozyme and tRNA (126–132). A variety of EPR techniques, such as cw-EPR, ENDOR (electron nuclear double resonance), and ESEEM (electron spin echo envelope modulation), have been used to obtain information on metal-binding sites, including locations, number of metal ions bound, affinity, ligand identity, and site geometry (125–132).

VII. Future Directions

SDSL studies of nucleic acids provide unique structural and dynamics information that compliments results obtained using other biophysical techniques. Advances in EPR spectroscopy and nitroxide labeling will make it possible to explore a number of important questions in the future. These include, but are not limited to, mapping global structures of nucleic acids and protein/nucleic acid complexes, monitoring dynamics in large nucleic acid assemblies, investigating site-specific electrostatic features, and investigating metal ion interactions.

Acknowledgments

We thank Gian G. Grant, Anna Popova, and Maria Frushicheva for advice and assistance in preparing the manuscript. Financial support is provided by the National Institutes of Health (R01 GM069557) and the National Science Foundation (MCB0546529). The authors also thank the William R. Wiley Environmental Molecular Sciences Laboratory for a pulse EPR instrumentation time award.

References

1. Altenbach, C., Flitsch, S. L., Khorana, H. G., and Hubbell, W. L. (1989). Structural studies on transmembrane proteins. 2. Spin labeling of bacteriorhodopsin mutants at unique cysteines. *Biochem.* **28**, 7806–7812.
2. Hubbell, W. L., and Altenbach, C. (1994). Investigation of structure and dynamics in membrane proteins using site-directed spin labeling. *Curr. Opin. Struct. Biol.* **4**, 566–573.
3. Hubbell, W. L., Cafiso, D. S., and Altenbach, C. (2000). Identifying conformational changes with site-directed spin labeling. *Nat. Struct. Biol.* **7**, 735–739.
4. Fajer, P. G. (2000). Electron spin resonance spectroscopy labeling in proteins and peptides analysis. *In* "Encyclopedia of Analytical Chemistry" (R. Meyers, Ed.), pp. 5725–5761. John Wiley & Sons, Chichester.
5. Borbat, P. P., Costa-Filho, A. J., Earle, K. A., Moscicki, J. K., and Freed, J. H. (2001). Electron spin resonance in studies of membranes and proteins. *Science* **291**, 266–269.
6. Columbus, L., and Hubbell, W. L. (2002). A new spin on protein dynamics. *Trends Biochem. Sci.* **27**, 288–295.
7. Fanucci, G. E., and Cafiso, D. S. (2006). Recent advances and applications of site-directed spin labeling. *Curr. Opin. Struct. Biol.* **16**, 644–653.
8. Klug, C. S., and Feix, J. B. (2008). Methods and applicants of site-directed spin labeling EPR spectroscopy. *Methods Cell Biol.* **84**, 617–658.
9. Qin, P. Z., and Dieckmann, T. (2004). Application of NMR and EPR methods to the study of RNA. *Curr. Opin. Struct. Biol.* **14**, 350–359.
10. Bobst, A. M., Sinha, T. K., and Pan, Y.-C. E. (1975). Electron spin resonance for detecting polyadenylate tracts in RNAs. *Science* **188**, 153–155.
11. Keyes, R. S., and Bobst, A. M. (1998). Spin-labeled nucleic acids. *In* "Biological Magnetic Resonance" (L. J. Berliner, Ed.), pp. 283–338. Plenum Press, New York.
12. Bondarev, G. N., Isaev-Ivanov, V. V., Isaeva-Ivanova, L. S., Kirillov, S. V., Kleiner, A. R., Lepekhin, A. F., Odinzov, V. B., and Fomichev, V. N. (1982). Study on conformational states of *Escherichia coli* tRNAPhe in solution by a modulation-free ESR-spectrometer. *Nucleic Acids Res.* **10**, 1113–1126.
13. Nothig-Laslo, V., Zivkovic, T., Kucan, Z., and Weygand-Durasevic, I. (1981). Binding of spermine to tRNATyr stabilizes the conformation of the anticodon loop and creates strong binding sites for divalent canons. *Eur. J. Biochem.* **117**, 263–267.
14. Robinson, B. H., Mailer, C., and Drobny, G. (1997). Site-specific dynamics in DNA: Experiments. *Annu. Rev. Biophys. Biomol. Struct.* **26**, 629–658.
15. Makinen, M. W., Mustafi, D., and Kasa, S. (1998). ENDOR of spin labels for structure determination: From small molecules to enzyme reaction intermediates. *In* "Biological Magnetic Resonance" (L. J. Berliner, Ed.), pp. 181–249. Springer-Verlag, New York, LLC.
16. Berliner, L. J. (1976). "Spin Labeling: Theory and Applications," p. 592. Academic Press, New York.
17. Marsh, D. (1981). Electron spin resonance: Spin labels. *Mol. Biol. Biochem. Biophys.* **31**, 51–142.
18. Eaton, G. R., Eaton, S. S., and Berliner, L. J. (2000). "Distance Measurements in Biological Systems by EPR."*Biol. Mag. Res. Vol. 19.* Kluwer, New York, NY.
19. Pake, G. E. (1948). Nuclear resonance absorption in hydrated crystals: Fine structure of the proton Line. *J. Chem. Phys.* **16**, 327–336.
20. Caruthers, M. H., Beaton, G., Wu, J. V., and Wiesler, W. (1992). Chemical synthesis of deoxyoligonucleotides and deoxyoligonucleotide analogs. *In* "Method. Enzymol" (D. M. J. Lilley and J. E. Dahlberg, Eds.), pp. 3–20. Elsevier Science & Technology Books, Academic Press, Inc., San Diego.

21. Usman, N., and Cedergren, R. (1992). Exploiting the chemical synthesis of RNA. *Trends Biochem. Sci.* **17**(9), 334–339.
22. Qin, P. Z., Butcher, S. E., Feigon, J., and Hubbell, W. L. (2001). Quantitative analysis of the GAAA tetraloop/receptor interaction in solution: A site-directed spin labeling study. *Biochem.* **40**, 6929–6936.
23. Edwards, T. E., Okonogi, T. M., Robinson, B. H., and Sigurdsson, S. T. (2001). Site-specific incorporation of nitroxide spin-labels into internal sites of the TAR RNA. Structure-dependent dynamics of RNA by EPR spectroscopy. *J. Am. Chem. Soc.* **123**, 1527–1528.
24. Kim, N., Murali, A., and DeRose, V. J. (2004). A distance ruler for RNA using EPR and site-directed spin labeling. *Chem. Biol.* **11**, 939–948.
25. Kao, S.-C., Polnaszek, C. F., Toppin, C. R., and Bobst, A. M. (1983). Internal motions in ribonucleic acid duplexes as determined by electron spin resonance with site-specifically spin-labeled uridines. *Biochem.* **22**, 5563–5568.
26. Kao, S. C., and Bobst, A. M. (1985). Local base dynamics and local structural features in RNA and DNA duplexes. *Biochem.* **24**, 5465–5469.
27. Spaltenstein, A., Robinson, B. H., and Hopkins, P. B. (1988). A rigid and nonperturbing probe for duplex DNA motion. *J. Am. Chem. Soc.* **110**(4), 1299–1301.
28. Miller, T. R., Alley, S. C., Reese, A. W., Solomon, M. S., McCallister, W. V., Mailer, C., Robinson, B. H., and Hopkins, P. B. (1995). A probe for sequence-dependent nucleic acid dynamics. *J. Am. Chem. Soc.* **117**(36), 9377–9378.
29. Keyes, R. S., and Bobst, A. M. (1995). Detection of internal and overall dynamics of a two-atom-tethered spin-labeled DNA. *Biochem.* **34**, 9265–9276.
30. Gannett, P. M., Darian, E., Powell, J., Johnson, E. M., Mundoma, C., Greenbaum, N. L., Ramsey, C. M., Dalal, N. S., and Budil, D. E. (2002). Probing triplex formation by EPR spectroscopy using a newly synthesized spin label for oligonucleotides. *Nucleic Acids Res.* **30**, 5328–5337.
31. Qin, P. Z., Hideg, K., Feigon, J., and Hubbell, W. L. (2003). Monitoring RNA base structure and dynamics using site-directed spin labeling. *Biochem.* **42**, 6772–6783.
32. Schiemann, O., Piton, N., Mu, Y., Stock, G., Engels, J. W., and Prisner, T. F. (2004). A PELDOR based nanometer distance ruler for oligonucleotides. *J. Am. Chem. Soc.* **126**, 5722–5729.
33. Barhate, N., Cekan, P., Massey, A. P., and Sigurdsson, S. Th. (2007). A nucleoside that contains a rigid nitroxide spin label: A fluorophore in disguise. *Angew. Chem. Int. Ed.* **46**(15), 2655–2658.
34. Piton, N., Mu, Y., Stock, G., Prisner, T. F., Schiemann, O., and Engels, J. W. (2007). Base-specific spin-labeling of RNA for structure determination. *Nucleic Acids Res.* **35**(9), 3128–3143.
35. Cai, Q., Kusnetzow, A. K., Hubbell, W. L., Haworth, I. S., Gacho, G. P., Van Eps, N., Hideg, K., Chambers, E. J., and Qin, P. Z. (2006). Site-directed spin labeling measurements of nanometer distances in nucleic acids using a sequence-independent nitroxide probe. *Nucleic Acids Res.* **34**, 4722–4734.
36. Cai, Q., Kusnetzow, A. K., Hideg, K., Price, E. A., Haworth, I. S., and Qin, P. Z. (2007). Nanometer distance measurements in RNA using site-directed spin labeling. *Biophys. J.* **93**(6), 2110–2117.
37. Qin, P. Z., Haworth, I. S., Cai, Q., Kusnetzow, A. K., Grant, G. P. G., Price, E. A., Sowa, G. Z., Popova, A., Herreros, B., and He, H. (2007). Measuring nanometer distances in nucleic acids using a sequence-independent nitroxide probe. *Nat. Protoc.* **2**(10), 2354–2365.
38. Nagahara, S., Murakami, A., and Makino, K. (1992). Spin-labeled oligonucleotides site specifically labeled at the internucleotide linkage. Separation of stereoisomeric probes and EPR spectroscopical detection of hybrid formation in solution. *Nucleosides Nucleotides* **11**, 889–901.

39. Price, E. A., Sutch, B. T., Cai, Q., Qin, P. Z., and Haworth, I. S. (2007). Computation of nitroxide-nitroxide distances for spin-labeled DNA duplexes. *Biopolymers* **87**, 40–50.
40. Schiemann, O., Weber, A., Edwards, T. E., Prisner, T. F., and Sigurdsson, S. T. (2003). Nanometer distance measurements on RNA using PELDOR. *J. Am. Chem. Soc.* **125**, 3334–3335.
41. Edwards, T. E., Okonogi, T. M., and Sigurdsson, S. T. (2002). Investigation of RNA-protein and RNA-metal ion interactions by electron paramagnetic resonance spectroscopy: The HIV TAR-Tat motif. *Chem. Biol.* **9**, 699–706.
42. Edwards, T. E., and Sigurdsson, S. T. (2002). Electron paramagnetic resonance dynamic signatures of TAR RNA-small molecule complexes provide insight into RNA structure and recognition. *Biochemistry* **41**, 14843–14847.
43. Edwards, T. E., and Sigurdsson, S. T. (2003). EPR spectroscopic analysis of TAR RNA-metal ion interactions. *Biochem. Biophys. Res. Commun.* **303**, 721–725.
44. Kim, N. K., Murali, A., and DeRose, V. J. (2005). Separate metal requirements for loop interactions and catalysis in the extended hammerhead ribozyme. *J. Am. Chem. Soc.* **127**, 14134–14135.
45. Edwards, T. E., and Sigurdsson, S. T. (2005). EPR spectroscopic analysis of U7 hammerhead ribozyme dynamics during metal ion induced folding. *Biochemistry* **44**, 12870–12878.
46. Ramos, A., and Varani, G. (1998). A new method to detect long-range protein-RNA contacts: NMR detection of electron-proton relaxation induced by nitroxide spin-labeled RNA. *J. Am. Chem. Soc.* **120**, 10992–10993.
47. Ramos, A., Bayer, P., and Varani, G. (1999). Determination of the structure of the RNA complex of a double-stranded RNA-binding domain from *Drosophila* Staufen protein. *Biopolymers* **52**, 181–196.
48. Ramos, A., Grünert, S., Adams, J., Micklem, D. R., Proctor, M. R., Freund, S., Bycroft, M., St Johnston, D., and Varani, G. (2000). RNA recognition by a Staufen double-stranded RNA-binding domain. *EMBO J.* **19**, 997–1009.
49. Varani, L., Gunderson, S. I., Mattaj, I. W., Kay, L. E., Neuhaus, D., and Varani, G. (2000). The NMR structure of the 38 kDa U1A protein–PIE RNA complex reveals the basis of cooperativity in regulation of polyadenylation by human U1A protein. *Nat. Struct. Biol.* **7**, 329–335.
50. Piton, N., Schiemann, O., Mu, Y., Stock, G., Prisner, T., and Engels, J. W. (2005). Synthesis of spin-labeled RNAs for long range distance measurements by peldor. *Nucleosides Nucleotides Nucleic Acids* **24**, 771–775.
51. Schiemann, O., Piton, N., Plackmeyer, J., Bode, B. E., Prisner, T. F., and Engels, J. W. (2007). Spin labeling of oligonucleotides with the nitroxide TPA and use of PELDOR, a pulse EPR method, to measure intramolecular distances. *Nat. Protoc.* **2**(4), 904–923.
52. Grant, G. P. G., and Qin, P. Z. (2007). A facile method for attaching nitroxide spin labels at the 5′-terminus of nucleic acids. *Nucleic Acids Res.* **35**(10), e77.
53. Macosko, J. C., Pio, M. S., Tinoco, I., Jr., and Shin, Y.-K. (1999). A novel 5′ displacement spin-labeling technique for electron paramagnetic resonance spectroscopy of RNA. *RNA* **5**, 1158–1166.
54. Dugas, H., and Caron, M. (1976). Specific spin-labeling of transfer ribonucleic acid molecules. *Nucleic Acids Res.* **3**, 19–34.
55. Toppin, C. R., Thomas, I. E., Bobst, E. V., and Bobst, A. M. (1983). Synthesis of spin labelled deoxynucleotide analogues and their incorporation with terminal deoxynucleotidyl transferase into DNA. *Int. J. Biol. Macromol.* **5**, 33–36.
56. Pauly, G. T., Thomas, I. E., and Bobst, A. M. (1987). Base dynamics of nitroxide-labeled thymidine analogs incorporated into (dA-dT)n by DNA polymerase I from *E. coli*. *Biochemistry* **26**, 7304–7310.

57. Strobel, O. K., Keyes, R. S., Sinden, R. R., and Bobst, A. M. (1995). Rigidity of a B–Z region incorporated into a plasmid as monitored by electron paramagnetic resonance. *Arch. Biochem. Biophys.* **324,** 357–366.
58. Bobst, A. M., Pauly, G. T., Keyes, R. S., and Bobst, E. V. (1988). Enzymatic sequence-specific spin labeling of a DNA fragment containing the recognition sequence of EcoRI endonuclease. *FEBS Lett.* **228,** 33–36.
59. Xiao, M., Martin, I., Yablonovitch, E., and Jiang, H. W. (2004). Electrical detection of the spin resonance of a single electron in a silicon field-effect transistor. *Nature* **430,** 435–439.
60. Hustedt, E. J., and Beth, A. H. (1999). Nitroxide spin-spin interactions: Applications to protein structure and dynamics. *Annu. Rev. Biophys. Biomol. Struct.* **28,** 129–153.
61. Hustedt, E. J., and Beth, A. H. (2000). Structural information from CW-EPR spectra of dipolar coupled nitroxide spin labels. *In* "Biological Magnetic Resonance: Distance Measurements in Biological Systems by EPR" (L. J. Berliner, G. R. Eaton, and S. S. Eaton, Eds.), pp. 155–184. Kluwer Academic, New York.
62. Hustedt, E. J., Smirnov, A. I., Laub, C. F., Cobb, C. E., and Beth, A. H. (1997). Molecular distances from dipolar coupled spin-labels: The global analysis of multifrequency continuous wave electron paramagnetic resonance data. *Biophys. J.* **72,** 1861–1877.
63. Mchaourab, H. S., Oh, K. J., Fang, C. J., and Hubbell, W. L. (1997). Conformation of T4 lysozyme in solution. Hinge-bending motion and the substrate-induced conformational transition studied by site-directed spin labeling. *Biochemistry* **36,** 307–316.
64. Rabenstein, M. D., and Shin, Y. K. (1995). Determination of the distance between two spin labels attached to a macromolecule. *Proc. Natl. Acad. Sci.* **92,** 8239–8243.
65. Xiao, W., and Shin, Y.-K. (2000). EPR spectroscopic ruler: The method and its applications. *In* "Biological Magnetic Resonance, Volume 19: Distance Measurements in Biological Systems by EPR" (L. J. Berliner, G. R. Eaton, and S. S. Eaton, Eds.), pp. 249–276. Kluwer Academic, New York.
66. Altenbach, C., Oh, K. J., Trabanino, R. J., Hideg, K., and Hubbell, W. L. (2001). Estimation of inter-residue distances in spin labeled proteins at physiological temperatures: Experimental strategies and practical limitations. *Biochemistry* **40,** 15471–15482.
67. Steinhoff, H. J., Radzwill, N., Thevis, W., Lenz, V., Brandenburg, D., Antson, A., Dodson, G., and Wollmer, A. (1997). Determination of interspin distances between spin labels attached to insulin: Comparison of electron paramagnetic resonance data with the X-ray structure. *Biophys. J.* **73,** 3287–3298.
68. Schweiger, A., and Jeschke, G. (2001). "Principles of Pulse Electron Paramagnetic Resonance." Oxford University Press, Oxford.
69. Borbat, P. P., and Freed, J. H. (2000). Double-quantum ESR and distance measurements. *In* "Biological Magnetic Resonance, Volume 19: Distance Measurements in Biological Systems by EPR" (L. J. Berliner, G. R. Eaton, and S. S. Eaton, Eds.), pp. 383–459. Kluwer Academic, New York.
70. Milov, A. D., Salikohov, K. M., and Shirov, M. D. (1981). Applications of ENDOR in electron-spin echo for paramagnetic center space distribution in solids. *Fiz. Tverd. Tela* **23,** 975–982.
71. Kurshev, V. V., Raitsimring, A. M., and Tsvetkov, Y. D. (1989). Selection of dipolar interaction by the "2 + 1" pulse train ESE *J. Magn. Reson. (1969)* **81**(3), 441–454.
72. Milov, A. D., Ponomarev, A. B., and Tsvetkov, Y. D. (1984). Electron-electron double resonance in electron spin echo: Model biradical systems and the sensitized photolysis of decalin. *Chem. Phys. Lett.* **110**(1), 67–72.
73. Larsen, R. G., and Singel, D. J. (1993). Double electron—electron resonance spin—echo modulation: Spectroscopic measurement of electron spin pair separations in orientationally disordered solids. *J. Chem. Phys.* **98**(7), 5134–5146.

74. Rakowsky, M. H., More, K. M., Kulikov, A. V., Eaton, G. R., and Eaton, S. S. (1995). Time-domain electron paramagnetic resonance as a probe of electron-electron spin-spin interaction in spin-labeled low-spin iron porphyrins. *J. Am. Chem. Soc.* **117**(7), 2049–2057.
75. Saxena, S., and Freed, J. H. (1996). Double quantum two-dimensional Fourier transform electron spin resonance: Distance measurements. *Chem. Phys. Lett.* **251**(1–2), 102–110.
76. Saxena, S., and Freed, J. H. (1997). Theory of double quantum two-dimensional electron spin resonance with application to distance measurements. *J. Chem. Phys.* **107**, 1317–1340.
77. Martin, R. E., Pannier, M., Diederich, F., Gramlich, V., Hubrich, M., and Spiess, H. W. (1998). Determination of end-to-rnd distances in a series of TEMPO diradicals of up to 2.8 nm length with a new four-pulse double electron electron resonance experiment. *Angew. Chem. Int. Ed.* **37**(20), 2834–2837.
78. Borbat, P. P., and Freed, J. H. (1999). Multiple-quantum ESR and distance measurements. *Chem. Phys. Lett.* **313**, 145–154.
79. Pannier, M., Veit, S., Godt, A., Jeschke, G., and Spiess, H. W. (2000). Dead-time free measurement of dipole-dipole interactions between electron spins. *J. Magn. Res.* **142**, 331–340.
80. Schiemann, O., and Prisner, T. F. (2007). Long-range distance determinations in biomacromolecules by EPR spectroscopy. *Q. Rev. Biophys.* **40**, 1–53.
81. Borbat, P. P., Davis, J. H., Butcher, S. E., and Freed, J. H. (2004). Measurement of large distances in biomolecules using double-quantum filtered refocused electron spin-echoes. *J. Am. Chem. Soc.* **126**, 7746–7747.
82. Grant, G. P. G., Popova, A., and Qin, P. Z. (2008). Diastereomer characterizations of nitroxide-labeled nucleic acids. *Biochem. Biophys. Res. Commun.* **371**, 451–455.
83. Ward, R., Keeble, D. J., El-Mkami, H., and Norman, D. G. (2007). Distance determination in heterogeneous DNA model systems by pulsed EPR. *Chem. Bio. Chem.* **8**, 1957–1964.
84. Jeschke, G., Pannier, M., and Spiess, H. W. (2000). Double electron-electron resonance: Methodical advances and application to disordered systems. In "Biological Magnetic Resonance, Volume 19: Distance Measurements in Biological Systems by EPR" (L. J. Berliner, G. R. Eaton, and S. S. Eaton, Eds.), pp. 493–512. Kluwer Academic, New York.
85. Jeschke, G., Panek, G., Godt, A., Bender, A., and Paulsen, H. (2004). Data analysis procedures for pulse ELDOR measurements of broad distance distributions. *Appl. Magn. Reson.* **26**, 223–244.
86. Bowman, M. K., Maryasov, A. G., Kim, N., and DeRose, V. J. (2004). Visulation of distance distribution from pulsed double electron-electron resonance data. *Appl. Magn. Reson.* **26**, 23–39.
87. Sale, K., Song, L., Liu, Y. S., Perozo, E., and Fajer, P. (2005). Explicit treatment of spin labels in modeling of distance constraints from dipolar EPR and DEER. *J. Am. Chem. Soc.* **127**, 9334–9335.
88. Darian, E., and Gannett, P. M. (2005). Application of molecular dynamics simulations to spin-labeled oligonucleotides. *J. Biomol. Struct. Dyn.* **22**, 579–593.
89. Griffith, H. O., and Jost, P. C. (1976). Lipid spin labels in biological membranes. In "Spin Labeling Theory and Application" (L. J. Berliner, Ed.), pp. 453–423. Academic Press, New York.
90. Mchaourab, H. S., Lietzow, M. A., Hideg, K., and Hubbell, W. L. (1996). Motion of spin-labeled side chains in T4 lysozyme. Correlation with protein structure and dynamics. *Biochemistry* **35**, 7692–7704.
91. Hubbell, W. L., Mchaourab, H. S., Altenbach, C., and Lietzow, M. A. (1996). Watching proteins move using site-directed spin labeling. *Structure* **4**, 779–783.
92. Robinson, B. H., Slusky, L. J., and Auteri, F. P. (1992). Direct simulation of continuous wave electron paramagnetic resonance spectra from Brownian dynamics trajectories. *J. Chem. Phys.* **96**, 2609–2616.

93. Steinhoff, H. J., and Hubbell, W. L. (1996). Calculation of electron paramagnetic resonance spectra from Brownian dynamics trajectories: Application to nitroxide side chains in proteins. *Biophys. J.* **71**, 2201–2212.
94. Budil, D. E., Sale, K. L., Khairy, K. A., and Fajer, P. G. (2006). Calculating slow-motional electron paramagnetic resonance spectra from molecular dynamics using a diffusion operator approach. *J Phys. Chem. A* **110**, 3703–3713.
95. Fajer, M. I., Li, H., Yang, W., and Fajer, P. G. (2007). Mapping electron paramagnetic resonance spin label conformations by the simulated scaling method. *J. Am. Chem. Soc.* **129**, 13840–13846.
96. Moro, G., and Freed, J. H. (1980). Efficient computation of magnetic resonance spectra and related correlation functions from stochastic Liouville equations. *J. Phys. Chem.* **84**, 2837–2840.
97. Hustedt, E. J., Spaltenstein, A., Kirchner, J. J., Hopkins, P. B., and Robinson, B. H. (1993). Motions of short DNA duplexes: An analysis of DNA dynamics using an EPR-active probe. *Biochemistry* **32**, 1774–1787.
98. Schneider, D. J., and Freed, J. H. (1989). Calculating slow motional magnetic resonance spectra: A user's guide. *In* "Spin Labeling: Theory and Applications" (L. J. Berliner, Ed.), pp. 1–76. Plenum Press, New York.
99. Budil, D. E., Lee, S., Saxena, S., and Freed, J. H. (1996). Nonlinear-least-squares analysis of slow-motion EPR spectra in one and two dimensions using a modified Levenberg-Marquardt algorithm. *J. Mag. Res. Series A* **120**, 155–189.
100. Liang, Z., and Freed, J. H. (1999). An assessment of the applicability of multifrequency ESR to study the complex dynamics of biomolecules. *J. Phys. Chem. B* **103**, 6384–6396.
101. Liang, Z., Freed, J. H., Keyes, R. S., and Bobst, A. M. (2000). An electron spin resonance study of DNA dynamics using the slowly relaxing local structure model. *J. Phys. Chem.* **104**, 5372–5381.
102. Qin, P. Z., Jennifer, I., and Oki, A. (2006). A model system for investigating lineshape/structure correlations in RNA site-directed spin labeling. *Biochem. Biophys. Res. Commun.* **343**, 117–124.
103. Columbus, L., Kalai, T., Jeko, J., Hideg, K., and Hubbell, W. L. (2001). Molecular motion of spin-labeled side chains in α-helices: Analysis by variation of side chain structure. *Biochemistry* **40**, 3828–3846.
104. Okonogi, T. M., Reese, A. W., Alley, S. C., Hopkins, P. B., and Robinson, B. H. (1999). Flexibility of duplex DNA on the submicrosecond timescale. *Biophys. J.* **77**, 3256–3276.
105. Okonogi, T. M., Alley, S. C., Reese, A. W., Hopkins, P. B., and Robinson, B. H. (2002). Sequence-dependent dynamics of duplex DNA: The applicability of a dinucleotide model. *Biophys. J.* **83**, 3446–3459.
106. Okonogi, T. M., Alley, S. C., Reese, A. W., Hopkins, P. B., and Robinson, B. H. (2000). Sequence-dependent dynamics in duplex DNA. *Biophys. J.* **78**, 2560–2571.
107. Cantor, C. R., and Schimmel, P. R. (1980). "Biophysical Chemistry," pp. 460–564. W.H. Freeman, San Francisco.
108. Hester, J. D., Bobst, E. V., Kryak, D. D., and Bobst, A. M. (2002). Identification of a single genome by electron paramagnetic resonance (EPR) with nitroxide-labeled oligonucleotide probes. *Free Radic. Res.* **36**, 491–498.
109. Keyes, R. S., Bobst, E. V., Cao, Y. Y., and Bobst, A. M. (1997). Overall and internal dynamics of DNA as monitored by five-atom-tethered spin labels. *Biophys. J.* **72**, 282–290.
110. Okonogi, T. M., Alley, S. C., Harwood, E. A., Hopkins, P. B., and Robinson, B. H. (2002). Phosphate backbone neutralization increases duplex DNA flexibility: A model for protein binding. *Proc. Natl. Acad. Sci.* **99**, 4156–4160.

111. Spaltenstein, A., Robinson, B. H., and Hopkins, P. B. (1989). Sequence- and structure-dependent DNA base dynamics: Synthesis, structure, and dynamics of site and sequence specifically spin-labeled DNA. *Biochemistry* **28**, 9484–9495.
112. Allain, F. H. T., and Varani, G. (1995). Structure of the P1 helix from group I self-splicing introns. *J. Mol. Biol.* **250**, 333–353.
113. Akke, M., Fiala, R., Jiang, F., Patel, D., and Palmer, A. G. (1997). Base dynamics in a UUCG tetraloop RNA hairpin characterized by 15N spin relaxation: Correlations with structure and stability. *RNA* **3**, 702–709.
114. Qin, P. Z., Feigon, J., and Hubbell, W. L. (2005). Site-directed spin labeling studies reveal solution conformational changes in a GAAA tetraloop receptor upon Mg^{2+}-dependent docking of a GAAA tetraloop. *J. Mol. Biol.* **351**, 1–8.
115. Cate, J. H., Gooding, A. R., Podell, E., Zhou, K., Golden, B. L., Kundrot, C. E., Cech, T. R., and Doudna, J. A. (1996). Crystal structure of a group I ribozyme domain: Principles of RNA packing. *Science* **273**(5282), 1678–1685.
116. Frankel, A. D., and Young, J. A. T. (1998). HIV-1: Fifteen proteins and an RNA. *Annu. Rev. Biochem.* **67**, 1–25.
117. Edwards, T. E., Robinson, B. H., and Sigurdsson, S. T. (2005). Identification of amino acids that promote specific and rigid TAR RNA-Tat protein complex formation. *Chem. Biol.* **12**, 329–337.
118. Blount, K. F., and Uhlenbeck, O. C. (2005). The structure-function dilemma of the hammerhead ribozyme. *Annu. Rev. Biophys. Biomol. Struct.* **34**, 415–440.
119. Shin, Y. K., and Hubbell, W. L. (1992). Determination of electrostatic potentials at biological interfaces using electron-electron double resonance. *Biophys. J.* **61**, 1443–1453.
120. Nielsen, R. D., Canaan, S., Gladden, J. A., Gelb, M. H., Mailer, C., and Robinson, B. H. (2004). Comparing continuous wave progressive saturation EPR and time domain saturation recovery EPR over the entire motional range of nitroxide spin labels. *J. Mag. Res.* **16**, 129–163.
121. Smirnova, T. I., Chadwick, T. G., Voinov, M. A., Poluektov, O., van Tol, J., Ozarowski, A., Schaaf, G., Ryan, M. M., and Bankaitis, V. A. (2007). Local polarity and hydrogen bonding inside the Sec14p phospholipid-binding cavity: High-field multi-frequency electron paramagnetic resonance studies. *Biophys. J.* **92**, 3686–3695.
122. Clore, G. M., Tang, C., and Iwahara, J. (2007). Elucidating transient macromolecular interactions using paramagnetic relaxation enhancement. *Curr. Opin. Struct. Biol.* **17**, 603–616.
123. Leulliot, N., Quevillon-Cheruel, S., Graille, M., van Tilbeurgh, H., Leeper, T. C., Godin, K. S., Edwards, T. E., Sigurdsson, S. T., Rozenkrants, N., Nagel, R. J., Ares, M., Jr., and Varani, G. (2004). A new alpha-helical extension promotes RNA binding by the dsRBD of Rnt1p RNAse III. *EMBO J.* **23**, 2468–2477.
124. DeRose, V. J. (2003). Metal ion binding to catalytic RNA molecules. *Curr. Opin. Struct. Biol.* **13**, 317–324.
125. Hoogstraten, C. G., Grant, C. V., Horton, T. E., DeRose, V. J., and Britt, R. D. (2002). Structural analysis of metal ion ligation to nucleotides and nucleic acids using pulsed EPR spectroscopy. *J. Am. Chem. Soc.* **124**, 834–842.
126. Horton, T. E., Clardy, D. R., and DeRose, V. J. (1998). Electron paramagnetic resonance spectroscopic measurement of Mn^{2+} binding affinities to the hammerhead ribozyme and correlation with cleavage activity. *Biochemistry* **37**, 18094–18101.
127. Morrissey, S. R., Horton, T. E., Grant, C. V., Hoogstraten, C. G., Britt, R. D., and DeRose, V. J. (1999). Mn^{2+}-nitrogen interactions in RNA probed by electron spin-echo envelope modulation spectroscopy: Application to the hammerhead ribozyme. *J. Am. Chem. Soc.* **121**, 9215–9218.

128. Morrissey, S. R., Horton, T. E., and DeRose, V. J. (2000). Mn^{2+} sites in the hammerhead ribozyme investigated by EPR and continuous-wave Q-band ENDOR spectroscopies. *J. Am. Chem. Soc.* **122**, 3473–3481.
129. Schiemann, O., Fritscher, J., Kisseleva, N., Sigurdsson, S. T., and Prisner, T. F. (2003). Structural investigation of a high-affinity MnII binding site in the hammerhead ribozyme by EPR spectroscopy and DFT calculations. Effects of neomycin B on metal-ion binding. *Chembiochem* **4**, 1057–1065.
130. Kisseleva, N., Khvorova, A., Westhof, E., and Schiemann, O. (2005). Binding of manganese(II) to a tertiary stabilized hammerhead ribozyme as studied by electron paramagnetic resonance spectroscopy. *RNA* **11**, 1–6.
131. Vogt, M., Lahiri, S., Hoogstraten, C. G., Britt, R. D., and DeRose, V. J. (2006). Coordination environment of a site-bound metal ion in the hammerhead ribozyme determined by 15N and 2H ESEEM spectroscopy. *J. Am. Chem. Soc.* **128**, 16764–16770.
132. Schiemann, O., Carmieli, R., and Goldfarb, D. (2007). W-band 31P-ENDOR on the high-affinity Mn^{2+} binding site in the minimal and tertiary stabilized hammerhead ribozymes. *Appl. Magn. Reson.* **31**, 543–552.
133. Butcher, S. E., Dieckmann, T., and Feigon, J. (1997). Solution structure of a GAAA tetraloop receptor RNA. *EMBO J.* **16**, 7490–7499.
134. LaConte, L. E. W., Voelz, V., Nelson, W., Enz, M., and Thomas, D. D. (2002). Molecular dynamics simulation of site-directed spin labeling: Experimental validation in muscle fibers. *Biophys. J.* **83**, 1854–1866.
135. Stoica, I. (2004). Using Molecular dynamics to simulate electronic spin resonance spectra of T4 lysozyme. *J. Phys. Chem. B* **108**, 1771–1782.
136. Beier, C., and Steinhoff, H.-J. (2006). A structure-based simulation approach for electron paramagnetic resonance spectra using molecular and stochastic dynamics simulations. *Biophys. J.* **91**, 2647–2664.
137. DeSensi, S. C., Rangel, D. P., Beth, A. H., Lybrand, T. P., and Hustedt, E. J. (2008). Simulation of nitroxide electron paramagnetic resonance spectra from Brownian trajectories and molecular dynamics simulations. *Biophys. J.* **94**, 3798–3809.

Molecular Computing with Deoxyribozymes

Milan N. Stojanovic

Department of Medicine, Columbia University

I. Introduction to Molecular Computing by Deoxyribozymes	199
II. Deoxyribozyme-Based Logic Gates	200
III. Deoxyribozyme-Based Circuits for Arithmetical Operations	204
IV. Deoxyribozyme-Based Automata: Circuits that Play Tic-Tac-Toe	206
V. Deoxyribozyme-Based Control of Downstream Elements	210
VI. Expanding Molecular Logic to Nanoparticles	213
VII. Other Approaches to Autonomous Computing with DNA	215
VIII. Conclusions and Future Visions	215
References	216

Silicomimetic molecular computing is based on an idea that individual molecules can perform basic logical operations and make simple decisions based on the presence or absence of multiple factors in solution. Using deoxyribozymes, nucleic acid catalysts made of DNA, and recognition regions for oligonucleotides, stem-loops, we constructed so-called full set of molecular logic gates, which allowed us to combine individual gates into more complex circuits. These gates and their circuits analyze sets of oligonucleotides as inputs and produce changed substrate oligonucleotides as outputs. Some examples of circuits are half- and full adders, which perform basic arithmetical operations, and molecular automata that can play game of tic-tac-toe. Our most advanced circuit integrates more than 100 nucleic acid logic gates to play a complete game of tic-tac-toe with 76 subgames. Both inputs and outputs of circuits can be coupled with upstream and downstream components, such as aptamers, for small molecule or protein release; this in turn, points to the possibility of autonomous therapeutic and diagnostic molecular devices.

I. Introduction to Molecular Computing by Deoxyribozymes

Adleman's seminal demonstration (1) that DNA may be suitable for massively parallel computing triggered a flurry of interest in DNA as an alternative to silicon-based computers, at least when it comes to solving certain types of problems. From the very beginning of our interest in this field, we were rather modest in our desire to compete with silicon using biomaterials. Instead, we decided to focus on something much simpler: the ability of molecules to make autonomous decisions. What does it mean when we say that molecules make decisions? It means that molecules interact with a set of other molecular or ionic species, called inputs, and based on interactions with each subset of inputs, produce or not produce another molecular species, called an output, as a "decision." It is preferable to have a definite decision (digital), in the form of YES or NO, or 1 or 0. The presence of each subset of inputs is tied to one value of the output (produced or not produced, YES, NO, 1, or 0); therefore, the rules of decision-making can be presented in the form of truth tables, similar to logical operations or Boolean calculations. Thus, these molecules can also be called molecular logic gates, and groups of molecules can be viewed as circuits, in analogy to modern electronic computers, in which logic gates are basic units that perform computing.

The calculations molecules perform are named NOT, AND, and OR based on their rules of operation being identical to their Boolean namesakes (2, 3). For example, a NOT gate, or inverter, uses one input molecule and produces the output only if that input is not present, while an AND gate, uses two molecules as inputs, providing an output if both inputs are present. OR gate uses also two inputs, but the output is produced if either or both are present. Similar to digital computing, we should be able to build more complex circuits from these basic computing elements (described in Section 4).

The behavior of any system in which recognition events turns off or on downstream events, such as structural or spectral changes, or catalytic activity, can be described through truth tables and reinterpreted as a logic gate-like behavior. This concept has led to some important advances in the fields of sensors, molecular electronics, and even quite complex logical operations have been demonstrated with this approach. Nevertheless, the ability to implement these logic gates in a plug-and-play fashion in more complex devices seems somewhat restricted at the moment, mostly by their incomplete modular designs and dissonant inputs and outputs.

II. Deoxyribozyme-Based Logic Gates

From the very beginning we wanted to enable the full modularity of our logic gates, which would allow us to use these gates for the analysis of large number of inputs, and in order to be able to arrange them readily in more complex molecular devices. Instead of using currents, as in digital computers, our logic gates use oligonucleotides as inputs and generate (calculate) new oligonucleotides as outputs based on a set of rules.

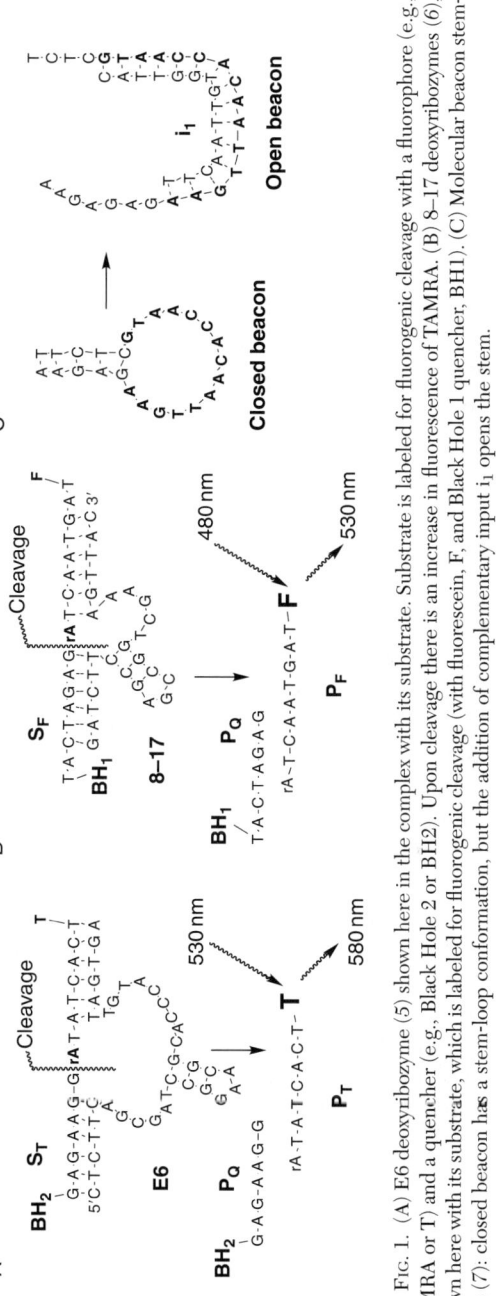

FIG. 1. (A) E6 deoxyribozyme (5) shown here in the complex with its substrate. Substrate is labeled for fluorogenic cleavage with a fluorophore (e.g., TAMRA or T) and a quencher (e.g., Black Hole 2 or BH2). (B) 8-17 deoxyribozymes (6), shown here with its substrate, which is labeled for fluorogenic cleavage (with fluorescein, F, and Black Hole 1 quencher, BH1). (C) Molecular beacon stem-loop (7): closed beacon has a stem-loop conformation, but the addition of complementary input i_1 opens the stem.

In order to achieve this behavior we use modular design (4) and combine two modules: a catalytic module (5, 6) that consists of deoxyribozymes (deoxyribonucleotide-based catalysts, Fig. 1A and 1B) and a recognition module, usually a stem-loop oligonucleotide inspired by molecular beacons (7) (Fig. 1C). The stem-loop establishes allosteric control over the activity of the deoxyribozyme, and conformational change (i.e., stem opening) in the stem-loop region influence the catalytic activity of the deoxyribozymes. We use two types of deoxyribozymes: phosphodiesterases, that cleave their substrates, and ligases, that put together two shorter substrates. The following discussion is applicable to both types of enzymes.

Our simplest design, called initially catalytic molecular beacon (8), and based on phosphodiesterase enzyme, uses positive allosteric regulation (Fig. 2A): a stem-loop region blocks access of the deoxyribozyme substrate to the substrate-recognition region. Upon binding of the complementary oligonucleotide to the loop, the stem will open, allowing substrate to bind and the catalytic reaction to proceed. This is an example of a detector or sensor gate, sometimes called by chemists a YES gate.

In another example we demonstrated negative allosteric regulation in a NOT gates (9) (Fig. 2B): a stem-loop region is embedded within the catalytic core of the deoxyribozyme and the binding of a complementary oligonucleotide distorts this core, turning off enzymatic activity. The drawback of this approach is that only certain enzymes can be used, that is, those deoxyribozymes that have a catalytic core that can be modified. For example, deoxyribozymes 8–17 cannot be used in this design, but only much slower E6 (Fig. 1A and B).

Multiple-input gates are constructed by the addition of two or three controlling modules to one deoxyribozyme. For instance, placement of two stem-loop controlling regions on the opposite ends of the substrate-recognition region will create an AND gate (Fig. 2C). This gate requires the presence of both inputs for enzymatic activity. Combination of sensor gate and NOT gate gives an ANDNOT gate (not shown), which is active in the presence of one input for as long as a second input is not present.

At this moment, we have a general approach to gates with up to three inputs. For example, a NOT gate-like element can be added to an AND gate in the ANDANDNOT gate (10) in Fig. 2D, to obtain a gate which is active in the presence of two inputs, while being shut down in the presence of the third.

The action of input oligonucleotides can be reversed by adding their complements (11), and we took advantage of this property to turn ANDANDNOT gate into ANDAND (Fig. 1E) and ANDNOTNOT gates (12). In these systems, individual gates are pre-complexed with inputs; for example, if we pre-complex a NOT loop in ANDANDNOT gate, the complex will behave as ANDAND gate. This inversion approach was suggested by an exceptional high-school student, Harvey Lederman, inspired by Yurke's machine (11).

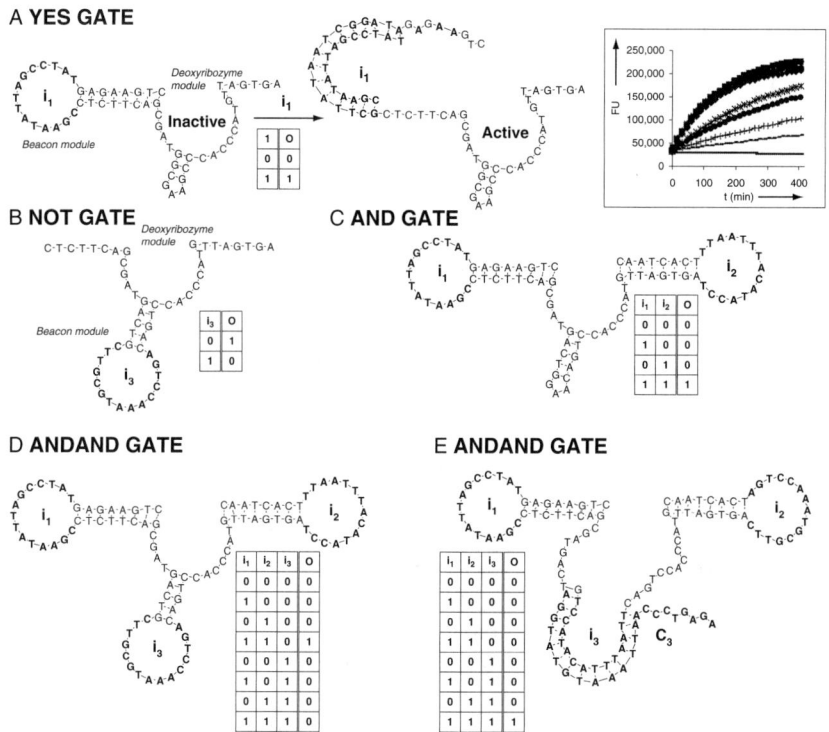

FIG. 2. A set of deoxyribozyme-based logic gates: (A) Catalytic molecular beacon or YES or sensor gate is constructed by attaching a beacon module to one of the substrate-recognition regions of the deoxyribozyme module. Upon the addition of an input (i_1), the gate is turned into its active form. The reaction can be monitored fluorogenically, and the graph in the insert shows increase in fluorescence with time upon addition of increasing amount of input. (B) NOT gate is constructed when a stem-loop is added to a catalytic core, in this case we use enzyme E6. Opening of the stem will distort the catalytic core and inhibit the reaction. (C) AND gate is constructed when two stem-loops are added to both substrate-recognition regions. Both stems have to open for a substrate to be cleaved, and both inputs have to be present. (D) A three-input gate (ANDANDNOT), which is active when two inputs (i_1 and i_2) are present, but not the third one (i_3), is a combination of an AND gate and a NOT gate. (E) The pre-complexation of stem-loops with a complementary oligonucleotide (c_3) leads to an inversion of function. An input complementary to c_3 has to be added to close the stem, and in this case activate the gate (ANDAND).

There are certain limitations on the structures of inputs that can be accepted by our gates. Some inputs have strong secondary structures, and cannot open up stems efficiently, while some loops favor alternative conformations with deoxyribozymes. Also, at room temperature, inputs different at only one position are not clearly differentiated by gates, except by minor differences in the initial reaction rates. The general Hamming distance (the difference in number of mismatches required for two inputs to be recognized as distinct inputs) is around three mismatches, with some exceptions (large number of G × T mismatches).

For both phosphodiesterase and ligase reactions, the formation of outputs can be monitored using gel electrophoresis. We usually take advantage of fluorogenic labeling of substrates (13) in order to monitor reactions of phosphodiesterases. If substrate is labeled with a 5′ fluorophore and a 3′ quencher (Fig. 1A), after its cleavage, the fluorophore is separated from the quencher and increase in fluorescence signal is observed.

III. Deoxyribozyme-Based Circuits for Arithmetical Operations

We discuss now parallel arrangement of molecular logic gates. In this approach, our gates do not directly communicate with each other, but they share common substrates. This is also called implicit OR arrangement, because even if only one gate is active, the substrate will be cleaved. For example, the simplest implicit OR circuit is made of two YES gates, with substrate being cleaved if either of these gates is active (9).

One challenge for any molecular computing is the ability to combine logic gates into more complex circuits that do something useful, for example, that perform basic arithmetic operations such as addition (14). In digital computers, this is achieved by cascading elements called adders. The first step in a design of oligonucleotide based adders is choosing a way to encode binary numbers in oligonucleotides. 1 at any digit can be encoded in the presence of a specific oligonucleotide, in which case 0 is defined by its absence. Thus, any number with n binary digits will be encoded by n oligonucleotides. Each number is encoded in a separate set of oligonucleotides, which is not very practical, but will suffice for these demonstrations. Output digits are also defined through separate catalytic activities, one each for the output bits.

A half-adder will take two binary digit inputs (bits) to produce two binary outputs—which are called a sum and a carry. Carry is used as an input for next element in a cascade, usually a full adder (see below). Any half adder, including molecular, should add in binary $00 + 00 = 00, 00 + 01 = 01, 01 + 00 = 01$, and $01 + 01 = 10$ (or in decimal $0 + 0 = 0, 0 + 1 = 1, 1 + 0 = 1$, and $1 + 1 = 2$).

In order to construct a deoxyribozyme-based half-adder (Fig. 3A), we need a system that could analyze the presence of two input molecules (for two bits, one each from our two numbers) and generate two different catalytic output activities within the prescribed set of rules for the adder behavior. Thus, the absence of both input molecules will leave the system as is, and neither of the outputs will be produced ($00 + 00 = 00$). The presence of any one (and only one)

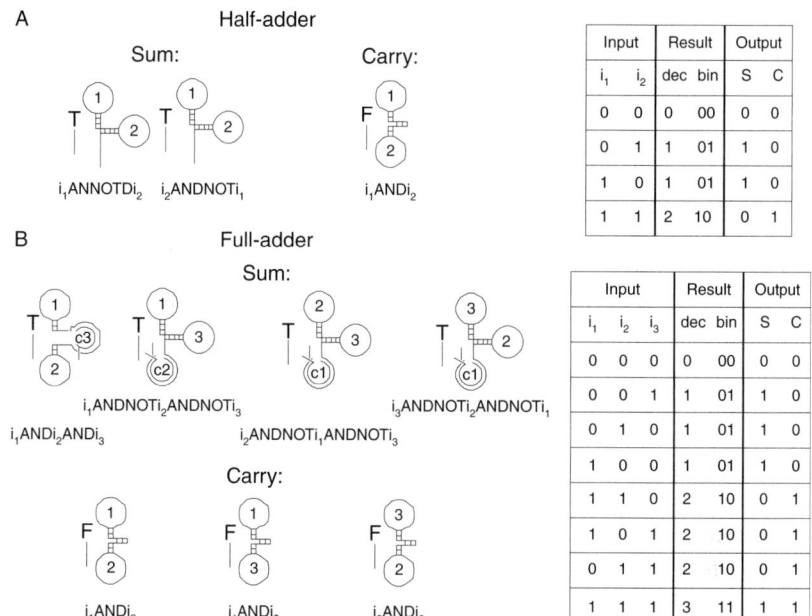

FIG. 3. Constructions of adders: the function of adders is monitored in two channels, sum as an increase in TAMRA fluorescence, while carry as an increase in fluorescein emission. (A) Deoxyribozyme-based half-adder is a mixture of three gates; two are ANDNOT gates based on E6 enzyme and cleave fluorogenic substrate with TAMRA (cf. Fig. 1A), while the third one is based on 8–17 and cleaves a fluorogenic substrate based on fluorescein. The former gates together are active only when only one of the inputs is present (that is, they behave as XOR gate required for sum), while the latter gate is active only when both inputs are present (i.e., AND gate, as required for carry). (B) Deoxyribozyme-based full adder is a mixture of seven gates; four gates cleave fluorogenic substrate with TAMRA and are based on E6 enzyme. These gates together behave as sum, and they are active if one and only one input is present or if all three inputs are present. Three gates based on 8–17 enzymes behave as carry: they are active when any two or all three inputs are present, and they cleave fluorescein-based fluorogenic substrate.

of the input molecules will trigger the activation of only the sum output (as in $00 + 01 = 01$ or $01 + 00 = 01$). Finally, in the presence of both inputs the half-adder should produce only the second output, while keeping the first one inactive (per $01 + 01 = 10$). The sum output was produced in our publication (15) with an implicit exclusive OR or implicit XOR gate built from two deoxyribozyme ANDNOT gates competing for the same substrate, and the carry output required an AND gate. We used combinations of the catalytic cores of deoxyribozymes E6 and 8–17, for sum and carry, respectively, in order for our logic gates to cleave two different oligonucleotide substrates, each labeled with different fluorophore/quencher pairs: fluorescein (F) and Black-Hole

Quencher 1 (BHQ1), and tetramethylrhodamine (TAMRA) and Black-Hole Quencher 2 (BHQ2). Thus, our deoxyribozyme-based half-adder needed a total of three deoxyribozyme logic gates (15).

A full adder will add three binary digits: two bits, one from each of the numbers, and a carry from a previous bit-addition step. This construct requires three-oligonucleotides as inputs. Full adder again produces two binary outputs, called a sum and a carry (Fig. 3B), requiring two separate catalytic outputs in deoxyribozyme design. Thus, a full adder is similar to a half-adder except that it processes an additional input, which is, in cascades of adders, usually a carry digit from a previous adder. The full adder produces two outputs according to the following rules: the absence of all three input molecules will leave the system as is, and neither of the outputs will be produced (00 + 00 + 00 = 00). The presence of any one (and only one) of the input molecules will activate only sum output (e.g., 01 + 00 + 00 = 01). The presence of any two inputs will activate only the carry (C) output (e.g., 01 + 01 + 00 = 10). The presence of all three inputs will activate both outputs (01 + 01 + 01 = 11).

Thanks to the general three-input gates, we could construct a full adder using implicit OR designs, rather than having to establish a serial connection (a cascade) between gates (12). Our molecular full adder, consisted of seven logic gates (Fig. 3B): three AND gates activating the fluorescein-labeled carry output (any two of the inputs or 01 + 00 + 00 = 10) and three ANDANDNOT and one ANDAND gates activating TAMRA-labeled sum output substrate. Three AND gates were also functioning as a carry in the presence of all three inputs, by being activated in parallel.

Two interesting design issues were identified and addressed during construction: (i) at higher gate concentrations (total concentrations >1 μM), the gates influenced each other, and therefore much lower concentrations were required to attain perfect full-adder behavior. It is likely that a judicious choice of oligonucleotides would allow the use of higher gate concentrations, although possibly at the expense of generality, (ii) we noticed that, despite our remarkable successes of modular design, we were not able to routinely rotate stem-loops in order to obtain all possible combinations of ANDNOTANDNOT gates. We expect that this apparent breakdown in modular design will likely be eliminated with the availability of improved oligonucleotide libraries that are specifically matched to the substrates and core enzymes sequences.

IV. Deoxyribozyme-Based Automata: Circuits that Play Tic-Tac-Toe

We can define molecular automata in several ways, for example, as sets of molecules that change states according to a series of inputs and a set of rules that determine the changes between each state. In a molecular world, we would like to approach the definition of automata a little more subjectively, as to differentiate them from ordinary circuits: molecular automata are molecular circuits that leave an observer with an impression of a meaningful response to a series of inputs. Essentially, humans translate their own challenges to automata in "language" molecules "understand," in our case input oligonucleotides, and obtains response over a sequential series of inputs. We have focused on the construction of molecular automata that play tic-tac-toe (known also as naughts-and-crosses) against a human player, because game playing is an unbiased test of the complexity that can be achieved by a new computation medium. Any result in this case is not something that we can arbitrarily argue whether it is a success or not, because our automata will either play or not play this game perfectly.

We chose tic-tac-toe because it is one of the simplest games of perfect information. There is also an element of tradition in this choice, because one of the first games played by computers was tic-tac-toe, over 50 years ago. Our first automaton, MAYA (an acronym derived from Molecular Array of YES and AND gates) (10), played a symmetry-pruned restricted game of tic-tac-toe, which allowed us to explore gate integration and decision making on a simplified scale. This design was restricted, at the time, by the lack of a general three-input logic gates. Our second automaton, MAYA-II (16), played an unrestricted game. The focus of MAYA-II project was to test the engineering limits of our deoxyribozyme-based approach to computing.

Games are played in 3 × 3 wells of 384 well plates (mimicking the nine fields of a tic-tac-toe game board), with individual wells sequentially numbered with #1–#9. In MAYA (shown in Fig. 4), the automaton always goes first in the middle well, and the human player is allowed to respond in only one of the corner wells or only one of the side wells (Fig. 3A) (10). This simplification reduces our engineering effort to programming the automata to play a total of 19 legal games (out of 76 possible, after the automaton's first move), 18 of which end in the automaton winning, because human is not playing perfectly, and one ending in draw, if both players play perfectly.

The automaton is turned on by adding Mg^{2+} or Zn^{2+} to all wells, activating a constitutively active deoxyribozyme in the middle well (#5). Human intentions are communicated to automaton through oligonucleotide inputs, which have to be added to all the wells. These inputs trigger response from gates, and the automaton's response is observed by monitoring a fluorescence increase in a response well. In MAYA, eight inputs are used encoding the position of the human move into wells #1–#4 and #6–#9 (#5 is already used by the automaton in the opening move). In order to move into well 1 the human player will add input i_1, and to move into well 4 the human player will add input i_4 to all wells. An array of 23 logic gates distributed in the remaining 8 wells calculates a response to the human player's input causing only one well to display an

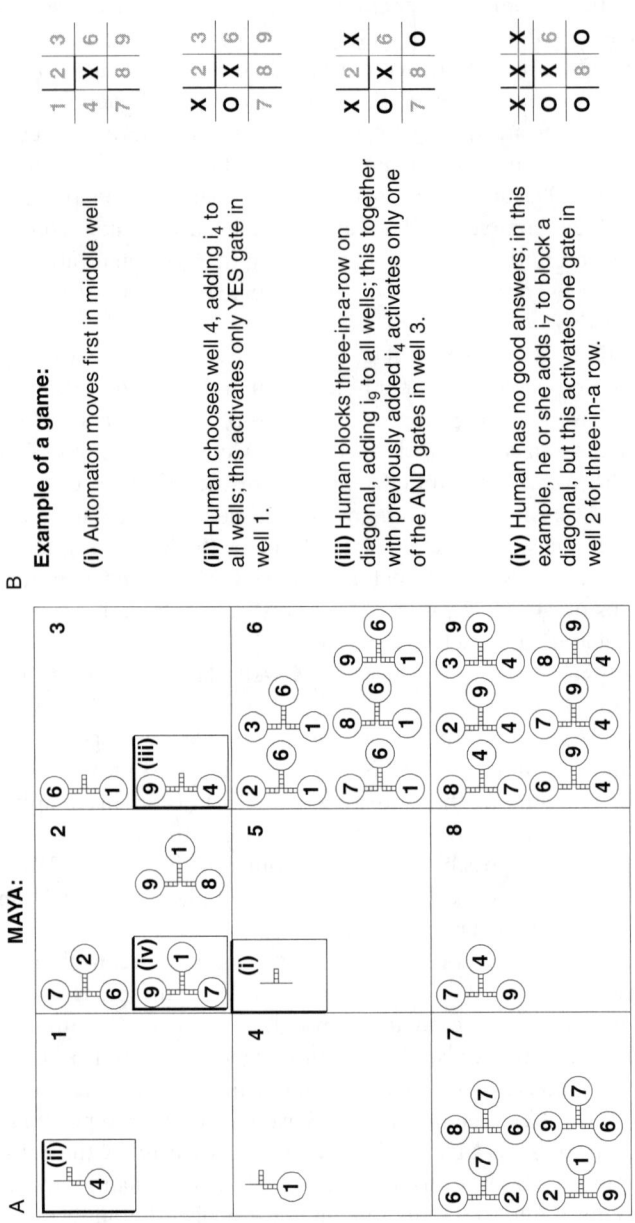

FIG. 4. MAYA, automaton that plays symmetry-pruned tic-tac-toe game: (A) Distribution of gates in wells. Central well (#5) contains constitutively active deoxyribozyme, while the other wells contain logic gates. Gates used in our example game are boxed. (B) An example of a game, in which human does not play perfectly and looses. There are total of 19 games encoded in this distribution of logic gates. Moves are labeled (i)–(iv).

increase in fluorescence, thus, indicating the automaton's chosen well. The cycle of human player input followed by automaton response continues until there is a draw or a victory for the automaton.

At the time, the combination of 23 logic gates (two YES and two AND gates, the rest being three-input gates) distributed in 8 wells was the test of the deoxyribozyme-based computing on the largest scale, and MAYA, in over 100 test games, truly excelled. While extensive testing indicated the lack of erroneous moves, we observed some background fluorescence increase in wrong wells due to imperfect digital behavior. For instance, upon addition of i_1 we noticed a strong and expected increase in well 4, but also a detectable increase in well 9 from the five partially activated ANDANDNOT gates. The digital behavior of individual gates can be perfected one gate at a time by increasing the lengths of inhibitory stems in the presence of specific inputs. This, however, suppresses the reaction rate in the presence of activating oligonucleotides leading to an increase in MAYA's "thinking time."

After the successful demonstration of MAYA, we decided to build a larger tic-tac-toe playing automaton, MAYA-II (16), in order to test the limits of integration of individual logic gates. While MAYA-II always played first in the middle well, the human player was free to choose the first move in the any of the remaining wells. Thus, the automaton plays all 76 permissible games. The automaton is also "user-friendly," with a two-color fluorogenic output system displaying both human and automaton moves. Human moves are displayed in a green channel (fluorescein) and automaton moves are displayed in a red channel (TAMRA), similar to our adders.

Translating a human move for MAYA-II required an increase from 8 to 32 input sequences. These new inputs coded both the well position and the order of the human move (their first, second, third, or fourth move into wells 1–8). The inputs were labeled i[NM] where N is the board position and M is the order of the humans move. For example, to play into well 7 for their first move, the human would add input i_{71}, but to play into well 7 on a third move the human would add input i_{73}. To calculate its next move and to display human moves, MAYA-II uses an array of 96 logic gates based on E6 deoxyribozymes. Human moves are displayed through an action of 32 YES gates based on the 8–17 deoxyribozymes. We tested (and retested) automaton, improving individual features, until it played perfectly, and clearly, with only minimal non-digital behavior.

The automaton move gates were immediately accepting almost all of the 32 input sequences but gates demonstrated wide variation in response strength, and we needed to optimize both gate structures and calibrate individual concentrations. This type of calibration resembles traditional circuit construction in silicon electronics, where circuit and switch calibration to high and low voltage signals is also required. Of course, it is preferable to create a system that minimizes the laborious trial and error of gate calibration, and we are currently constructing reference library of individual modules, for the immediate construction of reliable gates.

The success of MAYA-II, a mixture of 128 molecular logic gates, 32 input DNA sequences, and two fluorescent substrates for outputs, all in one solution, indicates the maturity of our deoxyribozyme-based logic gates as a modular, "plug and play" system. Based on its size, we can argue that MAYA-II represents the first medium scale integration of molecular logic gates in solution. The significance of this result is that it shows that large-scale, higher level computing using molecular logic gates is reasonable, with even larger molecular automata being feasible. For example, with the gates used in this paper, we can regularly detect down to about 10 nM gate concentrations, and the maximum total concentration of oligonucleotides in an individual well is estimated to be around 1 μM (with well-chosen libraries it is likely to be even more). Thus, at the minimum, we could operate up to 100 non-interfering gates in parallel in a single tube. In MAYA-II the maximum number of gates used per well was 18, significantly less than this number.

V. Deoxyribozyme-Based Control of Downstream Elements

While up to now we discussed only parallel arrangement of gates, serial arrangement or a cascade of elements is also possible. Serial arrangement can be based on a product of one "upstream" enzyme, serving as an input, and either inhibiting (product-inhibitor) or promoting (product-promoter) the other "downstream" enzyme. Alternatively, a substrate of an upstream enzyme can serve as an input and either inhibit or promote a downstream enzyme; the cleavage of this substrate can remove this allosteric regulation. So, the success of serial circuits is based on oligonucleotide motifs in a substrate and its products being sufficiently different, because either substrate or products have to be inactive in regulation of downstream elements. This represents a serious design problem; for example, if there is a motif in a product of phosphodiesterase enzyme that activates downstream element, the same motif is also always present in the originating substrate. With the ligase enzymes, the problem is opposite: motifs in shorter substrates are always present in longer products. One possible solution is to develop hidden motifs, through some form of conformational switching. Of note, these problems are avoided altogether in systems that undergo single turnover (17). However, we believe that the enzymatic module is crucial for building more complex circuits, because of a fan-out issue (the maximum number of downstream gates that a single upstream logic gate can feed) and our plans to have feedbacks. We now go through several published and unpublished examples in which a serial approach was established.

The most straightforward and the very first cascade (Fig. 5A) we reported used a ligase deoxyribozyme (18) module in an upstream gate, and a downstream phosphodiesterase YES gate (19). In this case, the product of a ligase gate reaction can bind to the stem-loop of a downstream phosphodiesterase gate (Fig. 5A), and is an example of a "product-promoter." Individually, the substrates are too short to activate downstream YES gates, although, interestingly, we discovered that synergistically they can activate downstream gates to some extent, leading to non-digital behavior. To avoid this problem, we had to introduce a single mismatch in one of the substrates (boxed G–G in YES gate in Fig. 5A). The ligase module requires chemically activated phosphoimidazole groups at the 3′ end of one of the substrates (PIM in Fig. 5A). Similar to the phosphodiesterase logic gate design we developed a full set of ligase logic gates (19), all of which can be cascaded.

In phosphodiesterase gates we could not, up to now, design practical substrate/product pairs, in which a product would *specifically* activate downstream gate, while substrate would not. An alternative design was more readily implemented: we used a substrate of an upstream logic gate as an inhibitor of a downstream NOT gate (thus the name substrate-inhibitor). In this way, only the active upstream gate can relieve this inhibition imposed on the downstream gate. Also, multiple cascade levels are possible, although the increase in time of computation becomes an issue. There are at least two factors slowing down this cascade: (i) very slow dissociation of substrate-inhibitor from its complex with the downstream logic gate, (ii) low concentration of free substrate-inhibitor, forcing the upstream enzyme to operate at substrate concentrations significantly below Km values. We have succeeded in increasing the rate of information transfer between layers of a cascade by introducing a mismatch between substrate-inhibitor and NOT loop in a gate, but even under the most favorable circumstances computing rates of several hours are needed. In principle, this design is general, and we have implemented in our unpublished work both AND and OR logic with it. We here provide an unpublished example of this approach: the coupling an upstream streptavidin sensor gate (20) to a downstream NOT gate (Fig. 5B).

The serial connection is not limited to inter-enzyme networks, and aptamers can be used for the same purpose. Aptamers are oligonucleotide-based recognition regions and it was immediately clear, at the intuitive level, at least, that we can use logic gates to control aptamers and their "binding states" (defined as bind, or 1 or not bind, or 0). We have developed a system that provides, through a cascade, logic gate control over both small molecules and proteins (21). This, we suggest, is an important step in the development of autonomous therapeutic and diagnostic devices.

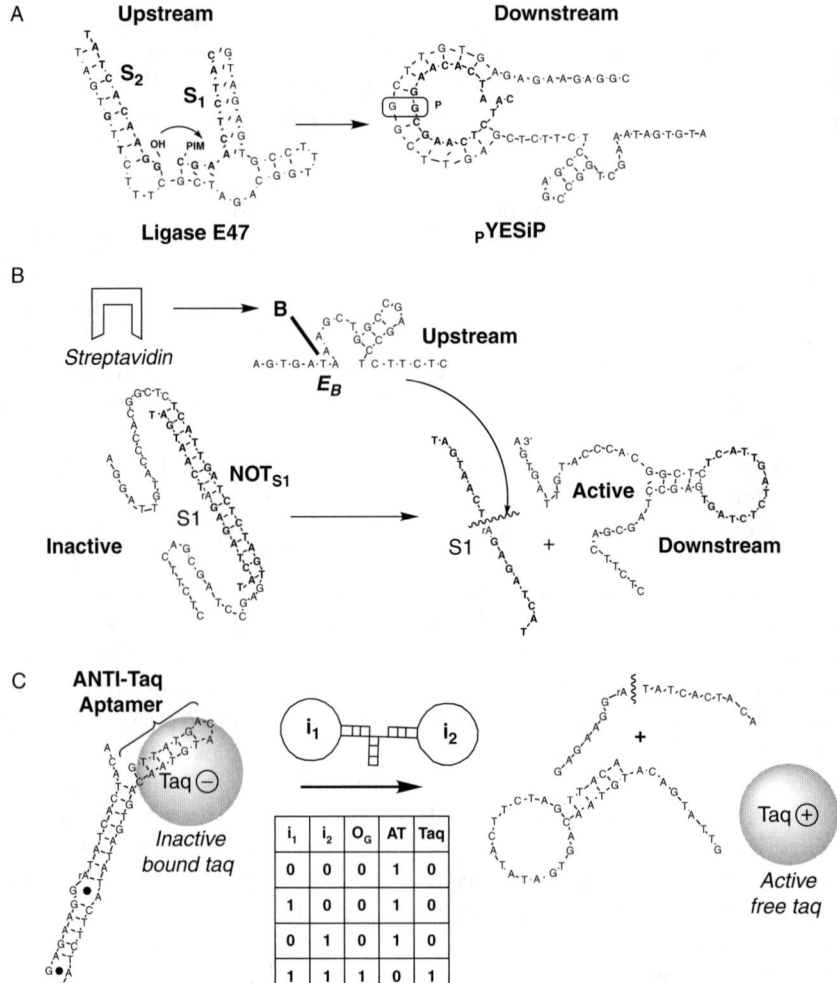

FIG. 5. Logic gates can be cascaded into downstream elements: (A) Ligase logic gates combine two short substrates into a longer product that feeds downstream phosphodiesterase gates. In this case, one of the substrates for ligase (S_1) is activated as phosphoimidazole (PIM). The boxed G–G in downstream gate is a mismatch introduced to avoid non-digital behavior in cascade. (B) Phosphodiesterases can cleave downstream logic gate inputs, leading to the release of downstream elements from inactive state. In this case the upstream phosphodiesterase Eb is itself biotinylated (B) and regulated by streptavidin. Binding of streptavidin to biotin distorts catalytic site and causes the catalytic cleavage to stop. (C) Logic gates can control aptameric binding events, such as the binding and release of active Taq DNA polymerase. An anti-Taq aptamer is kept in active state through complexation with a substrate for a logic gate. If logic gate is active, the substrate is removed, and the aptamer activity is turned off. This results in the activation of Taq polymerase.

The logic gate control of aptamers is similar to our phosphodiesterase–phosphodiesterase cascades: the substrate of an upstream logic gate is used to keep an aptamer engineered for structure-switching (22–24) in a ligand-binding state. The removal of substrate from solution switches the aptamer slowly off, and releases any bound ligand. In our first studies we adapted two aptamers for the control by a logic gate. The first one was malachite green aptamer, and we used decrease in fluorescence connected to the release of malachite green to monitor the information transfer between logic gate and aptamer. The second aptamer was a Taq DNA polymerase inhibitor, and we demonstrated logic gate control of Taq activity (Fig. 5C) (21).

The principles demonstrated in our initial studies indicate other aptamers can be integrated into molecular circuits. Through the process of *in vitro* selection and amplification (also known as SELEX) with either natural or unnatural nucleotides, aptamers that are responsive to almost any molecular or ionic species can be isolated. Thus, our demonstration indicates that molecular computation can be used to connect to almost any other event on a molecular scale.

VI. Expanding Molecular Logic to Nanoparticles

We now describe our initial efforts to integrate molecular computing devices with more traditional approaches to nanomedicine, such as those using nanoparticles for drug-delivery. This approach, if successful, would open possibilities to increase the functional complexity of delivery systems. We will describe one system, a three-layer cascade, in which microscopic particles coordinate their activity *without any direct physical contact*. More complex networks, including an AND hub, can be found in our initial publication (25).

The elementary unit of a network of microparticles is a single particle covered with a DNA computing or sensing element. Individual bead senses the presence of an input stimulus (or multiple stimuli) in solution, and, according to a set of rules encoded on this bead by computing elements, it releases an oligonucleotide signal as an output through a catalytic process. This can occur through one of the logic gates that are deposited on bead together with its substrate, and the released oligonucleotide is in this case one of the products. This signal can interact with another DNA element on a downstream bead, leading to the information transfer between beads and to a cascade. The communication between elements requires no physical contact and it occurs over the long range through diffusion of signaling molecules. The network activity can be monitored by multicolored flow cytometry, if we judiciously label individual oligonucleotides.

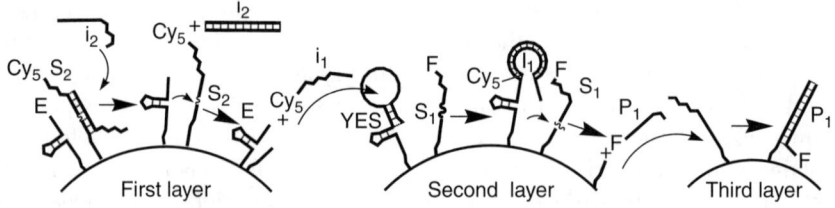

FIG. 6. Three-layer cascade with beads: the cascade starts with a first layer bead, which senses input i_2. This input releases free substrate S_2, which is then cleaved by enzyme (E). This process releases input i_1, which, in turn, activates YES gate on the second layer of beads. Activated YES gate cleaves co-deposited substrate S_1, releasing a product P_1. The product diffuses, being captured by its complement on the beads constituting the third layer. This and other cascades of particles can be monitored by flow cytometry, and in this case we labeled first substrate with Cy5, and the second substrate with fluorescein (F). After the cascade is triggered (i.e., when i_2 is added), on the first bead we will observe a decrease in Cy5 fluorescence, on the second bead, an increase in Cy5 and a decrease in fluorescein, while on the third bead, we will observe an increase in fluorescein intensity.

In one demonstrated three-layer cascade (Fig. 6) the middle (second) layer bead is coated homogenously with deoxyribozyme YES gate and its substrate S_1. In the presence of activating input i_1, the enzyme is 'turned on' cleaving the substrate and releasing one of the products from the bead in solution as an output. The product diffuses away from the bead, and is captured by the third layer (so called final acceptor bead), which is coated with its complement. Instead of manually adding input i_1, we can add new particle with another enzyme–substrate pair (E–S2), as the first layer. In order to demonstrate versatility of the elements, we blocked substrate with a complementary strand, and left enzyme E constitutively active. The addition of i_2 removes the blocking strand, allowing the enzyme to cleave its substrate S_2, releasing i_1 to activate YES gate in the middle layer. Omitting the middle layer produced no signal in the last layer, demonstrating that this is a true cascade.

The dynamic networks of beads are of interest as a new computing medium, but can also be useful in nanomedicine, for example, in molecular computing-controlled drug release. This approach can be used, we hope, to minimize the side effects of targeted immunotherapy whenever nanoparticles are used. Our results represent a step toward scenario in which networks will be used to assess the presence of multiple tissue markers, or they could assess additional information from the remote positions. In the current nanomedicine paradigm, more complex functions of microparticles are achieved by loading a single particle with additional functionalities, and in our approach, an increase in complexity of functions is achieved by forming networks of simpler particles.

VII. Other Approaches to Autonomous Computing with DNA

Deoxyribozymes are not the only approach used to autonomous computing with DNA, and we provide here selected examples. The first fully autonomous computing with so called DNA Seeman tiles (26) was proposed, and only later demonstrated, by Winfree (27) in the form of algorithmic self-assembly. Seeman tiles are rigid DNA complexes with four (or more) sticky ends. A mixture of such molecules with carefully designed complementary sticky ends, could give an asymmetric two dimensional crystal grown according to the set of rules ("program"). Winfree's group has more recently also reported an approach to molecular circuits based on DNA strand displacement (28). These are conceptually the simplest circuits and have no protein or ribozymes as components. These circuits can be several layers deep.

A different approach, based on an early proposal by Rothemund (29), but much improved, was implemented as finite automata by Shapiro's group (30, 31). In this system a double stranded DNA molecule with a sticky end serves as an input tape that is operated on by a restriction enzyme directed through a mixture of smaller DNA complexes with sticky ends ("program" or software). Various sequences within this DNA molecule are assigned states. Thus, the operation by a restriction enzyme changes states (i.e., different sticky ends are exposed) of the system while processing (moving over) the tape. This processing of DNA tape, that is, this cascade, depends on the mixture of external inputs, so, we can justly say that the cascade is programmable. Furthermore, DNA molecules constituting software can be made sensitive towards the presence of other RNA molecules (inputs), and this was used to obtain a "proof-of-concepts" demonstration of an autonomous therapeutic device for mRNA analysis.

VIII. Conclusions and Future Visions

Over the past several years, we have established arbitrary logic calculations in solution, limited only by some cross-reactivity and concentration effects of various elements, and our patience to wait for our computations to be observable. We are intrigued by opportunities to build increasingly complex networks of molecules, networks that will perhaps be trainable, that could be subjected to evolutionary processes, and that could show emergent properties. For example, we hope to develop over the next several years completely amorphous mixtures of computing molecules, containing mixtures logic gates, which can be trained to play different strategies of tic-tac-toe by playing the desired strategies with them.

However, our biggest long-term challenge is to adapt these systems to *in vivo* therapeutic applications. For example, the therapy for hematological neoplasms using immunotoxin conjugates has been limited by concurrent development of toxic effects in normal lymphocytes. Yet, lymphocytes display a set of surface markers characteristic for their lineage, that is, a position on a hematopoietic tree. These surface markers can be used as inputs for molecular logic, leading to the decisions to kill the cell or not.

Acknowledgment

The authors are grateful for funding from Leukemia and Lymphona Society, NSF, NASA, and Searle.

References

1. Adleman, L. M. (1994). Molecular computation of solutions to combinatorial problems. *Science* **266**(5187), 1021–1024.
2. de Silva, A. P., Leydet, Y., Lincheneau, C., and McClenaghan, N. D. (2006). Chemical approaches to nanometre-scale logic gates. *J. Phys. Condens. Matter* **18**(33), S1847–S1872.
3. Credi, A. (2007). Molecules that make decisions. **46**(29), 5472–5475.
4. Breaker, R. R. (2002). Engineered allosteric ribozymes as biosensor components. *Curr. Opin. Biotechnol.* **13**, 31–39.
5. Breaker, R. R., and Joyce, G. F. (1995). A DNA enzyme with Mg(2+)-dependent RNA phosphoesterase activity. *Chem. Biol.* **2**(10), 655–660.
6. Santoro, S. W., and Joyce, G. F. (1997). *Proc. Natl. Acad. Sci. USA* **94**, 4262–4266.
7. Tyagi, S., and Krammer, F. R. (1996). Molecular beacons: Probes that fluoresce upon hybridization. *Nat. Biotechnol.* **14**, 303–309.
8. Stojanovic, M. N., de Prada, P., and Landry, D. W. (2001). Catalytic molecular beacons. *ChemBioChem* **2**(6), 411–415.
9. Stojanovic, M. N., Mitchell, T. E., and Stefanovic, D. (2002). Deoxyribozyme-based logic gates. *J. Am. Chem. Soc.* **124**(14), 3555–3561.
10. Stojanovic, M. N., and Stefanovic, D. (2003). A deoxyribozyme-based molecular automaton. *Nat. Biotechnol.* **21**(9), 1069–1074.
11. Yurke, B., Turberfield, A. J., Mills, Jr., A. P., Simmel, F. C., and Neumann, J. L. (2000). A DNA-fuelled molecular machine made of DNA. *Nature* **406**(6796), 605–608.
12. Lederman, H., Macdonald, J., Stefanovic, D., and Stojanovic, M. N. (2006). Deoxyribozyme-based three-input logic gates and construction of a molecular full adder. *Biochemistry* **45**(4), 1194–1199.
13. Macdonald, J., Stefanovic, D., and Stojanovic, M. N. (2006). Solution-phase molecular-scale computation with deoxyribozyme-based logic gates and fluorescent readouts. *In* "Fluorescent Energy Transfer Nucleic Acid Probes: Designs and Protocols" (V. Didenko, Ed.), **335**, pp. 343–363. Humana Press, Totowa, NJ.
14. McCluskey, E. J. (1986). "Logic Design Principles: With Emphasis on Testable Semicustom Circuits." Prentice Hall, Englewood Cliffs, NJ.

15. Stojanovic, M. N., and Stefanovic, D. (2003). Deoxyribozyme-based half adder. *J. Am. Chem. Soc.* **125**, 6673–6676.
16. Macdonald, J., Sutovic, M., Lederman, H., Pendri, K., Lu, W., Andrews, B. L., Stefanovic, D., and Stojanovic, M. N. (2006). Medium scale integration of molecular logic gates in an automaton. *Nano Lett.* **6**(11), 2598–2603.
17. Penchovsky, R., and Breaker, R. R. (2005). *Nat. Biotechnol.* **23**, 1424–1433.
18. Cuenoud, B., and Szostak, J. W. (1995). *Nature* **375**, 611–614.
19. Stojanovic, M. N., Semova, S., Kolpashchikov, D., Macdonald, J., Morgan, C., and Stefanovic, D. (2005). Deoxyribozyme-based ligase logic gates and their initial circuits. *J. Am. Chem. Soc.* **127**, 6914–6915.
20. Stojanovic, M. N., de Prada, P., and Landry, D. W. (2000). *Nucleic Acids Res.* **28**, 2915–2918.
21. Kolpashchikov, D. M., and Stojanovic, M. N. (2005). Boolean control of aptamer binding States. *J. Am. Chem. Soc.* **127**(32), 11348–11351.
22. Nutiu, R., and Li, Y. (2003). Structure-switching signaling aptamers. *J. Am. Chem. Soc.* **125**(16), 4771–4778.
23. Dittmer, W. U., Reuter, A., and Simmel, F. C. (2004). A DNA-based machine that can cyclically bind and release thrombin. *Angew. Chem. Int. Ed. Engl.* **43**(27), 3550–3553.
24. Dirks, R. M., and Pierce, N. A. (2004). Triggered amplification by hybridization chain reaction. *Proc. Natl. Acad. Sci. USA* **101**(43), 15275–15278.
25. Ruslan, Y, Sergei, R, and Milan, N. S. (2007). Networking particles over distance using oligonucleotide-based devices. *J. Am. Chem. Soc.* **129**(50), 15581–15584.
26. Carbone, A., and Seeman, N. C. (2002). Circuits and programmable self-assembling DNA structures. *Proc. Natl. Acad. Sci.* **99**, 12577–12582.
27. Rothemund, P. W. K., Papadakis, N., and Winfree, E. (2004). Algorithmic self-assembly on DNA sierpinski triangles. *PLoS Biol.* **2**, e424.
28. Georg, S., David, S., and David, Yu Z. (2006). Erik winfree enzyme-free nucleic acid logic circuits. *Science* **314**(5805), 1585–1588.
29. Rothemund, P. W. K. (1995). A DNA and restriction enzyme implementation of Turing machines. In DNA Based Computers, Proceedings of a DIMACS Workshop, April 1995, pp. 75–119.
30. Benenson, Y., Paz-Elizur, T., Adar, R., Keinan, E., Livneh, Z., and Shapiro, E. (2001). Programmable and autonomous computing machine made of biomolecules. *Nature* **414**, 430–434.
31. Benenson, Y., Gil, B., Ben-Dor, U., Adar, R., and Shapiro, E. (2004). An autonomous molecular computer for logical control of gene expression. *Nature* **429**, 423–429.

Molecular Colony Technique: A New Tool for Biomedical Research and Clinical Practice

ALEXANDER B. CHETVERIN AND
HELENA V. CHETVERINA

Institute of Protein Research of the Russian Academy of Sciences, Pushchino, Moscow Region, Russia 142290

I. Introduction 220
II. Detection of Airborne RNAs 221
    A. A Mysterious Enzyme 221
    B. The Puzzle of Spontaneous RNA Synthesis 222
    C. Invention of the Molecular Colony Technique 223
III. Monitoring Reactions Between Single Molecules 225
    A. Cell-Free System for RNA Recombination 225
    B. Spontaneous Rearrangements in RNA Sequences 226
    C. Possible Role of Spontaneous RNA Recombination 227
IV. Cell-Free Gene Cloning 228
    A. Gene Cloning and Expression in Molecular Colonies 228
    B. Advantages of the MCT Cloning 230
    C. Comparison to Other *In Vitro* Cloning Systems 231
V. Molecular Colonies as a Precellular Form of Life 233
VI. Molecular Colony Diagnostics 235
    A. MCT Assay of DNA and RNA Targets 235
    B. Visualizing Molecular Colonies 236
    C. Detection of DNA and RNA Targets in Clinical Samples 238
    D. Quantitative *In Situ* Assay 239
VII. Gene and Gene Expression Analysis 241
    A. Sequencing on Molecular Colonies 241
    B. Single-Nucleotide Polymorphism 242
    C. Alternative Splicing 244
    D. Allele and mRNA Copy Number 245
VIII. Opportunities Provided by the Molecular Colony Technology 247
    References 249

Having emerged as a tool for answering a purely academic question about the origin of spontaneous RNA synthesis in cell-free reactions, the molecular colony technique (MCT) has become an elaborated technology with unique capabilities and a great biotechnological potential. Reviewed here are the history of invention of MCT and its numerous applications, including studies on chemical reactions between single molecules, cloning and screening of genes *in vitro*, molecular diagnostics, and gene structure and expression analysis.

I. Introduction

Molecular colony technique (MCT) comprises the amplification of nucleic acids in immobilized media, such as in a gel. Because the gel matrix prevents convection of the entrapped liquid and restricts diffusion of nucleic acids, the amplification products do not spread throughout the medium. Rather, they form more or less compact molecular colonies, which are reminiscent of bacterial colonies growing on the surface of a nutrient agar. If amplification is carried out in a thin gel layer, a 2-D pattern of molecular colonies is generated, each colony comprising many copies (a clone) of a single starting RNA or DNA template molecule.

We conceived the idea of MCT in 1989 while attempting to find the source of the mysterious "spontaneous" RNA synthesis manifested by $Q\beta$ replicase (RNA-directed RNA polymerase of phage $Q\beta$) in the absence of any added template. Accordingly, the first version of MCT comprised growing RNA colonies in an agarose gel containing $Q\beta$ replicase and its substrates, ribonucleoside 5'-triphosphates (rNTPs) (*1*).

After demonstration of the feasibility of the $Q\beta$ replicase version of MCT, it became obvious that nucleic acid colonies can be obtained with the use of any *in vitro* exponential amplification system. Among them are polymerase chain reaction (PCR) (*2*) and isothermal (i.e., occurring at a constant temperature) amplification reactions, such as self-sustained sequence replication (3SR) (*3*), nucleic acid sequence-based amplification (NASBA) (*4*), strand displacement amplification (*5*), rolling circle amplification (*6*), and loop-mediated DNA amplification (*7*). In addition to the $Q\beta$ replicase version, 3SR and PCR versions of MCT were disclosed in a series of our patent applications filed in 1992 (*8–13*). The first Russian patent on MCT was granted in 1995 (*8*), and the first US patent in 1997 (*9*). Those patents, as well as our paper published in 1993 (*14*), teach using a heat-resistant matrix, such as a polyacrylamide gel, for the PCR version of MCT.

For many years, MCT was exclusively developed in our laboratory in Russia. In addition to investigation of the origin of spontaneous RNA synthesis, we used the $Q\beta$ replicase version for detecting rare events of recombinations (sequence rearrangements) in RNA molecules (*15, 16*). The situation changed at the end of 1999, when Mitra and Church of the Harvard Medical School published their first paper on the PCR version of MCT (17), whose only difference from our published protocol (8, 9) was that one of the PCR primers was immobilized on the gel matrix with the acrydite chemistry (18, 19). They termed the method "polony" (for *polymerase colony*) technology, and developed several important MCT applications, including single-nucleotide polymorphism (SNP) genotyping (20), *in situ* sequencing on DNA colonies (21), and studies on alternative pre-mRNA splicing (22). Regretfully, they failed to accurately disclose the MCT authorship. Even more, in one interview, Mitra claimed: "It was Church's idea to amplify single molecules in acrylamide" (23). Hopefully, this will change and all laboratories working on MCT, whatever term they prefer to use for its name, will combine their efforts to further develop this promising technology.

In this chapter, we discuss various scientific and practical applications of MCT developed to date in different laboratories in Russia, United States, and China. In addition to already mentioned applications, these include *in vitro* gene cloning and molecular diagnostics.

II. Detection of Airborne RNAs

A. A Mysterious Enzyme

Qβ replicase is a unique RNA-directed RNA polymerase capable of producing up to 10^{10} copies of a single starting template molecule in only 10 min at room temperature (24) and this is the absolute record of the rate of *in vitro* nucleic acid amplification. Like DNA amplification in PCR, the RNA synthesis occurs exponentially as long as the enzyme is in molar excess over template (25). Because both the original RNA and its complementary copy can serve as replicase templates, the number of RNA molecules increases by a factor of two in each round of replication. Importantly, the template and its complementary copy do not form duplex and are released single-stranded after completion of a replication round (26). Therefore, unlike PCR which requires that the reaction medium is periodically heated to melt the double-stranded DNA product, Qβ replicase reaction is isothermal, and each next round of replication begins immediately after completion of the previous one.

Qβ replicase is a relatively simple enzyme. All reactions leading to the exponential amplification of RNA are performed by a three-subunit protein complex consisting of a 65,317 Da polypeptide encoded by the viral genome and two *Escherichia coli* proteins that usually participate in the mRNA translation on the ribosome, elongation factors EF-Tu и EF-Ts (27). In contrast to shorter satellite RNAs, the genomic Qβ phage RNA requires two more host proteins for its replication; these are ribosomal protein S1 and RNA-binding protein Hfq (earlier referred to as HF, a host factor), which solely participate in the recognition of the Qβ RNA plus strand. The only obvious function of the latter two proteins in the phage replication seems not to allow Qβ replicase to recognize the genomic RNAs of the phages from other serological groups.

Yet, despite its simple structure, Qβ replicase is a sort of black box that keeps most of its secrets after more than 40 years of studies. It is still unclear what polypeptide elongation factors EF-Tu и EF-Ts do in the RNA synthesis reaction (28)? Why does Qβ replicase spend at the termination step up to 90% of the overall replication time (29)? What is the purpose of the 3′-terminal adenylylation of each product RNA strand (30, 31)? What is the mechanism of maintaining the single strandedness of the replication intermediate in the absence of single strand-binding proteins and helicases, and without spending any energy for unwinding the double helix that must be formed during template-directed synthesis of the complementary nascent strand (26)? What does ensure the high template selectivity in the absence of sequence-specific primers and promoters (24)?

B. The Puzzle of Spontaneous RNA Synthesis

Probably, the most mysterious feature of Qβ replicase was its ability to spontaneously synthesize RNAs possessing nonrandom nucleotide sequences. Incubation of a highly purified replicase in the absence of template with all four rNTPs inevitably resulted in the generation of large amounts of a variety of RNA species whose length ranged from several tens to several hundreds of nucleotides (24). This fact seemed even more amazing in view of the reluctance of Qβ replicase to amplify any cellular RNAs and even the genomic RNAs of closely related phages (32).

In 1975, Sumper and Luce of the laboratory of Manfred Eigen, a Nobel Prize Laureate, succeeded in the purification of Qβ replicase to a state when the addition of only five molecules of a replicable RNA to a 200-μl reaction stimulated RNA synthesis above the spontaneous level. It followed that each aliquot of replicase they used for the reactions might have been contaminated by no more than a few templates molecules. Yet, after the authors managed to reduce the incubation volume 10,000-fold (down to 20 nl), they still continued to observe spontaneously synthesized RNAs in each individual sample (33). This result gave rise to a hypothesis that efficient Qβ replicase templates are generated *de novo* by virtue of random nucleotide condensation followed by evolution of fortuitously formed replicable polyribonucleotides into rapidly amplifiable RNA species (34, 35). This hypothesis dominated until 1991, when Qβ replicase version of MCT allowed us to reveal the source of spontaneously generated templates (1).

The authors of the hypothesis treated it very seriously, referring to the *de novo* RNA synthesis as to "a primary model system for evolution at the molecular level" that "gives clear answers as to how information is formed by evolutionary events" (*36*). They published on this subject dozens of papers furnished with elaborated mathematical models. Indeed, the problem is of a great scientific importance, insomuch as it directly relates to the question about the upper limit of the rate of molecular evolution.

Selection experiments demonstrated that only a few of the initial diversity of 10^{12} unique sequences of 50–77 nt in length were replicable. Yet, the selected RNAs replicated far less efficiently than products of the spontaneous synthesis, even though all the sequences subjected to selection were deliberately provided with terminal clusters (5'-GGG и CCCA-3') characteristic of the natural Qβ replicase templates (*37*). These findings suggested that the probability of generation of a replicable RNA by chance is very low. Furthermore, since the postulated products of random nucleotide condensation were never detected, the number of sequences subjected to evolution in a spontaneous reaction must be even smaller than 10^{12} used in the selection experiments (which corresponds to ≈40 ng of a 70-nt-long RNA, a readily detectable amount). Hence, the rate of molecular evolution in a tube must be literally of an astronomical scale for very efficient Qβ replicase templates not only be created but also amplified to a detectable amount within the span of a spontaneous reaction [20–40 min (*33*), with the duration of a single replication round for optimized templates being about 20 s].

C. Invention of the Molecular Colony Technique

It is noteworthy that with some variations, the theme of "spontaneous generation" repeatedly arises in scientific discussions. Almost one and a half century ago, Louis Pasteur disproved the doctrine of spontaneous generation of life from nonliving organic matter by demonstrating that no life could arise in a boiled meat broth unless solid particles heavier than air were allowed to enter. Those experiments also convincingly demonstrated that microorganisms are everywhere—even in the air (*38*). After such a vivid lesson, one should expect that a scholar would thoroughly rule out all other possibilities before claiming something to be "spontaneously generating."

Surprisingly, the Pasteur's teaching was ignored, although there were a number of indications that the apparently spontaneous RNA synthesis is caused by sample cross-contamination. The most important observation was a striking similarity of RNAs synthesized in different experiments and laboratories (*24, 33*). Given the ability of microorganisms to be disseminated through air, why not to expect the same of much smaller RNA molecules? However, such a possibility could not be explored with the Pasteur's approach: the reaction mixtures containing Qβ replicase cannot be boiled or otherwise sterilized to destroy "germs" (RNA templates) without also completely inactivating the enzyme.

In this situation, we made use of a strategy developed by Robert Koch, a Pasteur's contemporary and rival (39), who succeeded in the isolation of tuberculosis bacteria by propagating them on the surface of a gel. In this format, each bacterial cell gave rise to a colony comprising multiple copies of the original cell (a clone) (40). One advantage of the Koch's approach as compared with the cell propagation in a broth is that it provides for direct determination of the number of bacterial cells in a sample—by counting the number of colonies.

Likewise, in order to detect and count RNA molecules that could be amplified by Qβ replicase, we decided to amplify them in a gel containing the replicase and rNTPs. Figure 1 shows the result of the first experiment on RNA amplification in an agarose gel performed in January 1989 (41). A melted low gelling temperature agarose was mixed at 40 °C with Qβ replicase, rNTPs, and buffer, and poured into a 4-cm-diameter petri dish. The dish was placed on ice and, after solidification of agarose, incubated at 37 °C for 1 h. Then the agarose was stained with ethidium bromide and inspected under ultraviolet light. As expected, the incubation resulted in a "spontaneous" synthesis of RNA forming a fluorescing complex with ethidium bromide. However, besides a few dozens of bright spots comprising RNA colonies, there were regions of a low fluorescence and even not fluorescing at all. The latter disagreed with the claim of Sumper and Luce that spontaneous RNA synthesis occurs in every smallest aliquot of the reaction medium (33) and, hence, the entire gel would have been colored more or less evenly. Our result suggested that RNA colonies resulted from a limited number of replicable molecules that had been entrapped by the agarose before the onset of RNA amplification.

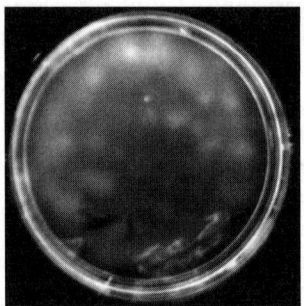

FIG. 1. RNA colonies that have spontaneously grown in a 2% ultra-low gelling temperature agarose (type IX, Sigma), containing 80 mM Tris–HCl (pH 8.0), 13 mM $MgCl_2$, 1 mM EDTA, 10% glycerol, 5 mM phosphoenol pyruvate, pyruvate kinase (20 μg/ml), 400 μM ATP, CTP, GTP, and UTP, and Qβ replicase (50 μg/ml). For other details, see Section II.C. Reproduced from reference (41) with permission.

To ascertain what the source of these molecules was, the experimental design was slightly modified. Two layers of agarose were used, one of which contained rNTPs and the other, cast atop the first layer, contained Qβ replicase. In this format, the reaction was controlled by the diffusion of rNTPs into the enzyme layer and RNA amplification occurred at the interface between the layers. Dishes were kept on a bench for 1 h before casting the enzyme layer, with some of the dishes being covered whereas the others left uncovered. All open dishes exhibited more RNA colonies and higher overall fluorescence intensity than did the covered dishes (*1*). The least number of colonies grew in experiments that were carried out in a room where no experiments with replicable RNAs had been previously carried out (*14*).

These experiments have clearly demonstrated that RNA molecules are present in the laboratory air and can invade amplification reactions, giving rise to the impression of a template-free reaction. Thus, everything has fallen into place: there is no need anymore in postulating the incredibly high rates of molecular evolution; a number of mystic features of the spontaneous synthesis such as close sequence similarity of RNAs generated in separate reactions have received rational explanations, and finally, the central dogma of molecular biology (*42*) has retained its validity because the postulate of "-creation of genetic information by a protein enzyme" (*43*) is no longer needed to explain the appearance of specific RNA sequences in the absence of added templates.

III. Monitoring Reactions Between Single Molecules

A. Cell-Free System for RNA Recombination

The ability of MCT to detect single molecules makes it a unique tool for studying extremely rare reactions such as recombinations between RNA molecules. After the discovery of genetic recombination at the level of DNA, indications began to accumulate that a similar process may occur at the level of RNA as well. From their studies on an exchange of genetic markers between related strains of poliovirus, whose genome consists of RNA, Hirst (*44*) and Ledinko (*45*) concluded that RNA molecules can probably recombine. Later, Agol *et al.* showed that poliovirus recombination results in the formation of a chimeric protein and, hence, must comprise a rearrangement of the sequence of the RNA template (*46*). Finally, King *et al.* demonstrated the formation of recombinant viral RNAs by direct sequencing (*47*). However, inasmuch as all experiments on RNA recombination were performed either in living cells or in crude cell lysates (*48*), a possibility could not be excluded that the process included a reverse transcription step. Hence, recombination might have occurred between viral cDNAs, rather than between RNAs, and recombinant RNAs might have been the products of transcription of recombinant DNAs.

The Qβ replicase version of MCT allowed the RNA recombination to be studied in a purified cell-free system that was devoid of any DNA, dNTPs (2′deoxynucleoside 5′-triphpsphates), reverse transcriptases, as well as enzymes needed for DNA recombination and transcription and, hence, the involvement of DNA intermediates could be excluded (15). Mutually supplementing 5′ and 3′ fragments of a natural Qβ replicase template served as recombination substrates. When separate, such RNA fragments were not capable of replication, but if they were rejoined in one molecule in a proper order, a replicable RNA might form that would be amplified by Qβ replicase. This provided for the positive selection of recombinant molecules. Formation of recombinant RNAs was monitored by the appearance of RNA colonies whose number reflected the number of recombination events. In early experiments, a mixture of fragments was distributed over a Qβ replicase-containing agarose layer, which was then covered by a nylon membrane impregnated with rNTPs (15). In addition to serving as a substitute for the second agarose layer, the membrane retarded the diffusion of RNA molecules by reversibly interacting with them. This reduced the size of RNA colonies and, hence, increased the resolving power of MCT (14).

These experiments showed that recombinant RNAs can indeed be generated in the purified system, and this was the first direct proof of the existence of intermolecular recombination at the RNA level (15). Moreover, in accord with the results of *in vivo* studies on genetic recombination in RNA bacteriophages (49), recombination in this *in vitro* system was nonhomologous and occurred at a similar frequency, $\approx 10^{-5}$ per nucleotide (50, 51). This provided an indirect evidence that in bacterial cells too, recombinant RNAs were formed by reactions between RNA molecules, without an involvement of DNA intermediates.

B. Spontaneous Rearrangements in RNA Sequences

Further studies showed that RNA recombination in the presence of Qβ replicase can only occur if there are free hydroxyl groups at the 3′-terminus of the 5′ fragment (15). Recombination was completely suppressed if the 5′ fragment was oxidized with sodium periodate that converted the *cis*-glycol (2′- and 3′-hydroxyl groups) at the 3′-end of RNA into a dialdehyde (52). However, recombination was completely restored when the oxidized fragment was treated with aniline that eliminated the dialdehyde producing a one nucleotide shorter 3′-phosphoryl RNA (53), followed by dephosphorylation with a phosphatase (15). These observations suggested the way how to separate the recombination reaction itself from the colony amplification of the recombination products (MCT assay), and hence to ascertain if RNA recombination can occur without replicase.

To this end, a mixture of the 5' and 3' fragments was incubated under chosen conditions and, before the MCT assay, the fragments were oxidized with periodate to prevent recombination between them in the Qβ replicase-containing agarose. In this experimental setting, RNA colonies could only appear if recombination between the fragments had occurred before the oxidation step (16).

It turned out that RNA fragments can recombine in the absence of both Qβ replicase and rNTPs. The only requirement is the presence of divalent cations, such as Mg^{2+}. However, such a "spontaneous" recombination occurs at a several orders of magnitude lower rate than in the presence of Qβ replicase. Also no free hydroxyl groups are needed indicating that a quite different chemistry is employed. Further, spontaneous RNA rearrangements can occur within one molecule (in cis), resulting in deletions of its internal regions. Importantly, the crossover sites are randomly distributed along the nucleotide sequence of the reacting fragments. This suggests that no special (ribozyme-like) structures are required for the reaction and, hence, that the capability of spontaneous sequence rearrangements is an intrinsic chemical property of polyribonucleotides (16).

The mechanism of spontaneous RNA recombination likely includes a nonhydrolytic cleavage of the sugar-phosphate backbone of RNA yielding 2',3'-cyclic phosphate and 5'-hydroxyl termini, followed by a reverse reaction in which the cleavage products are *trans*-ligated. As far as such a mechanism requires the participation of 2'-hydroxyls, the spontaneous sequence rearrangements seem to be an exclusive property of RNA, and should not occur between DNA molecules. An alternative mechanism that includes intermediate formation of branched structures (lariats) (50) is unlikely because inclusion in the reaction mixture of debranching endonuclease that specifically cleaves the 2', 5' internucleotide bonds (54) does not decrease the yield of recombinant molecules (our unpublished observations).

C. Possible Role of Spontaneous RNA Recombination

At first glance, the rate of spontaneous RNA recombination, $\approx 10^{-9} \text{ h}^{-1}$ per internucleotide bond at 37 °C (16), is extremely low. The products of this reaction can hardly be detected without MCT, even more so on the huge background of the unrecombined fragments. Yet, spontaneous sequence rearrangements may play an important role in the evolution of genomes, composed of either RNA or DNA. Even being not promoted by cellular RNA-binding proteins, the spontaneous recombination should result in a new recombinant RNA molecule in each eukaryotic cell every minute. This means that there may occur up to 10^{20} such events during the life span of the human body (16). Reverse transcription and integration of even a minute fraction of the generated sequences would significantly change the genome and should be considered among major factors of genome instability and oncogenic transformation (55).

Alexander Spirin have proposed that spontaneous RNA recombination might have played an important role in the origin and evolution of the RNA world (56–58). In particular, spontaneous sequence rearrangements might have been the main source of new RNA variants subjected to the natural selection and might have been the first mechanism of the generation of long polyribonucleotides capable of folding into globules possessing specific ligand-binding and catalytic activities. The cotemporal mechanism of RNA synthesis through consecutive template-directed additions of mononucleotides might have emerged later, after RNA polymerase ribozymes were created as a result of the sequence rearrangement events. We would like to add that the capability of spontaneous recombination may serve as another argument in favor of antecedency of RNA to DNA.

The experimental approach developed for monitoring spontaneous rearrangements was later used to explore the mechanisms of RNA recombination by viral RNA-directed RNA polymerases (59). In this case, the incubation mixture, in addition to being oxidized, was deproteinized by phenol extraction. It was found that viral RNA polymerases use diverse mechanisms to produce recombinant molecules. For example, protein 3Dpol (poliovirus RNA polymerase) uses primer extension, which is a part of the template switch mechanism. However, primer extension is rejected by $Q\beta$ replicase, which instead uses still undefined mechanism, in which important role belongs to the 3′-terminal hydroxyls of the 5′ recombination substrate.

The above consideration shows that the ability to detect and enumerate nucleic acid molecules makes MCT a unique research tool that permits chemical reactions between single molecules to be studied.

IV. Cell-Free Gene Cloning

A. Gene Cloning and Expression in Molecular Colonies

When the density of molecular colonies is not exceedingly high, each of them comprises a homogenous progeny (clone) of one initial molecule. It was therefore tempting to use MCT for *in vitro* gene cloning. Initially, we tried the

Qβ replicase version of MCT for *in vitro* cloning of mRNAs embedded in a natural Qβ replicase template (24). Unfortunately, when loaded with mRNAs, such RNA vectors lost their capacity of exponential amplification (60). Even worse, they underwent spontaneous deletions of most of the mRNA inserts (16), resulting in an exponential amplification of shorter RNAs that rapidly outgrew the original recombinant molecules (60). Therefore, we switched to other enzymatic reaction that provide for the exponential amplification of nucleic acids, each of which can be utilized by MCT (8, 9, 14).

The most promising results were obtained with the PCR version of MCT (PCR-MCT). As far as PCR involves repeated sample heating, temperature-resistant media, such as a polyacrylamide gel, should be used instead of agarose. Originally, we prepared amplification gels by polymerizing acrylamide in the presence of all the PCR reagents including dNTPs and DNA polymerase (8–13), and this protocol was reproduced by Mitra and Church (17). Later, we found that more consistent results can be obtained by first preparing an "empty" gel, washing it to remove any soluble substances, drying, and soaking the dehydrated gel in a complete PCR cocktail (61). Compared to the original protocol, this ensured a better preservation of the DNA polymerase activity. Also, this permitted many gels to be prepared for future use, thus eliminating the need for repeating this cumbersome procedure in every MCT experiment. This modification was then taken over by Church *et al.* (22) without a reference to our work.

After having optimized the gel porosity by varying the concentration of acrylamide and of the cross-linking reagent (N,N'-methylenebisacrylamide), we succeeded in obtaining colonies that contained up to 10^8 copies of long (up to 2 Kbp) DNA fragments that had enough capacity to accommodate an entire protein-coding sequence flanked by upstream and downstream untranslated regions containing a transcription promoter and translation enhancers (62). Within statistical scatter, the number of colonies was equal to the number of template molecules seeded on the gel. Hence, MCT allows up to 100% elements of a genetic library to be cloned and tested, compared to only 0.0001–0.01% when cloning is performed in living cells. The poor efficiency of *in vivo* cloning is due to a low yield of each of the steps of insert ligation into a cloning vector and of cell transformation by the ligation product (63).

This protocol enabled us to isolate multiple cDNA clones from a portion of library that corresponded to the RNA content of a single cell (62). Such efficiency was never obtained with the *in vivo* cloning techniques.

What is most important is that cDNA colonies can be expressed *in situ*, thereby providing for screening the clones for the encoded functions. For *in situ* transcription, the PCR gel is dried and reswollen in a solution containing the transcription buffer, a DNA-directed RNA polymerase, and rNTPs. Incubation of the gel during 2 h at 37 °C results in the synthesis of at least 10 copies of RNA of up to 1700 nucleotides in length per a DNA copy contained in a colony (62). Transcription in molecular colonies provides for cloning and direct selection of RNA molecules with desired catalytic or binding properties, such ribozymes and aptamers.

However, the above procedure proved not optimal for *in situ* translation because the PCR components, especially the alkaline buffer, are potent inhibitors of protein synthesis in the wheat germ extract used in these experiments. The problem was solved by soaking the gel in a specially formulated saline alcohol solution, which solubilized and washed away any inhibitors of translation, but precipitated DNA and hence preserved the colony pattern. Then the gel was dried and reswollen in a solution containing all the components necessary for a combined transcription–translation. With green fluorescent protein as an example, gene expression in molecular colonies was shown to result in the synthesis of an active (and, hence, correctly folded) polypeptide. On average, each colony produced $\approx 10^9$ protein molecules, which is sufficient for its detection by binding with antibodies or ligands, as well as by the protein's enzyme activity (62).

The synthesis of an active protein in molecular colonies was found to require a higher concentration of thiol compounds than does protein synthesis in a test tube or in a cell-free reactor (64). The most plausible explanation is that molecular colonies are expressed in a thin gel layer, in which case the surface-to-volume ratio of the reaction medium is higher. As a result, the reaction medium becomes more susceptible to oxidation by the atmospheric oxygen. This may affect either the components of the cell-free system or the nascent protein (e.g., by inducing the formation of incorrect S–S bond during its folding), or both.

B. Advantages of the MCT Cloning

Cloning in molecular colonies is a true molecular cloning, in contradistinction to the common gene cloning *in vivo* which, although often being referred to as "molecular cloning" (65), is, in fact, the cloning of cells (or viruses infecting cells) rather than of molecules. Compared with cloning *in vivo*, gene cloning in molecular colonies has a number of advantages. Cloning and screening becomes possible without natural selection and in the presence of unnatural nucleotides and amino acids. Genes can be directly screened by the properties of their expression products because there are no cell walls or membranes. Clones can be analyzed under conditions different from PCR amplification, transcription, or translation because the reaction medium can easily be changed by soaking the gel with an appropriate solution. A molecular colony comprises genetically pure DNA, which can directly be used for genetic manipulations. Finally, insomuch as each gene and its expression product are located in the same colony, MCT can function as a molecular display. Unlike phage display or another form of genetic display, linking of a polypeptide to its gene is achieved here without modifying the protein with a tag sequence or fusing it with another protein and, hence, without affecting its structure and function. In other words, with the cloning in molecular colonies, one can do everything achievable with cloning in living cells, and much more.

C. Comparison to Other *In Vitro* Cloning Systems

After the invention of MCT, several alternative methods, each based on the exhausting dilution principle, were proposed for the *in vitro* gene cloning. Seemingly, the feasibility of such principle was for the first time demonstrated in Russia by Lukyanov *et al.* (66), who diluted a mixture of restriction DNA fragments to the average of 1 molecule per PCR tube. In approximately 30% of tubes, gel electrophoresis of the PCR products revealed a single DNA band whose mobility matched that of one of the mixed fragments. Three years later, a similar approach was published by Vogelstein and Kinzler of the John Hopkins University under the name of "digital PCR" (67), regretfully, without a reference to the Russian work. The same principle is used in methods in which DNA mixtures are sorted on either free microbeads (68) or microbeads encapsulated within water-in-oil emulsion droplets (69, 70). Although each of these methods can, in principle, result in the isolation of DNA clones, clonal purity of the preparations obtained needs to be verified by independent means, such as by cloning in bacterial cells. The following consideration illustrates the difference in the efficiencies of gene cloning by dilution and by growing colonies, whether cellular or molecular.

At cloning by dilution, DNA molecules are randomly distributed among compartments such as among wells of a PCR plate. A clone will be isolated if there occurs just one DNA molecule in a well. A probability that there is more than one molecule in a well equals

$$P_{>1} = 1 - \sum_{m=0}^{1} \frac{\lambda^m}{m!} e^{-\lambda},$$

where λ is the mean population of wells and m is the number of molecules in a well (the Poisson distribution). Hence, if DNA preparation is diluted to the average of 1 molecule per well ($\lambda = 1$), then 26.4% of wells will receive 2 or more molecules, which is 41.8% of all populated wells (63.2%). In other words, in almost a half of the wells in which specific PCR products accumulate, they are not clones. A good enough probability of obtaining a clone (99.5%) is expected at a mean population of wells of no more than 0.01 (i.e., 1 molecule per 100 wells).

However, 99% of wells will remain empty in this case. This means that as few as four clones can be isolated on a 384-well PCR plate provided that a sample distributed among the 384 wells contains four DNA molecules.

In case of colonies, there are no discrete compartments. If more than one molecule (or cell) is seeded on a gel, there is a probability that two or more colonies will overlap. However, if they overlap only partially, then in many instances such a dual colony can be discriminated against for its shape being not round. This is in contrast to a well, of which it is difficult to say if it is populated by the progeny of one molecule or of two or more molecules, especially if the properties of these molecules are similar. A dual colony can be discriminated against if the centers of the constituent colonies are apart $\geq 20\%$ of their diameter (Fig. 2). Therefore, we may assume that a dual colony will be mistaken for a clone if the center of a second colony will fall within a "critical circle" (cocentered with the first colony) whose diameter is 40% of the colony diameter (Fig. 2). Hence, the probability that the isolated material is not a clone is $\approx f(n-1)$, where f is a fraction of the gel occupied by the critical circle and n is the number of molecules seeded on the gel.

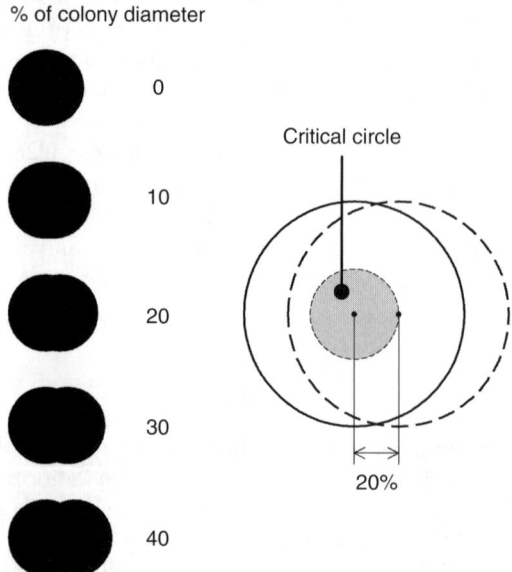

FIG. 2. Scheme of overlapping colonies. For details, see Section IV.C. Reproduced from reference (*41*) with permission.

In our MCT cloning experiments [colony diameter ≈0.5 mm, gel diameter 14 mm (62)] this probability is 0.02% or 0.5% when 2 or 25 molecules are seeded, respectively. Thus, a single 14-mm gel allows as many pure clones to be obtained (at a 99.5% warranty) as do 2500 wells (more than six 384-well plates) when cloning is performed by dilution. Furthermore, it is possible to significantly reduce the colony size and, hence, to increase the MCT resolving power. For example, using a denser gel allows the diameter of DNA colonies to be reduced to less than 10 μm (17). This permits thousands of clones to be resolved on a 14-mm gel.

V. Molecular Colonies as a Precellular Form of Life

As it was discussed in Section II.C, there are no facts supporting the hypothesis that the rate of RNA evolution can be so high as to provide for the *de novo* generation of replicable RNAs within a few hours. However, it is still quite possible that replicable RNAs as well as their replicases might have emerged on the Earth or on another planet during a much longer time period, giving rise to the RNA world (71). It seems hardly probable that the question of how the life had originated will be one day answered with a certainty. Yet, one could attempt to generate, on the basis of experimentally proven facts, a likely concept of how it might have occurred.

Assuming that the life has originated in a natural course of events, one inevitably becomes confronted with the problem of a mechanism of evolution before the first cell has emerged. Indeed, even the most primitive cells that could be imagined would be too complex to emerge by chance as a single-step event. Discussions of this problem in the literature have led to a conclusion that evolution might not occur in a pool filled with individual molecules, because there must be certain segregation and compartmentalization of molecules for the natural selection to operate (56, 72, 73).

First of all, compartmentalization is needed to link a gene to its product for natural selection be able to identify the gene that makes a better product (72). Here is the essence of a genetic display. Even if there is no clear distinction between a "gene" and its "product" (as with self-encoding ribozyme), one molecule is still not sufficient. In the extreme example of a self-replicating ribozyme, at least three molecules are to be involved in the replication process and must colocalize: a template strand (gene), its complementary copy that directs the synthesis of new templates, and the molecule that acts as a replicase ribozyme (gene product).

Second, segregation is needed for selective amplification of better ribozyme variants that are generated by virtue of mutations and sequence rearrangements; otherwise, all molecules present in the pool will be amplified equally well (57, 73). Furthermore, if a mutant RNA replicase emerged, with superior efficiency or accuracy, it would better replicate other RNA replicases but would have no selective advantage for itself (73). Hence, a common pool, like any commune, levels down everything to a mediocrity, and gives no advantage to rarely emerging molecules with outstanding abilities.

Finally, compartmentalization increases the local concentration (74), which favors interactions between molecules even if their affinity towards each other is low (75).

According to Alexander Oparin, compartmentalization at early steps of evolution was due to formation of coacervates, microscopic droplets of condensed colloids composed of oppositely charged polypeptides and polysaccharides (74). However, now it is generally accepted that polypeptides and polysaccharides had arisen much later than the RNA world. Alternatively, RNA molecules could be compartmentalized by encapsulation within lipid vesicles (75, 76). However, this would produce another problem. In addition to compartmentalizing macromolecules, the lipid membrane would create an impermeable barrier for the exchange between the compartment and the environment, in particular, for the influx of low molecular weight substances, such as RNA precursors, for the efflux of the products of metabolism, and also for the horizontal transfer of genetic information that is required for a rapid evolution at the population level (77). Transport of hydrophilic molecules across lipid membranes could occur through specific pores, molecular pumps, or other complex mechanisms, but these could only emerge at relatively late steps of evolution.

Jack Szostak was the first who noted that molecular colonies, similar to those produced as a result of RNA amplification in the $Q\beta$ replicase-containing agarose, might have served as a precellular form of compartmentalization in the RNA world (73). Instead of encapsulation with a lipid membrane, compartmentalization is achieved here due to a relatively low diffusion rates of macromolecules as compared with low molecular weight substances. RNA colony might grow in porous minerals (73) or in moist clays (56). This idea was further elaborated by Spirin who suggested that first inhabitants of the RNA world might have been mixed colonies containing three types of RNA molecules: ligand-binding RNAs responsible for a selective adsorption and accumulation of nutrients from the environment, ribozymes catalyzing the metabolic reactions of nucleotide synthesis, and ribozymes catalyzing the complementary replication of the RNAs making up a colony. Compartmentalization of RNA molecules in colonies would trigger the mechanism of natural selection. Individual colonies whose RNA molecules possessed better activities and better supplemented each other functions would overgrow other colonies with worse characteristics. In the absence of enveloping membranes, a population of colonies might benefit from a free exchange of genetic material, which would provide for a high rate of evolution. For example, RNA molecules could be disseminated through air, as they did in our experiments (1).

Finally, the demonstration of the possibility of *in situ* translation (62) suggests that RNA colonies might have created a translation apparatus and become selectable according to functions of the synthesized proteins even before they had acquired a membrane envelope. In other words, molecular colonies display all the basic features that must be possessed by precellular organisms. Thus, molecular colonies, which were initially devised as a means to ascertain the nature of spontaneous RNA synthesis, had probably been created by the nature and become the first inhabitants of the Earth long before they were invented by man.

VI. Molecular Colony Diagnostics

A. MCT Assay of DNA and RNA Targets

Detection of replicable RNAs in the laboratory air (Section II.C) has been, in fact, the first diagnostic application of MCT. Those experiments have shown that MCT is able to detect single "infectious" molecules and to enumerate them. Unfortunately, the Qβ replicase version of MCT turned out a poor diagnostic tool (61) because of the structural requirements Qβ replicase imposes upon its templates (24, 78) and because of the high rate of RNA recombination, especially in the presence of Qβ replicase (15, 59), which results in numerous target-specific RNA colonies due to probe recombination irrespective of the presence of target in the analyzed sample.

Primer-dependent systems, particularly PCR (2, 79), are more suitable for diagnostic assays. Carrying out PCR in a gel (PCR-MCT, see Section IV.A) allows both DNA and RNA targets to be assayed as DNA colonies; in the latter case, RNA must first be converted into cDNA by reverse transcription with an RNA-directed DNA polymerase.

In early diagnostic experiments, we used Tth DNA polymerase for assaying both DNA and RNA targets, taking advantage of the ability of this enzyme to function both as an RNA-directed DNA polymerase (reverse transcriptase) and as a DNA-directed DNA polymerase in a bicine/Mn^{2+} buffer system (80, 81). A dry polyacrylamide gel was loaded with a PCR cocktail (Section IV.A) containing this enzyme as well as DNA and/or RNA targets by soaking for 30 min at ≈60 °C, which also provided for the reverse transcription of RNA targets, and then PCR cycles were performed. This one-step, one-enzyme protocol allowed 100% of DNA molecules and 13–16% of RNA molecules to be detected (61). The lower recovery of RNA targets reflected the yield of reverse transcription by Tth DNA polymerase.

However, later we found that this protocol failed to detect RNA targets in clinical samples containing large amounts of nontarget nucleic acids. We discovered that DNA or RNA, if present in the reaction mixture at a concentration of ≥ 5 ng/μl, selectively inhibited the reverse transcriptase activity of Tth DNA polymerase without affecting its DNA-dependent polymerase activity (82). These results show that despite its apparent attractiveness, the single-enzyme RT-PCR assay, in which Tth DNA polymerase serves for both reverse transcription and PCR (83–85), is of a limited diagnostic utility. It can be safely used only if the expected titer of an RNA target is at least 10 molecules per 1 μl of blood, while at a lower burden of an RNA target, the use of the Tth RT-PCR system may lead to false negative results.

The best results were obtained when a mutant M-MLV reverse transcriptase lacking the RNase H activity (86) was used for reverse transcription followed by in-gel PCR with Taq DNA polymerase. Large amounts of nontarget nucleic acids do not inhibit the M-MLV enzyme; moreover, they stabilize the enzyme at the elevated temperature that increases the primer selectivity. After reverse transcription, the components of M-MLV reaction buffer that are inhibitory to Taq DNA polymerase (such as dithiothreitol required for maintaining the reverse transcriptase activity) should be removed from the sample (by, e.g., ethanol/acetone precipitation or gel-filtration though a spin column). With this protocol, up to 50% of RNA molecules can be detected even if the concentration of nontarget nucleic in the reaction mixture exceeds 1 μg/μl (82).

B. Visualizing Molecular Colonies

DNA or RNA colonies are invisible by themselves, but they can be made visible in a number of ways, such as by staining with fluorescent intercalating dyes ethidium bromide (1, 9, 14) or SYBR Green I (17), by incorporation of radioactively labeled nucleotides (9, 14) or by in-gel hybridization with an unlabeled oligonucleotide that is then labeled by *in situ* incorporation of fluorescent nucleotides [(20, 21); also, see Section VII.A].

Conveniently, molecular colonies can be detected by hybridization with sequence-specific oligonucleotide probes on a nylon membrane to which the colony contents are transferred by blotting during or after amplification (14, 15, 61). Hybridization on a blotting membrane has a number of advantages compared to in-gel hybridization. First, double-stranded nucleic acids can be easily melted on the membrane by denaturation with formamide (RNA) or alkali (DNA) and then cross-linked to the membrane in a single-stranded form by UV irradiation. This eliminates any need in the immobilization of PCR primers on the gel matrix (17) and in the partial depolymerization of the matrix or in the use of electrophoresis to facilitate the removal of unattached product strand (20). Also, unhybridized probes can be washed away much easier from a membrane than from a gel. Finally, mechanical properties of a nylon membrane allows virtually unlimited number of reprobing cycles to be performed, whereas polyacrylamide gel becomes detached from the glass slide after 10–14 cycles of *in situ* hybridization (21).

Initially, molecular colonies were detected on blotting membranes by hybridizing with radioactively labeled probes (14, 15, 61). More convenient fluorescently labeled probes were not used because of a belief that the high intrinsic fluorescence of nylon (87, 88) will not allow the relatively low signal from colonies to be detected. Unexpectedly, in a recent study, we have found that molecular colonies can be readily detected on a nylon membrane by hybridization with fluorescent probes. Moreover, the signal-to-background ratio is high enough to allow molecular colonies to be detected even without washing away the unhybridized probe (89). Altogether, the suggested improvements greatly reduce the duration and simplify the hybridization procedure.

The main drawback of all of the above methods for the detection of molecular colonies is that they require that the gel is opened after PCR amplification. This may result in the environment being heavily contaminated with the amplification products, which may interfere with, or even preclude, subsequent analyses.

The latter problem has been overcome by our recent finding (90, 91) that PCR colonies can be visualized *in situ* with any of the common homogeneous fluorescence detection systems, including fluorescent intercalating dyes, hybridization probes, or combinations thereof, which are introduced in the reaction medium before PCR and whose fluorescence changes (usually increases) upon interaction with the amplification products. Figure 3 illustrates the detection of growing DNA colonies with a pair of adjacently hybridizing probes whose proximal ends are provided with fluorophores capable of fluorescent resonance energy transfer (FRET) (92). Another example of a homogeneous detection system is presented by molecular beacon, an oligonucleotide probe capable of folding into a hairpin whose loop is complementary to a target sequence, and whose proximal ends (held together by the stem) carry a fluorophore and its quencher. Fluorescence of the fluorophore is efficiently quenched when the molecular beacon is in the hairpin (closed) conformation and is not quenched when it is in the stretched (open) conformation attained when the probe hybridizes to its target (93).

With homogeneous detection systems, no postamplification steps such as opening the gel, DNA blotting, hybridization, and removal of unbound substrates, dyes, or probes are needed, and the growth of DNA colonies can be monitored in

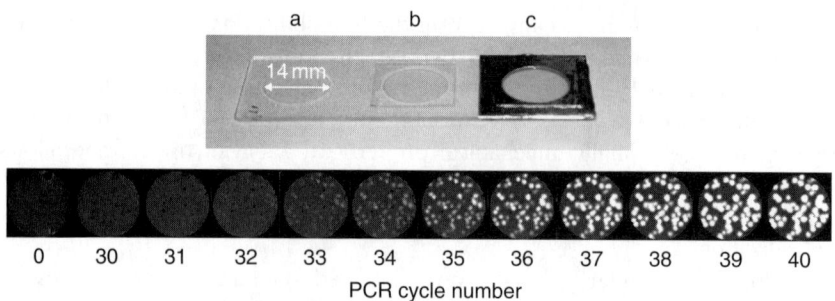

FIG. 3. Real-time monitoring of DNA colonies growing in a polyacrylamide gel in the presence of adjacently hybridizing FRET probes labeled with fluorophores FAM (energy donor) and Cy5 (energy acceptor) (91). Top: A homemade slide for in-gel PCR in which 0.4-mm-deep reaction wells were made by drilling. (a) Reaction well with a film of dry polyacrylamide gel attached to its bottom; (b) gel soaked in a solution containing all the PCR components and the FRET probes; (c) reaction well sealed with a piece of adhesive foil in which a round window was made to enable monitoring of the gel fluorescence. Bottom: images of a gel obtained with a microchip scanner after the indicated PCR cycles. For further explanations, see Section VI.B.

real time. This enhances the performance, simplifies, and reduces the time and cost of the MCT assays, as well as the risk of sample cross-contamination. Altogether, these features greatly increase the diagnostic power of MCT.

C. Detection of DNA and RNA Targets in Clinical Samples

It is generally believed that PCR (or RT-PCR) can detect single DNA or RNA molecules. However, this is only true when purified targets are assayed. A real sensitivity of PCR assays of clinical samples is 2–3 orders of magnitude lower. For example, in a blinded study, the lower sensitivity limit of the detection of the variola virus DNA by the real time PCR was found to be ≈500 molecules (94). The main reason for the sensitivity loss is the competing amplification of nontarget DNA that occurs because of a limited specificity of primer hybridization. Given the complexity of the human genome and a huge excess of human nucleic acids over the virus DNA in an analyzed sample, it is not surprising that PCR primers sometimes anneal and prime on human sequences, even when there are mismatched or unmatched bases. At low virus loads, the nonspecific DNA synthesis may completely hide the signal generated by the target. If that happens, the PCR assay will produce a false-negative result.

In the MCT format, competition between amplicons is greatly reduced because the respective colonies are spatially separated. In a specially designed experiment, it was shown that the number of colonies produced by 300 molecules of human immunodeficiency virus type 1 (HIV-1) RNA was not changed when up to 1,000,000,000 molecules of human hepatitis B virus DNA were simultaneously amplified in the same gel (61). The competition from nonspecific synthesis is also virtually eliminated. Neither the number nor the size of virus-specific DNA colonies were affected in presence of up to a trillion-fold excess of nucleic acids isolated from the whole human blood (61, 82).

The success of a clinical assay is to a great extent determined by the steps preceding PCR, such as storage of clinical samples and nucleic acid isolation. Although blood is commonly used for clinical diagnostics, it is rich in nucleases that rapidly destroy DNA and especially RNA, and contains large amounts of proteins as well as other biopolymers and low molecular weight compounds that may interfere with reverse transcription and DNA amplification.

Recently, it was found that RNA and DNA retain their integrity for at least 3 days at room temperature, no less than 2 weeks at +4 °C, and more than a year at –20 °C when whole blood samples are stored as lysates containing 4 M guanidine thiocyanate. Storage time at room temperature can be substantially prolonged if nucleic acids are precipitated by two volumes of isopropanol (95). This method of sample preservation provides for a safe storage and transportation at ambient temperature and is compatible with the subsequent procedure of nucleic acid isolation.

We also found that popular nucleic acid isolation protocols based on the binding of DNA and RNA to silica or glass surfaces (96) give low yields and fail to completely remove reverse transcription and PCR inhibitors (82). Best results are obtained with phenol extraction, either after proteinase K digestion in the presence of sodium dodecylsulfate for the isolation of DNA (61) or after sample lysis with guanidinium thiocyanate (GTC) for the isolation of RNA (97, 98). Both DNA and RNA can be isolated from a GTC-lysed blood at a nearly 100% yield by extraction with a neutral phenol, after most proteins have been removed from the sample by isopropanol precipitation (82).

With the above methods for sample preservation and nucleic acid isolation, the MCT assay can reliably detect one DNA target molecule and two RNA molecules in 100-μl aliquots of the whole blood (82). This is the highest sensitivity ever achieved in molecular diagnostics.

D. Quantitative *In Situ* Assay

All colonies in the real-time detection experiment of Fig. 3 became visible almost synchronously. This was because each colony originated from a single template molecule. However, when a colony originates from a cluster of molecules, it becomes visible earlier because a lesser number of PCR cycles are needed to produce detectable amount of DNA at that location (note that such a multimolecular colony differs from a colony produced by one molecule in that it does not, in general, represent a molecular clone). Hence, the initial number of molecules in each colony can be estimated using the principle utilized in the solution real-time PCR (also termed "quantitative PCR") to estimate the copy number of a target in a test tube, by noting the cycle number at which the fluorescence signal from a colony reaches certain (threshold) level. This provides for another application termed "quantitative *in situ* assay" (QISA) (*91*).

For example, QISA can be used for quantifying specific DNA or RNA sequences in individual cells. To this end, cells are entrapped in a layer of polyacrylamide gel. The gel is then soaked with guanidine thiocyanate or detergents to lyse the cells. The lysing reagents are washed away with a saline alcohol solution (similar to one used to remove from a gel the PCR components, see Section IV.A) that causes precipitation and hence preserves the location of nucleic acids. Then the gel is dried and processed using primers and fluorescent probes specific to the DNA or RNA target to be assayed, and the time of appearance of DNA colonies is registered (*91*).

At the first glance, QISA resembles the well-known "*in situ* PCR" (*99, 100*), but there are important differences. In the *in situ* PCR, cells are treated to make their walls semipermeable, preferably permeable for PCR reaction components, including dNTPs, primers, and DNA polymerase, but not permeable for template DNA or RNA and for the amplification products. Amplification is carried out in solution, in which the cells are suspended or which covers cells attached to the surface of a slide. Because it is not possible to simultaneously make the cell membrane readily permeable for a DNA polymerase and prevent any leakage of DNA, a certain (sometimes substantial) efflux of PCR products from the cells and their migration to other locations cannot be excluded. This may result in errors in the target localization and makes quantitative target assays hardly possible.

In QISA, cells entrapped in a gel matrix are completely lysed. Therefore, target molecules are freely accessible to the components of an amplification system. Yet, due to immobilization of the medium, target molecules and their copies remain at the locations formerly occupied by their home cells. Also, the growing molecular colonies can be detected in real time allowing the target DNA or RNA molecules in each cell to be quantified, which is not possible with the *in situ* PCR. Many cells can be analyzed in parallel on a single gel, for example, for the presence and copy number of a particular mRNA. This would provide information on the level of expression of the corresponding gene in each of these cells, which can be useful for both research and diagnostic purposes.

The same approach could provide for the detection and quantitation of pathogenic microorganisms in clinical samples, such as in blood or stool. A similar procedure could be used for the determination of localization and local concentration of mRNAs in tissue slices. After having analyzed a series of slices, it would be possible to create the 3-D expression map of a particular gene in a tissue, in an organ, or even in an entire organism, such as in an embryo's body (91).

VII. Gene and Gene Expression Analysis

A. Sequencing on Molecular Colonies

In 2003, Church et al. have demonstrated that molecular colonies can be sequenced directly in the PCR gel, without extracting and handling their DNA contents (21). To this end, they prepared a library of short DNA fragments flanked by two constant regions containing primer-binding sites, and used appropriate primers for in-gel amplification of the fragments. As one of the PCR primers was immobilized on the gel matrix, its extension produced a strand that was covalently attached to the gel. The complementary unattached strands were removed from the gel by electrophoresis upon denaturation of DNA colonies with formamide.

DNA sequencing was performed on all colonies simultaneously, by hybridizing the gel-attached strands with a PCR primer, which is complementary to their 3′-terminal constant region, followed by a series of consecutive template-directed single-base extensions of the primer using a DNA polymerase and a fluorescently labeled nucleotide, a technology earlier developed at Affymax for a SNP analysis (101). After washing off any unincorporated label, the gel was scanned to ascertain which colonies became labeled; this revealed the identity of the template base immediately upstream of the annealed primer in those colonies. To provide for further extensions, the nucleotide possessed a free 3′-hydroxyl group. Also, to distinguish between nucleotides incorporated at consecutive extensions, a fluorescent label was attached to the nucleotide via a disulfide bond (102) or via a photocleavable linker (103) and, after gel scanning, the fluorophore was removed by treatment with concentrated β-mercaptoethanol or by UV irradiation, respectively. The extension/scanning/cleavage cycles were repeated with other labeled nucleotides. Theoretically, this reiterative procedure can be continued indefinitely long, revealing the template sequence lying upstream of the primer-binding site in each colony.

The authors estimated the yield of an individual extension step to be 99.8%, and predicted that the primer can be extended 150 bases without significant dephasing of the sequence information. In practice, however, the procedure was limited to 10–14 cycles because the gel detached from the glass slide, but this is seemingly a mere technical problem that can be solved in a number of ways. For example, the sequencing reactions could be performed on a blotting membrane (Section VI.B).

Another problem is the presence in a template of homonucleotide tracts resulting in the incorporation per cycle of many nucleotides, whose number cannot be accurately determined from the fluorescence measurements. One solution to this problem could be the use of fluorescently labeled nucleotide analogues with a reversibly blocked 3′-hydroxyl group, acting as strand extension terminators (104–107). This allows only one nucleotide to be added in each cycle irrespective of the length of a homonucleotide tract.

Instead of many consecutive extensions with labeled mononucleotides, DNA sequences adjacent to constant template regions can be determined by *in situ* ligating a primer, hybridized to such a region, with labeled oligonucleotides. In a recent work (108), the feasibility of this approach was demonstrated by ligating the primer with a pool of degenerate nonamers labeled with fluorescent dyes. As many pools of nonamers were prepared as there were positions that were to be queried (in this work, up to seven), and nonamers in the pool for querying a particular position were each colored with one of four fluorescent dyes according to the identity of the base at that position. After template-directed ligation by a DNA ligase followed by four-color imaging of the gel, the ligated primer–nonamer pairs were stripped away from their templates and a new ligation-imaging cycle was performed using a pool of nonamers labeled according to a different query position. It was shown that a template nucleotide can be accurately identified when the query position is six to seven bases from the ligation junction.

Since many thousands of DNA colonies can be resolved in one gel (17, 109), the *in situ* sequencing can potentially become a very high-throughput method for data acquisition. As far as this method produces a list of relatively short oligonucleotide runs present in a DNA (and sometimes linked groups of such runs), it is best suited for the resequencing of known genomes in order to refine the existing data and to determine individual sequence variations.

Alternatively, material from an individual DNA colony can be electrophoretically injected into capillary, and then sequenced using the conventional capillary electrophoresis technology (110).

B. Single-Nucleotide Polymorphism

Individual genetic variations in humans, as well as in other biological species, are mostly represented by single nucleotide substitutions, also termed single nucleotide polymorphisms (SNPs). It is generally believed that mapping SNPs can help to identify genes involved in genetic disorders, diagnostic markers, and predictors of individual drug response [see, e.g., reference (101)]. Recent studies suggest that in many instances, there is not one SNP that produces effect, but, instead, multiple SNPs interact to alter the expression of a gene or the function of the encoded protein (111). Often, these alterations only occur when certain combinations of SNPs are present *in cis*, that is, within the same DNA molecule, and hence, it is the haplotype (a set of SNPs within an entire chromosome or its large segment encompassing one or more genes) which may be of a diagnostic or predictive value. Furthermore, it was shown that SNPs are inherited as large haplotype blocks spanning more than 100 Kb along the chromosome (average size of a gene) with relatively few variants of each block in a population (*112–114*), and that the haplotype block structure is conserved among mammals (*115*).

Obtaining haplotype information is technically challenging, primarily because of the difficulty of separating sister chromosomes of a diploid genome. The first human haplotype map with ten 500-kilobase regions of a fine SNP structure has been recently obtained as a result of efforts of a big consortium (*116*) for a cost of US$100 million (*117*).

Church *et al.* (*20*) were the first who showed that MCT allows a long-range haplotype information to be easily obtained at a low cost and without the need in separating the chromosomes. This is because amplification of two or more loci will produce overlapping DNA colonies if those loci are linked by being present within the same chromosome, but it will usually produce separate colonies if the loci are unlinked. SNPs were detected in the amplified fragments with the primer extension approach (Section VII.A), which is particularly suitable for the determination of the identity of single bases.

Recently, Church *et al.* managed to link SNPs spanning a 153-Mb region of human chromosome 7 (*118*). To this end, they embedded in a polyacrylamide gel condensed (protein-bound) chromosomes having a size of less than 1 μm, and then removed proteins by proteinase K digestion and ruptured the chromosomal DNA into relatively small pieces by freezing and thawing the gel. Rupturing DNA increased the yield of PCR product in a colony but did not unlink SNPs because the pieces were entrapped in the gel matrix. In one gel, the authors were able to discern SNP clusters from 414 chromosomes. They also demonstrated that correct haplotypes can be inferred even by analyzing a highly heterogeneous sample representing a mixture of chromosomes obtained from several tens of individuals and argued that a 14-loci haplotyping experiment on 50 individuals using a single amplification gel is equivalent to 2600 allele-specific PCR assays, meaning a more than three orders of magnitude improvement in throughput. Finally, they noted that single-cell meiotic recombinations and even rare mitotic recombinations and chromosome translocations in cancer cells can be detected and characterized using this long-range haplotyping assay.

Another example of the use of MCT for determining sets of linked SNPs is presented by the recent work of a group from Duke University on the detection of minor drug-resistant populations of HIV in infected individuals (*109*). The authors performed sequencing on molecular colonies by the single base extension method and showed that this allows a minor population carrying a set of mutations responsible for HIV drug resistance to be detected even if it comprises as low as 0.01% of the total viral RNA preparation. This is 2000 times better than the sensitivity achievable by a direct sequencing of the RNA pool (20%) and provides for a more accurate prediction of the treatment outcome. The authors noted that insomuch as single viral cDNA molecules become embedded and separately amplified in the gel, the artifact sequences that are commonly generated at conventional PCR through recombination are eliminated.

A similar approach was applied for the quantitative monitoring of known drug-resistant mutations in the fusion BCR-ABL tyrosine kinase gene whose expression is a cause of the chronic myeloid leukemia (119). Nine mutations in the BCR-ABL kinase domain account for more than 90% of mutations resulting in the resistance to the kinase inhibitor imatinib, which is widely used for chemotherapy (120). With MCT, it is possible to detect one drug-resistant leukemic cell among as many as 10^4 nonresistant cells (121).

MCT has been used to quantitatively analyze the methylation pattern of gene P16 in stomach tumor cells (122). Pretreatment of DNA with bisulfite resulted in the conversion of unmodified cytosines into uracyls, whereas methylated cytosines remain unchanged. The resulting SNPs were detected by hybridization with fluorescent probes. In this work, the isothermal hyperbranched rolling circle amplification, rather PCR, was used to produce DNA colonies.

C. Alternative Splicing

One more example of multiple linked traits is presented by alternatively spliced mRNAs that contain various combinations of a set of exons. There can be a huge number of alternatively spliced sequences. Thus, the presence or absence of 1 exon produces 2 variants, whereas 10 exons may produce up to 2^{10}, or 1024 combinations, in which each exon is either present or absent. At standard analysis of alternatively spliced mRNAs with the solution RT-PCR, some minor species may remain undetected and cumbersome analysis is needed to identify all splicing variants and to determine their proportion. Also, false "splicing variants" may arise due to sequence recombinations during both reverse transcription and PCR steps.

As was demonstrated by Church et al., MCT helps to overcome all these problems (22). After in-gel amplification of cDNA using primers matching the outermost exons to ensure that every splicing variant is amplified, the colonies were interrogated with exon-specific probes. This indicated which exons were present and which were absent in each of the colonies, resulting both in an unambiguous detection of all splicing variants in one gel and in a precise determination of the copy number of each of the variants. To identify exons, the colonies were hybridized with unlabeled oligonucleotide probes that were then labeled *in situ* by extension with fluorescent nucleotides as discussed in Section VII.A. It should be noted, however, that the same information could be easily obtained by colony hybridization with prelabeled fluorescent oligonucleotides, preferably on a nylon membrane (89), which also provides for a fast repetitive reprobing (see Section VI.B).

The MCT exon profiling was used to analyze mRNA variants of the murine CD44 gene possessing 10 variable exons (22). After having analyzed ≈9000 colonies, 69 distinct splicing isoforms were detected, which was more than twice the number previously identified. Exon quantifications performed on normal and transformed cells showed that cancer transformation resulted in statistically very significant changes in the proportion of these isoforms (in each case, $p<10^{-10}$), although the observed changes of individual isoforms were relatively modest (1.3- to 2.9-fold) and would likely be missed by a nondigital assay such as the solution RT-PCR. The results suggest that splicing isoforms can serve as tumorigenesis markers, but only when a digital assay, such as MCT, is used for their quantification. In another study, MCT assay was successfully used to monitor the relative expression levels of two splicing variants of the oncogene K-*ras*, K-RAS2A and K-RAS2B (123). A similar approach was recently used to demonstrate statistically significant Alzheimer's disease-associated changes in the exon pattern of the mRNA coding for protein tau (a microtubule-associated protein that is important for establishing and maintaining the neuronal morphology), which are not detectable with a solution-based quantitative PCR (124).

D. Allele and mRNA Copy Number

As far as MCT allows the number of nucleic acid molecules to be directly assessed by counting the number of colonies, it is perfectly suited for any kind of quantitative analysis of DNA and RNA targets. This analysis is similar to diagnostic assays discussed above (Section VI) and can utilize any format of colony detection, including either probe hybridization on a blotting membrane or *in situ* staining with a homogeneous fluorescent detection system (Section VI.B), or a single base extension of an unlabeled probe with a fluorescent nucleotide analogue (Section VII.A).

In one of the first reports of this sort, MCT was used to determine the proportion of individual yeast strains in a competition growth experiment in which eight strains carrying different plasmids coding for a mutated phosphoglycerate kinase gene were pooled and grown in a single culture. Plasmid DNA was extracted, DNA colonies were grown, and single base extension reactions were performed to identify the colonies whose DNA carries a particular nucleotide substitution in that gene (*125*). The same approach was later used to quantitatively analyze the competition in a pool of as many as 54 yeast strains producing various mutant variants of human glucose-6-phosphate dehydrogenase (*126*).

In another work, MCT was used to detect gene deletions resulting in a loss of heterozygosity (LOH) (*127*). When LOH occurs in a tumor-suppressor gene region, this may result in cancer transformation of the cell if the remaining copy of the gene happens to be inactivated by, for example, a point mutation. Therefore, LOH in tumor-suppressor genes is used as a common marker of cancer cells. In this work, MCT analysis of a pancreatic cell line, Panc-1, showed that there were approximately twice as many K-*ras*2 colonies as there were p53 colonies, indicating LOH in the p53 gene.

One of the most promising applications of MCT is the digital monitoring of gene expression. This presents a greatly improved assay sensitive enough for identifying and quantifying small but biologically important and statistically relevant changes in gene expression over time. The potential of this approach was demonstrated in a work in which 14 genes involved in galactose metabolism in *Saccharomyces cerevisiae* were analyzed for their expression levels in glucose and galactose minimal media (*128*). The authors conclude that MCT assay is likely to have profound implications in the field of functional genomics because the gene expression measurements are digital in nature and therefore more accurate than any other technologies. Also they noted that this technique has a larger dynamic range than do other expression profiling techniques. Namely, they were able to quantitatively measure gene expression at very low levels with similar precision as to measurements at very high expression levels. In a recent work, MCT assay was used to accurately measure the copy number of several individual mRNAs in a few embryonic stem cells (*129*).

Finally, MCT provides for accurate quantifications of different alleles of the same gene and their expression levels. After DNA colonies had grown, those produced by two alleles of a gene or by their respective mRNAs can be differentially labeled by performing *in situ* single base addition assay. A simple counting of the differentially labeled colonies allows the relative expression levels of the two alleles to be precisely determined. To validate this technique, the relative expression levels of the protein kinase D2 gene in a family of heterozygous patients bearing the 4208G/A SNP were examined and compared to the literature (*130*). The authors conclude that MCT has at least two major advantages in measuring the allelic variations of gene expression. First, the measurements are digital and hence more accurate than the "analogue" approaches for quantifying allelic variation. In addition, the MCT approach provides an absolute measure of the relative expression of the two alleles, whereas other approaches are normalized to the population average, which is assumed to be 1:1 [see also reference (20)].

VIII. Opportunities Provided by the Molecular Colony Technology

Various applications of MCT stem from its unique ability of compartmentalized amplification and expression of individual DNA or RNA molecules. Three immediate derivatives of this basic ability are exploited in the applications developed to date: (1) isolation of molecular clones, (2) detection and enumeration of individual molecules, and (3) identification of structural cis-elements, that is, elements covalently linked by being present in the same DNA or RNA molecule.

Cloning is the most straightforward application of MCT because each individual colony represents a molecular clone (i.e., multiple copies of a single DNA or RNA molecule) by its origin. As discussed in Section IV.C, MCT is a much more efficient cloning method than are methods based on exhaustive dilution of a genetic material. Being combined with *in situ* expression (Section IV.A), MCT cloning can be used for establishing the relationship between a gene and its encoded function (performed by either protein or RNA), for the identification of groups of genes encoding regulatory networks, interacting proteins and/or RNAs, and as a versatile tool for protein and aptamer engineering and drug discovery.

The ability to detect and enumerate individual nucleic acid molecules makes MCT a powerful tool for a number of analytical applications such as single-molecule chemistry (Section III); absolute quantification of copies of individual genes, alleles, and RNA species including in single cells (Section VII.D); and, of course, any type of quantitative molecular diagnostics (Section VI), including assays for infectious agents, cancer diagnostics, environmental monitoring, forensic tests, as well as assays for genetically modified organisms. Unlike standard assays based on the amplification of nucleic acids in solution, MCT assay does not require measurements of signal intensity, only the presence or absence of a colony needs to be determined irrespective of the number of constituent molecules. This means that MCT is a digital technology in its nature, which greatly increases reliability of the assay.

Another important feature of MCT is that it spatially separates amplification reactions initiated by different template molecules present in a sample. This suppresses the competition between simultaneously amplified targets in multiplex assays, and also virtually eliminates the interference from nonspecific synthesis that limits the sensitivity of solution assays (Section VI.C). Therefore, MCT assays manifest the unsurpassed sensitivity, permitting as low as one DNA or two RNA molecules to be detected on a trillion-fold larger background of nontarget nucleic acids. Also, template separation greatly reduces the probability of amplification artifacts that commonly arise during PCR due to both homologous and nonhomologous recombinations between DNA species present in the analyzed samples (2, *131–137*).

It should be noted that MCT assay can potentially be used for detecting single molecules of substances other than DNA or RNA. For example, in the approach termed proximity ligation assay [reviewed in (*138*)], a protein target triggers ligation of two DNA pieces linked to affinity probes that can simultaneously bind to two different sites of the same protein molecule. After binding, the two DNA pieces come close and can be joined by ligase, resulting in a recombinant DNA strand that serves as a surrogate marker for the protein target to be detected. Single molecules of the recombinant DNA can then be detected and enumerated with the MCT assay. The role of protein-specific reagents can be played by antibodies, protein-binding DNA or RNA aptamers, specific protein ligands, such as enzyme substrate or cofactor analogues. Obviously, similar approaches could be developed for assaying a variety of molecules. In principle, this sort of MCT assays could provide for the detection of single molecules of any substance that is complex enough to form on its surface at least two high-affinity binding determinants. Such MCT assays can be used for diagnostic purposes and for the detection of rare products or intermediates of a chemical reaction.

Finally, the ability to identify *cis*-elements that give rise to overlapping colonies makes MCT a unique tool for long-range genotyping, haplotyping, determining linked exons, and for *in situ* sequencing (Section VII).

To date, the MCT potential has been explored in a number of fields, such as single-molecule chemistry, cell-free cloning and screening of genes, molecular diagnostics, as well as gene expression and sequence analysis, and in each of these fields, MCT has demonstrated a superior performance. This permits us to conclude that MCT is becoming a developed technology with unique capabilities and high potentials in basic and applied research, many areas of biotechnology, and clinical practice.

Acknowledgments

The development of various aspects of MCT in Russia has been made possible owing to invaluable contribution of our collaborators Timur Samatov, Marina Falaleeva, and Victor Ugarov, and long-term financial support from the International Science Foundation (grants MTO000 and MTO300), Howard Hughes Medical Institute (grants 75195-544701 and INTNL55000302),

Russian Foundation for Basic Research (grants 93-04-06550, 96-04-48329, 96-04-48331, 99-04-48672, 02-04-48320, 05-04-48897, 07-04-01114 and 08-04-01660), INTAS (International Association for the promotion of cooperation with scientists from the New Independent States of the former Soviet Union, grants 95-1365, 97-10348, and 01-2012), and from programs "Molecular and Cell Biology" of the Russian Academy of Sciences and "Living Systems" of the Russian Federal Agency for Science and Innovations.

References

1. Chetverin, A. B., Chetverina, H. V., and Munishkin, A. V. (1991). On the nature of spontaneous RNA synthesis by Qβ replicase. *J. Mol. Biol.* **222**, 3–9.
2. Saiki, R. K., Gelfand, D. H., Stoffel, S., Scharf, S. J., Higuchi, R., Horn, G. T., Mullis, K. B., and Erlich, H. A. (1988). Primer-directed enzymatic amplification of DNA with a thermostable DNA polymerase. *Science* **239**, 487–491.
3. Guatelli, J. C., Whitfield, K. M., Kwoh, D. Y., Barringer, K. J., Richman, D. D., and Gingeras, T. R. (1990). Isothermal, *in vitro* amplification of nucleic acids by a multienzyme reaction modeled after retroviral replication. *Proc. Natl. Acad. Sci. USA* **87**, 1874–1878.
4. Compton, J. (1991). Nucleic acid sequence-based amplification. *Nature* **350**, 91–92.
5. Walker, G. T., Little, M. C., Nadeau, J. G., and Shank, D. D. (1992). Isothermal *in vitro* amplification of DNA by a restriction enzyme/DNA polymerase system. *Proc. Natl. Acad. Sci. USA* **89**, 392–396.
6. Fire, A., and Xu, S.-Q. (1995). Rolling replication of short DNA circles. *Proc. Natl. Acad. Sci. USA* **92**, 4641–4645.
7. Notomi, T., Okayama, H., Masubuchi, H., Yonekawa, T., Watanabe, K., Amino, N., and Hase, T. (2000). Loop-mediated isothermal amplification of DNA. *Nucleic Acids Res.* **28**, e63.
8. Chetverin, A. B., and Chetverina, H. V. (1995). Method of nucleic acid copying, method of their expression and a medium for their realization. *Russian Federation Patent* 2,048,522.
9. Chetverin, A. B., and Chetverina, H. V. (1997). Method for amplification of nucleic acids in solid media. *U.S. Patent* 5,616,478.
10. Chetverin, A. B., and Chetverina, H. V. (1998). Method of nucleic acid cloning. *Russian Federation Patent* 2,114,175.
11. Chetverin, A. B., and Chetverina, H. V. (1998). Method of identification of nucleic acid. *Russian Federation Patent* 2,114,915.
12. Chetverin, A. B., and Chetverina, H. V. (1999). Method for amplification of nucleic acids in solid media and its application for nucleic acid cloning and diagnostics. *U.S. Patent* 5,958,698.
13. Chetverin, A. B., and Chetverina, H. V. (1999). Solid medium for amplification and expression of nucleic acids as colonies. *U.S. Patent* 6,001,568.
14. Chetverina, H. V., and Chetverin, A. B. (1993). Cloning of RNA molecules *in vitro*. *Nucleic Acids Res.* **21**, 2349–2353.
15. Chetverin, A. B., Chetverina, H. V., Demidenko, A. A., and Ugarov, V. I. (1997). Nonhomologous RNA recombination in a cell-free system: Evidence for a transesterification mechanism guided by secondary structure. *Cell* **88**, 503–513.
16. Chetverina, H. V., Demidenko, A. A., Ugarov, V. I., and Chetverin, A. B. (1999). Spontaneous rearrangements in RNA sequences. *FEBS Lett.* **450**, 89–94.
17. Mitra, R. D., and Church, G. M. (1999). *In situ* localized amplification and contact replication of many individual DNA molecules. *Nucleic Acids Res.* **27**, e34.
18. Boles, T. C., Kron, S. J., and Adams, C. P. (1999). Nucleic acid-containing polymerizable complex. *U.S. Patent* 5,932,711.

19. Rehman, F. N., Audeh, M., Abrams, E. S., Hammond, P. W., Kenney, M., and Boles, T. C. (1999). Immobilization of acrylamide-modified oligonucleotides by co-polymerization. *Nucleic Acids Res.* **27,** 649–655.
20. Mitra, R. D., Butty, V. L., Shendure, J., Williams, B. R., Housman, D. E., and Church, G. M. (2003). Digital genotyping and haplotyping with polymerase colonies. *Proc. Natl. Acad. Sci. USA* **100,** 5926–5931.
21. Mitra, R. D., Shendure, J., Olejnik, J., Olejnik, E. K., and Church, G. M. (2003). Fluorescent *in situ* sequencing on polymerase colonies. *Anal. Biochem.* **320,** 55–65.
22. Zhu, J., Shendure, J., Mitra, R. D., and Church, G. M. (2003). Single molecule profiling of alternative pre-mRNA splicing. *Science* **301,** 836–838.
23. Marx, V. (2004). Helping SNPs to speak up louder than before. *Genomics & Proteomics* (Reed Elsevier, Inc.), March Issue.
24. Chetverin, A. B., and Spirin, A. S. (1995). Replicable RNA vectors: Prospects for cell-free gene amplification, expression and cloning. *Prog. Nucleic Acid Res. Mol. Biol.* **51,** 225–270.
25. Haruna, I., and Spiegelman, S. (1965). Autocatalytic synthesis of a viral RNA *in vitro. Science* **150,** 884–886.
26. Weissmann, C., Feix, G., and Slor, H. (1968). *In vitro* synthesis of phage RNA: The nature of the intermediates. *Cold Spring Harb. Symp. Quant. Biol.* **33,** 83–100.
27. Carmichael, G. G., Landers, T. A., and Weber, K. (1976). Immunochemical analysis of the functions of the subunits of phage Qβ ribonucleic acid replicase. *J. Biol. Chem.* **251,** 2744–2748.
28. Blumenthal, T., and Carmichael, G. (1979). RNA replication: Function and structure of Qβ replicase. *Annu. Rev. Biochem.* **48,** 525–545.
29. Biebricher, C. K., Eigen, M., and Gardiner, W. C. (1983). Kinetics of RNA replication. *Biochemistry* **22,** 2544–2559.
30. Rensing, U., and August, J. T. (1969). The 3′-terminus and the replication of phage RNA. *Nature* **224,** 853–856.
31. Weber, H., and Weissmann, C. (1970). The 3′-termini of bacteriophage Qβ plus and minus strands. *J. Mol. Biol.* **51,** 215–224.
32. Haruna, I., and Spiegelman, S. (1965). Specific template requirements of RNA replicases. *Proc. Natl. Acad. Sci. USA* **54,** 579–587.
33. Sumper, M., and Luce, R. (1975). Evidence for *de novo* production of self-replicating and environmentally adapted RNA structures by bacteriophage Qβ replicase. *Proc. Natl. Acad. Sci USA* **72,** 162–166.
34. Biebricher, C. K., Eigen, M., and Luce, R. (1981). Product analysis of RNA generated *de novo* by Qβ replicase. *J. Mol. Biol.* **148,** 369–390.
35. Biebricher, C. K., Eigen, M., and Luce, R. (1986). Template-free RNA synthesis by Qβ replicase. *Nature* **321,** 89–91.
36. Biebricher, C. K., Eigen, M., and McCaskill, J. S. (1993). Template-directed and template-free RNA synthesis by Qβ replicase. *J. Mol. Biol.* **231,** 175–179.
37. Brown, D., and Gold, L. (1995). Selection and characterization of RNAs replicated by Qβ replicase. *Biochemistry* **34,** 14775–14782.
38. Pasteur, L. (1860). Expériences relatives aux générations dites spontanées. *C. R. Acad. Sci.* **50,** 303–307.
39. Weiss, R. A. (2005). Robert Koch: The Grandfather of Cloning. *Cell* **123,** 539–542.
40. Koch, R. (1881). Zur Untersuchung von pathogenen Organismen. *Mitteilungen aus dem Kaiserlichen Gesundheitsamte* **1,** 1–48.
41. Chetverin, A. B., and Chetverina, H. V. (2007). Scientific and practical applications of molecular colonies. *Mol. Biol. (Mosk.)* **41,** 250–261.
42. Crick, F. H. (1958). On protein synthesis. *Symp. Soc. Exp. Biol.* **12,** 138–163.

43. McCaskill, J. S., and Bauer, G. J. (1993). Images of evolution: Origin of spontaneous RNA replication waves. *Proc. Natl. Acad. Sci. USA* **90**, 4191–4195.
44. Hirst, G. K. (1962). Genetic recombination with Newcastle disease virus, poliovirus and influenza virus. *Cold Spring Harbor Symp. Quant. Biol.* **27**, 303–309.
45. Ledinko, N. (1963). Genetic recombination with poliovirus type 1: Studies of crosses between a normal horse serum-resistant mutant and several guanidine-resistant mutants of the same strain. *Virology* **180**, 107–119.
46. Romanova, L. I., Tolskaya, E. A., Kolesnikova, M. S., and Agol, V. I. (1980). Biochemical evidence for intertypic genetic recombination of polioviruses. *FEBS Lett.* **118**, 109–112.
47. King, A. M. Q., McCahon, D., Slade, W. R., and Newman, J. W. I. (1982). Recombination in RNA. *Cell* **29**, 921–928.
48. Jarvis, T. C., and Kirkegaard, K. (1992). Poliovirus RNA recombination: Mechanistic studies in the absence of selection. *EMBO J.* **11**, 3135–3145.
49. Olsthoorn, R. C. L., and van Duin, J. (1996). Random removal of inserts from an RNA genome: Selection against single-stranded RNA. *J. Virol.* **70**, 729–736.
50. Chetverin, A. B. (1999). A new look at recombination of RNA. *Mol. Biol. (Mosk.)* **33**, 985–996.
51. Chetverin, A. B. (1999). The puzzle of RNA recombination. *FEBS Lett.* **460**, 1–5.
52. Steinschneider, A., and Fraenkel-Conrat, H. (1966). Studies of nucleotide sequences in tobacco mosaic virus ribonucleic acid. III. Periodate oxidation and semicarbazone formation. *Biochemistry* **5**, 2729–2734.
53. Steinschneider, A., and Fraenkel-Conrat, H. (1966). Studies of nucleotide sequences in tobacco mosaic virus ribonucleic acid. IV. Use of aniline in stepwise degradation. *Biochemistry* **5**, 2735–2743.
54. Nam, K., Hudson, R. H. E., Chapman, K. B., Ganeshan, K., Damha, M. J., and Boeke, J. D. (1994). Yeast lariat debranching enzyme: Substrate and sequence specificity. *J. Biol. Chem.* **269**, 20613–20621.
55. Chetverin, A. B. (2004). Replicable and recombinogenic RNAs. *FEBS Lett.* **567**, 35–41.
56. Spirin, A. S. (2002). Omnipotent RNA. *FEBS Lett.* **530**, 4–8.
57. Spirin, A. S. (2005). RNA world and its evolution. *Mol. Biol. (Mosk.)* **39**, 466–472.
58. Spirin, A. S. (2005). Origin, possible forms of being, and size of the primeval organisms. *Paleontol. J. (Mosk.)* **39**, 364–371.
59. Chetverin, A. B., Kopein, D. S., Chetverina, H. V., Demidenko, A. A., and Ugarov, V. I. (2005). Viral RNA-directed RNA polymerases use diverse mechanisms to promote recombination between RNA molecules. *J. Biol. Chem.* **280**, 8748–8755.
60. Morozov, I. Y., Ugarov, V. I., Chetverin, A. B., and Spirin, A. S. (1993). Synergism in replication and translation of messenger RNA in a cell-free system. *Proc. Natl. Acad. Sci. USA* **90**, 9325–9329.
61. Chetverina, H. V., Samatov, T. R., Ugarov, V. I., and Chetverin, A. B. (2002). Molecular colony diagnostics: Detection and quantitation of viral nucleic acids by in-gel PCR. *BioTechniques* **33**, 150–156.
62. Samatov, T. R., Chetverina, H. V., and Chetverin, A. B. (2005). Expressible molecular colonies. *Nucleic Acids Res.* **33**, e145.
63. Roberts, R. W., and Ja, W. W. (1999). In vitro selection of nucleic acids and proteins: What are we learning? *Curr. Opin. Struct. Biol.* **9**, 521–529.
64. Chetverin, A. B., Samatov, T. R., and Chetverina, H. V. (2008). Gene cloning and expression in molecular colonies. In "Cell-Free Protein Synthesis: Methods and Protocols" (A. S. Spirin and J. R. Swartz, Eds.), pp. 191–206. Wiley-VCH, Weinheim.
65. Sambrook, J., and Russell, D. W. (2001). "Molecular Cloning." 3rd edn. Cold Spring Harbor Laboratory Press, Cold Spring Harbor, NY.

66. Lukyanov, K. A., Matz, M. V., Bogdanova, E. A., Gurskaya, N. G., and Lukyanov, S. A. (1996). Molecule by molecule PCR amplification of complex DNA mixtures for direct sequencing: An approach to *in vitro* cloning. *Nucleic Acids Res.* **24,** 2194–2195.
67. Vogelstein, B., and Kinzler, K. W. (1999). Digital PCR. *Proc. Natl. Acad. Sci. USA* **96,** 9236–9241.
68. Brenner, S., Williams, S. R., Vermaas, E. H., Storck, T., Moon, K., McCollum, C., Mao, J. I., Luo, S., Kirchner, J. J., Eletr, S., DuBridge, R. B., Burcham, T. *et al.* (2000). In vitro cloning of complex mixtures of DNA on microbeads: Physical separation of differentially expressed cDNAs. *Proc. Natl. Acad. Sci. USA* **97,** 1665–1670.
69. Sepp, A., Tawfik, D. S., and Griffiths, A. D. (2002). Microbead display by *in vitro* compartmentalisation: Selection for binding using flow cytometry. *FEBS Lett.* **532,** 455–458.
70. Dressman, D., Yan, H., Traverso, G., Kinzler, K. W., and Vogelstein, B. (2003). Transforming single DNA molecules into fluorescent magnetic particles for detection and enumeration of genetic variations. *Proc. Natl. Acad. Sci. USA* **100,** 8817–8822.
71. Gilbert, W. (1986). Origin of life: The RNA world. *Nature* **319,** 618.
72. Gilbert, W., and de Souza, S. J. (1999). Introns and the RNA world. In "The RNA World" (R. F. Gesteland, T. R. Cech, and J. F. Atkins, Eds.), 2nd edn., pp. 221–231. Cold Spring Harbor Laboratory Press, Cold Spring Harbor, NY.
73. Szostak, J. W. (1999). Constrains on the sizes of the earliest cells. In "Size Limits of Very Small Microorganisms: Proceedings of a Workshop." pp. 120–125. National Academy Press, Washington, DC.
74. Oparin, A. I. (1941). "The Origins of Life on the Earth." 2nd edn., USSR Academy of Sciences, Moscow.
75. Jay, D. G., and Gilbert, W. (1987). Basic protein enhances the incorporation of DNA into lipid vesicles: Model for the formation of primordial cells. *Proc. Natl. Acad. Sci. USA* **84,** 1978–1980.
76. Hanczyc, M. M., Fujikawa, S. M., and Szostak, J. W. (2003). Experimental models of primitive cellular compartments: Encapsulation, growth, and division. *Science* **302,** 618–622.
77. Woese, C. R. (1998). The universal ancestor. *Proc. Natl. Acad. Sci. USA* **95,** 6854–6859.
78. Axelrod, V. D., Brown, E., Priano, C., and Mills, D. R. (1991). Coliphage Qβ RNA replication: RNA catalytic for single-strand release. *Virology* **184,** 595–608.
79. Saiki, R. K., Scharf, S., Faloona, F., Mullis, K. B., Horn, G. T., Erlich, H. A., and Arnheim, N. (1985). Enzymatic amplification of β-globin genomic sequences and restriction site analysis for diagnosis of sickle cell anemia. *Science* **230,** 1350–1354.
80. Myers, T. W., and Gelfand, D. H. (1991). Reverse transcription and DNA amplification by a *Thermus thermophilus* DNA polymerase. *Biochemistry* **30,** 7661–7666.
81. Myers, T. W., and Sigua, C. L. (1995). Amplification of RNA: High temperature reverse transcription and DNA amplification with *Thermus thermophilus* DNA polymerase. In "PCR Strategies" (M. A. Innis, D. H. Gelfand, and J. J. Sninsky, Eds.), pp. 58–68. Academic Press, San Diego.
82. Chetverina, H. V., Falaleeva, M. V., and Chetverin, A. B. (2004). Simultaneous assay of DNA and RNA targets in the whole blood using novel isolation procedure and molecular colony amplification. *Analyt. Biochem.* **334,** 376–381.
83. Young, K. K. Y., Resnick, R. M., and Myers, T. W. (1993). Detection of hepatitis C virus RNA by a combined reverse transcription-polymerase chain reaction assay. *J. Clin. Microbiol.* **31,** 882–886.
84. Mulder, J., McKinney, N., Christopherson, C., Sninsky, J., Greenfield, L., and Kwok, S. (1994). Rapid and simple PCR assay for quantitation of human immunodeficiency virus type 1 RNA in plasma: Application to acute retroviral infection. *J. Clin. Microbiol.* **32,** 292–300.

85. Young, K. K. Y., Archer, J. J., Yokosuka, O., Omata, M., and Resnick, R. M. (1995). Detection of hepatitis C virus RNA by a combined reverse transcription PCR assay: Comparison with nested amplification and antibody testing. *J. Clin. Microbiol.* **33**, 654–657.
86. Kotewicz, M. L., Sampson, C. M., D'Alessio, J. M., and Gerard, G. F. (1988). Isolation of cloned Moloney murine leukemia virus reverse transcriptase lacking ribonuclease H activity. *Nucleic Acids Res.* **16**, 265–277.
87. Weiss, R. B., Kimball, A. W., Gesteland, R. F., Ferguson, F. M., Dunn, D. M., Di Sera, L. J., and Cherry, J. L. (1995). Automated hybridization/imaging device for fluorescent multiplex DNA sequencing. *US Patent* 5,470,710.
88. Guerasimova, A., Ivanov, I., and Lehrach, H. (1999). A method of one-step enzyme labelling of short oligonucleotide probes for filter hybridization. *Nucleic Acids Res.* **27**, 703–705.
89. Chetverina, E. V., Kravchenko, A. V., Falaleeva, M. V., and Chetverin, A. B. (2007). Express hybridization of molecular colonies with fluorescent probes. *Russian J. Bioorg. Chem.* **33**, 423–430.
90. Samatov, T. R., Chetverina, H. V., and Chetverin, A. B. (2006). Real-time monitoring of DNA colonies growing in a polyacrylamide gel. *Anal. Biochem.* **356**, 300–302.
91. Chetverin, A. B., Samatov, T. R., and Chetverina, H. V. (2007). Non-invasive molecular colony methods, kits and apparatus *PCT* Publication W. 2007111639.
92. Cardullo, R. A., Agrawal, S., Flores, C., Zamecnik, P. C., and Wolf, D. E. (1988). Detection of nucleic acid hybridization by nonradiative fluorescence resonance energy transfer. *Proc. Natl. Acad. Sci. USA* **85**, 8790–8794.
93. Tyagi, S., and Kramer, F. R. (1996). Molecular beacons—probes that fluoresce upon hybridization. *Nat. Biotechnol.* **14**, 303–308.
94. LeDuc, J. W., Damon, I., Meegan, J. M., Relman, D. A., Huggins, J., and Jahrling, P. B. (2002). Smallpox research activities: US Interagency Collaboration, 2001. *Emerg. Infect. Dis.* **8**, 743–745.
95. Kravchenko, A. V., Chetverina, E. V., and Chetverin, A. B. (2006). Preservation of nucleic acid integrity in guanidine thiocyanate lysates of whole blood. *Russian J. Bioorg. Chem.* **32**, 547–551.
96. Boom, R., Sol, C. J., Salimans, M. M., Jansen, C. L., Wertheim-van Dillen, P. M., and van der Noordaa, J. (1990). Rapid and simple method for purification of nucleic acids. *J. Clin. Microbiol.* **28**, 495–503.
97. Chirgwin, J. M., Przybyla, A. E., MacDonald, R. J., and Rutter, W. J. (1979). Isolation of biologically active ribonucleic acid from sources enriched in ribonuclease. *Biochemistry* **18**, 5294–5299.
98. Chomczynski, P., and Sacchi, N. (1987). Single-step method of RNA isolation by acid guanidinium thiocyanate-phenol-chloroform extraction. *Anal. Biochem.* **162**, 156–159.
99. Haase, A. T., Retzel, E. F., and Staskus, K. A. (1990). Amplification and detection of lentiviral DNA inside cells. *Proc. Natl. Acad. Sci. USA* **87**, 4971–4975.
100. Uhlmann, V., Silva, I., Luttich, K., Picton, S., and O'Leary, J. J. (1998). In cell amplification. *Mol. Pathol.* **51**, 119–130.
101. Shapero, M. H., Leuther, K. K., Nguyen, A., Scott, M., and Jones, K. W. (2001). SNP genotyping by multiplexed solid-phase amplification and fluorescent minisequencing. *Genome Res.* **11**, 1926–1934.
102. Shimkus, M. L., Guaglianone, P., and Herman, T. M. (1986). Synthesis and characterization of biotin-labeled nucleotide analogs. *DNA* **5**, 247–255.
103. Olejnik, J., Krzymanska-Olejnik, E., and Rothschild, K. J. (1998). Photocleavable affinity tags for isolation and detection of biomolecules. *Methods Enzymol.* **291**, 135–154.

104. Welch, M. B., and Burgess, K. (1999). Synthesis of fluorescent, photolabile 3'-O-protected nucleoside triphosphates for the base addition sequencing scheme. *Nucleosides Nucleotides* **18**, 197–201.
105. Ruparel, H., Bi, L., Li, Z., Bai, X., Kim, D. H., Turro, N. J., and Ju, J. (2005). Design and synthesis of a 3'-O-allyl photocleavable fluorescent nucleotide as a reversible terminator for DNA sequencing by synthesis. *Proc. Natl. Acad. Sci. USA* **102**, 5932–5937.
106. Ju, J., Kim, D. H., Bi, L., Meng, Q., Bai, X., Li, Z., Li, X., Marma, M. S., Shi, S., Wu, J., Edwards, J. R., Romu, A., *et al.* (2006). Four-color DNA sequencing by synthesis using cleavable fluorescent nucleotide reversible terminators. *Proc. Natl. Acad. Sci. USA* **103**, 19635–19640.
107. Turcatti, G., Romieu, A., Fedurco, M., and Tairi, A. P. (2008). A new class of cleavable fluorescent nucleotides: Synthesis and optimization as reversible terminators for DNA sequencing by synthesis. *Nucleic Acids Res.*, published online February 7, 2008, doi:10.1093/nar/gkn021.
108. Shendure, J., Porreca, G. J., Reppas, N. B., Lin, X., McCutcheon, J. P., Rosenbaum, A. M., Wang, M. D., Zhang, K., Mitra, R. D., and Church, G. M. (2005). Accurate multiplex polony sequencing of an evolved bacterial genome. *Science* **309**, 1728–1732.
109. Cai, F., Chen, H., Hicks, C. B., Bartlett, J. A., Zhu, J., and Gao, F. (2007). Detection of minor drug-resistant populations by parallel allele-specific sequencing. *Nat. Methods* **4**, 123–125.
110. Kosobokova, O., Gavrilov, D. N., Khozikov, V., Stepukhovich, A., Tsupryk, A., Pan'kov, S., Somova, O., Abanshin, N., Gudkov, G., Tcherevishnik, M., and Gorfinkel, V. (2007). Electrokinetic injection of DNA from gel micropads: Basis for coupling polony technology with CE separation. *Electrophoresis* **28**, 3890–3900.
111. Davidson, S. (2000). Research suggests importance of haplotypes over SNPs. *Nat. Biotechnol.* **18**, 1134–1135.
112. Daly, M. J., Rioux, J. D., Schaffner, S. F., Hudson, T. J., and Lander, E. S. (2001). High-resolution haplotype structure in the human genome. *Nat. Genet.* **29**, 229–232.
113. Patil, N., Berno, A. J., Hinds, D. A., Barrett, W. A., Doshi, J. M., Hacker, C. R., Kautzer, C. R., Lee, D. H., Marjoribanks, C., McDonough, D. P., Nguyen, B. T., Norris, M. C., *et al.* (2001). Blocks of limited haplotype diversity revealed by high-resolution scanning of human chromosome 21. *Science* **294**, 1719–1723.
114. Gabriel, S. B., Schaffner, S. F., Nguyen, H., Moore, J. M., Roy, J., Blumenstiel, B., Higgins, J., DeFelice, M., Lochner, A., Faggart, M., Liu-Cordero, S. N., Rotimi, C., *et al.* (2002). The structure of haplotype blocks in the human genome. *Science* **296**, 2225–2229.
115. Guryev, V., Smits, B. M. G., van de Belt, J., Verheul, M., Hubner, N., and Cuppen, E. (2006). Haplotype block structure is conserved across mammals. *PLoS Genet.* **2**, e121.
116. International HapMap Consortium (2005). A haplotype map of the human genome. *Nature* **437**, 1299–1320.
117. Couzin, J. (2002). HapMap launched with pledges of $100 million. *Science* **298**, 941–942.
118. Zhang, K., Zhu, J., Shendure, J., Porreca, G. J., Aach, J. D., Mitra, R. D., and Church, G. M. (2006). Long-range polony haplotyping of individual human chromosome molecules. *Nat. Genet.* **38**, 382–387.
119. Daley, G. Q., Van Etten, R. A., and Baltimore, D. (1990). Induction of chronic myelogenous leukemia in mice by the P210BCR/ABL gene of the Philadelphia chromosome. *Science* **247**, 824–830.
120. Hughes, T., Deininger, M., Hochhaus, A., Branford, S., Radich, J., Kaeda, J., Baccarani, M., Cortes, J., Cross, N. C., Druker, B. J., Gabert, J., Grimwade, D., *et al.* (2006). Monitoring CML patients responding to treatment withtyrosine kinase inhibitors: Review and recommendations for harmonizing current methodology for detecting BCR-ABL transcripts and kinase domain mutations and for expressing results. *Blood* **108**, 28–37.

121. Nardi, V., Raz, T., Cao, X., Wu, C. J., Stone, R. M., Cortes, J., Deininger, M. W., Church, G., Zhu, J., and Daley, G. Q. (2008). Quantitative monitoring by polymerase colony assay of known mutations resistant to ABL kinase inhibitors. *Oncogene* **27**, 775–782.
122. Zhou, D., Zhang, R., Fang, R., Cheng, L., Xiao, P., and Lu, Z. (2008). Methylation pattern analysis using high-throughput microarray of solid-phase hyperbranched rolling circle amplification products. *Electrophoresis* **29**, 626–633.
123. Butz, J. A., Roberts, K. G., and Edwards, J. S. (2004). Detecting changes in the relative expression of KRAS2 splice variants using polymerase colonies. *Biotechnol. Prog.* **20**, 1836–1839.
124. Conrad, C., Zhu, J., Conrad, C., Schoenfeld, D., Fang, Z., Ingelsson, M., Stamm, S., Church, G., and Hyman, B. T. (2007). Single molecule profiling of tau gene expression in Alzheimer's disease. *J. Neurochem.* **103**, 1228–1236.
125. Merritt, J., DiTonno, J. R., Mitra, R. D., Church, G. M., and Edwards, J. S. (2003). Parallel competition analysis of *Saccharomyces cerevisiae* strains differing by a single base using polymerase colonies. *Nucleic Acids Res.* **31**, e84.
126. Merritt, J., Butz, J. A., Ogunnaike, B. A., and Edwards, J. S. (2005). Parallel analysis of mutant human glucose 6-phosphate dehydrogenase in yeast using PCR colonies. *Biotechnol. Bioeng.* **92**, 519–531.
127. Butz, J., Wickstrom, E., and Edwards, J. S. (2003). Characterization of mutations and LOH of p53 and K-ras2 in pancreatic cancer cell lines by immobilized PCR. *BMC Biotechnol.* **3**, 11.
128. Mikkilineni, V., Mitra, R. D., Merritt, J., DiTonno, J. R., Church, G. M., Ogunnaike, B., and Edwards, J. S. (2004). Digital quantitative measurements of gene expression. *Biotechnol. Bioeng.* **86**, 117–124.
129. Rieger, C., Poppino, R., Sheridan, R., Moley, K., Mitra, R., and Gottlieb, D. (2007). Polony analysis of gene expression in ES cells and blastocysts. *Nucleic Acids Res.* **35**, e151.
130. Butz, J. A., Yan, H., Mikkilineni, V., and Edwards, J. S. (2004). Detection of allelic variations of human gene expression by polymerase colonies. *BMC Genet.* **5**, 3.
131. Meyerhans, A., Vartanian, J.-P., and Wain-Hobson, S. (1990). DNA recombination during PCR. *Nucleic Acids Res.* **18**, 1687–1691.
132. Pääbo, S., Irwin, D. M., and Wilson, A. C. (1990). DNA damage promotes jumping between templates during enzymatic amplification. *J. Biol. Chem.* **265**, 4718–4721.
133. Jansen, R., and Ledley, F. D. (1990). Disruption of phase during PCR amplification and cloning of heterozygous target sequences. *Nucleic Acids Res.* **18**, 5153–5156.
134. Marton, A., Delbecchi, L., and Bourgaux, P. (1991). DNA nicking favors PCR recombination. *Nucleic Acids Res.* **19**, 2423–2426.
135. Odelberg, S. J., Weiss, R. B., Hata, A., and White, R. (1995). Template-switching during DNA synthesis by *Thermus aquaticus* DNA polymerase I. *Nucleic Acids Res.* **23**, 2049–2057.
136. Zaphiropoulos, P. G. (1998). Non-homologous recombination mediated by *Thermus aquaticus* DNA polymerase. I. Evidence supporting a copy choice mechanism. *Nucleic Acids Res.* **26**, 2843–2848.
137. Yu, W., Rusterholtz, K. J., Krummel, A. T., and Lehman, N. (2006). Detection of high levels of recombination generated during PCR amplification of RNA templates. *Biotechniques* **40**, 499–507.
138. Gustafsdottir, S. M., Schallmeiner, E., Fredriksson, S., Gullberg, M., Soderberg, O., Jarvius, M., Jarvius, J., Howell, M., and Landegren, U. (2005). Proximity ligation assays for sensitive and specific protein analyses. *Anal. Biochem.* **345**, 2–9.

Index

A

Amino-modifier C6 dT, 62
Aptamers, 211–213
Arithmetic operations, deoxyribozyme-based
   full adder circuits, 205–206
   half-adder circuits, 204–205
α−synuclein, 2
Avalanche photodiodes (APDs), 63

B

Base excision repair (BER), 127
Beak-induced replication (BIR), 130
Biopolymers
   conformational dynamics
      auto-correlation function, 46–49
      cross-correlation function, 47, 49, 52–53
      DNA hairpins, 57–58
      dsDNA, 59
      flow and relaxation methods, 36–37
      nucleosomes, 58
      fluorophore–quencher and FRET
         binding of nucleic acid, 45–46
         DNA hairpin, 49–50
         nucleosomes wrapping–unwrapping, 51
Biotin–streptavidin interactions, 55

C

5-carboxytetramethylrhodamine (TMR), 43
Catalytic molecular beacon, 202–203
Cell-free gene cloning
   dilution principle
      DNA molecule isolation, 231
      dual colony, 232
   molecular colony technique
      advantages, 230–231
      PCR technique, 229–230

Centromeres and sister chromatid
   cohesion, 125–126
CHRomatin Accessibility Complex (CHRAC)
   centromeres and sister chromatid
      cohesion, 125–126
   Dpb4/p17, 123–125
   Drosophila melanogaster, 123
   telomere-proximal silencing, 122
Chromatin remodeling complexes
   centromeres and sister chromatid
      cohesion, 125–126
   Dpb4/p17, 123–125
   Drosophila, 122–123
Chromatin states regulation
   gene silencing, 121
   telomere-proximal silencing, 122
   and telomeres, 122
Circular permuted RNAs (cpRNAs), 81–82
Coarse-grained simulation, 92
Coaxial stacking, 73
Comparative sequence analysis, 92
Conformational dynamics, nucleic acids
   auto-correlation decays
      proximity factor, 54–55
      simultaneous measurement, 52–53
      in single-stranded DNA, 48
      spatially offset laser beams, 53–54
      two-state system, 46–47
   cross-correlation decays
      in nucleosomes, 47, 49
      simultaneous measurement, 52–53
      spatially offset laser beams, 53–54
   DNA
      condensation and packaging, 59–60
      hairpins, 57–58
      mobility and flexibility, 59
   dsDNA, 59
   flow and relaxation methods, 36–37
   immobilization strategies, 55–56
   nucleosomes, 58
   quencher and FRET

Conformational dynamics, nucleic acids (*cont.*)
  DNA hairpin, 49–50
  nucleosomes wrapping–unwrapping, 51
Continuous wave EPR spectrum, 152, 172–173
Cryogenic atomic force microscopy (cryo-AFM), 86
Crystal cell, 10
Cyclopamine, 6–7

D

Deoxynucleoside triphosphate (dNTP), 110–112
Deoxyribozyme
  arithmetic operations
    full adder circuits, 205–206
    half-adder circuits, 204–205
  downstream elements control, 210–213
  logic gates
    allosteric regulation and molecular beacon, 202–203
    catalytic and recognition module, 201–202
    implicit OR arrangement, 204
  molecular automata
    MAYA, 207–209
    MAYA II, 209–210
  nanoparticles and three-layer cascade, 213–214
  oligonucleotides, 200
Dipolar coupling, 154
DNA
  autonomous computing, 215
  binding, 113
  duplexes
    dynamic behaviour, 180–182
    electrostatic potential distributions, 188
DNA hairpin
  auto and cross correlation functions
    proximity factor, 55
    spatially offset laser beams, 53–54
  dual-beam approach, 57
  fluorophore–quencher
    oxazine derivative, 48
    rhodamine 6G-dabcyl, 49–50
DNA polymerase ε (DNA pol ε)
  biochemical properties
    DNA binding, 113
    exonuclease, 112
    fidelity, 113–114
    polymerization, 110–112
  checkpoint control
    cell cycle progression and replication, 118–120
    Dpb2 and cell cycle, 120
  chromatin remodeling complexes
    centromeres and sister chromatid cohesion, 125–126
    Dpb4/p17, 123–125
    Drosophila melanogaster, 122–123
  chromatin states regulation
    gene silencing, 121
    telomere-proximal silencing, 122
    and telomeres, 122
  diverse cellular processes, 103
  DNA damages
    base excision repair, 127
    mismatch repair, 128–129
    nucleotide excision repair, 127–128
  gene names, 108
  physical and functional interactions, 109–110
  recombinational repair
    break-induced replication (BIR), 130
    double-strand break repair, 129
    gene conversion, 129
    mammalian recombination complex, 130
    rDNA recombination, 130–131
  replication of DNA
    fork formation, 116–118
    polymerase, 115–116
  Schizosaccharomyces pombe
    accessory subunits, 132
    catalytic subunit, 131–132
  structure
    catalytic subunit, 102–106
    holoenzyme, 106–109
  xenopus replication system, 133
Double electron-electron resonance (DEER) spectroscopy, 165–167
Double-strand breaks (DSB), 129
Dpb2
  and cell cycle, 120
  pol ε holoenzyme, 106–109
Drosophila melanogaster
  blood cell lines, 10–11
  in cancer research
    cell migration, 5–6
    mosaic analysis method, 4–5
    signaling pathways, 5

INDEX

therapeutic agents, 6–7
genes related to human diseases, 11–23
genetic model organism, 1–2
hypoxic response delineation
  hypoxia inducible factor 1, 8–9
  mitotic cycles arrest, 7
  innate immunity, 9–10
  neurological disease model
    Alzheimer's disease, 3
    Parkinson's disease, 2
    polyglutamine tract repeat
      disorders, 3–4
dsDNA, breathing fluctuations, 59

E

Energy transfer, 35
EPR lineshape analysis
  DNA duplexes, 180–182
  RNA molecules
    hammerhead ribozyme, 187
    HIV trans-activation responsive
      type, 184–186
    TLR-type containing GAAA tetraloop
      receptor, 182–184
  τ_R effect, 180–181
Evolution of RNA
  encapsulation, 234
  linking and segregation, 233–234
  translation, 235
Exonucleolytic activity, 112

F

Fenton reaction, 78–80
Fluorescein
  calibration method, 64
  triplet dynamics, 45
Fluorescence
  emission properties, 35
  spontaneous relaxation processes, 34
Fluorescence correlation spectroscopy (FCS)
  biopolymers, conformational dynamics
    DNA hairpin, 49–50, 57–58
    dsDNA, 59
    nucleosomes, 58
  fluctuation quantification
    auto-correlation functions, 42
    in detector setup, 40–42
    intensity fluctuations, 39
    measurement, 38–39
    normalized correlation functions, 40
    on immobilized molecules, 56
  instrumentation, 62–64
  photoisomerizations, 45
  quenching and FRET, 45–46
  signal-to-noise ratio, 64–65
  signal translational diffusion
    autocorrelation decay, 43–44
    observation volume, 42–43
    residence time, 44
  triplet dynamics, 44–45
  two-photon excitation, 65
Fluorescence resonance energy transfer
  molecular colony detection, 237
  RNA folding, 85
  and SDSL, 162–163
Fluorophore–quencher
  binding of DNA, 45–46
  deoxyribozyme-based half-adder, 205
  hairpin dynamics of DNA, 49–50
  wrapping–unwrapping of DNA, 51
Fluorophores
  chemical structures, 61–62
  photochemical stability, 60–61
  spectrum absorption, 60
Förster resonance energy transfer
  (FRET)
  applications, 36
  biopolymer labeling, 45–46
  energy transfer rate, 35
  nucleosomes wrapping–unwrapping, 50

G

GAAA tetraloop
  binding with RNA receptor, 181
  conformational changes, 185
  TLR RNA, 182–184
  τ_R effect, 180
Gene conversion, 129
Gene expression, molecular colony
  technique
  allele and mRNA copy number, 245–246
  alternative splicing, 244–245
Gene silencing, 121
Genetic mosaicism, 5

H

Hairpin ribozyme, 75
Hammerhead ribozyme, 75, 187
hedgehog gene, 6
Hepatitis delta virus ribozyme, 75
Hill equation, 89
Histone-fold motifs (HFMs), 107
HIV trans-activation responsive RNA, 184–186
Holoenzyme, 106–109
Homologous RNA modeling, 91–92
Homophila database, 2
Hydroxyl radical footprinting, 78–80
hypnos gene, 9
Hypoxia inducible factor-1 (HIF-1), 8–9

I

Implicit OR arrangement, 204
Internitroxide distances interpretation
  geometry-based modeling, 172
  molecular dynamics (MD)
    simulations, 170–172
  NASNOX discrete conformer
    search, 170–171
Interstrand cross-links (ICL), 128
Introns, 76

L

Lamellocytes, 10–11
Ligase logic gate, 211–212

M

MANIP program, 91
MAYA (Molecular Array of YES
  and AND gates)
  limitations test, 209–210
  well plates, 207–209
MAYA II, 209–210
Mdm2 oncoprotein, 110
M-DNA formation, 60
MicroRNAs (miRNAs), 77
Microscopic ordered macroscopic disordered
  (MOMD) model, 176
Molecular automata
  MAYA, 207–209
  MAYA II, 209–210
Molecular colony technique (MCT)
  applications, 247–248
  cell-free gene cloning
    advantages, 230–231
    dilution principle, 231–232
    PCR technique, 229–230
  DNA and RNA target detection
    in clinical samples, 238–239
    polymerase, 235–236
    reverse transcriptase, 236
  evolution and compartmentalization
    linking of gene, 233
    segregation of gene, 234
  gene expression analysis
    allele and mRNA copy number,
      245–246
    alternative splicing, 244–245
    quantitative *in situ* assay, 239–241
  RNA recombination study, 225–226
  sequencing of DNA and RNA
    PCR primer, 241
    in situ ligation, 242
  SNPs determination
    drug-resistant mutation, 244
    haplotype information, 243
  spontaneous RNA synthesis
    Koch's approach, 223–225
    recombination, 228–229
  visualization
    fluorescence detection systems,
      237–238
    hybridization method, 236–237
Molecular dynamics (MD)
  simulations, 170–172
Molecular interactions monitoring, τ_R effect
Molecular logic gates
  allosteric regulation and molecular
    beacon, 202–203
  catalytic and recognition module, 201–202
  deoxyribozyme-based
    full adder circuits, 205–206
    half-adder circuits, 204–205
  implicit OR arrangement, 204
  limitations, 204
  nanoparticles and three-layer
    cascade, 213–214
  oligonucleotides, 200
Mosaic analysis method, 4–5

INDEX

N

NASNOX discrete conformer search, 169–170
Negative allosteric regulation, 202–203
Nitroxide dynamics
  and EPR spectral lineshape, 173–174
  and spectral simulations
    MD simulation, 172–173
    MOMD model, 176
    SRLS model, 176
    stochastic Liouville equation (SLE), 175–176
Nitroxide-labeled macromolecules tumbling. See τ_R effect
Nitroxide labeling
  enzymatic methods
    nitroxide-modified nucleotides, 161–162
    at nucleic acid terminus, 161
  future developments, 162
  solid-phase chemical synthesis
    internal phosphate site, 154–159
    at 2′ position of sugar, 159
Nitroxide motion correlation
  quantitative spectral simulations, 178–179
  semiempirical nitroxide mobility, 177–178
Nitroxide scanning, 178
Nitroxide-solvent accessibility, 187–188
Non-coding RNAs, 72
Nuclear magnetic resonance (NMR) spectroscopy, 83
Nucleic acid DataBase (NDB), 82
Nucleic acid labeling, 62
Nucleosomes, conformational dynamics
  auto-correlation decays, 47, 49
  sequence positioning, 58
  wrapping–unwrapping, 50
Nucleotide excision repair (NER), 127–128
Nucleotide triphosphates (NTPs), 161

P

Paramagnetic metal ions probing, 189
Paramagnetic relaxation enhancement (PRE), 188–189
parkin gene, 2
Phosphodiesterase logic gate, 211–212
Photocross-linking approach, 81
Photo-induced electron transfer (PET), 46
Photoisomerizations, 45
Photomultipliers, 63
Photon counting histogram (PCH), 57
Plasmatocytes, 9–10
Polarity effects, 189
Pol ε. See DNA polymerase ε
Polymerization, 110–112
Presenilin gene, 3
Primary messenger RNA (pre-mRNA), 77
Proximity factor, 54–55

Q

$Q\beta$ replicase
  gene cloning, 229–230
  as MCT version, 220
  recombination of RNA
    cell-free system, 226
    spontaneous rearrangements, 227
    spontaneous recombination, 228–229
  RNA synthesis
    Koch's approach, 224–225
    nonrandom nucleotide sequences, 222–223
    protein complex and elongation factors, 221–222
Quantitative *in situ* assay (QISA), 239–241

R

Rhodamine 6G
  calibration standard, 64
  triplet dynamics, 45
Rhodamine 6G-dabcyl quencher modification, 50
Ribose zipper, 73
Ribosomal RNA (rRNA), 77
Ribozyme
  introns, 76
  rRNA and miRNA, 77–78
  types, 75
RNA
  ab initio modeling, 90
  evolution and compartmentalization
    encapsulation, 234
    linking and segregation, 233–234
    translation, 235
  folding and molecular dynamics
    free energy, 87–88

RNA (cont.)
　ionic environment, 88–89
　recombination
　　cell-free system, 226
　　in genome evolution, 228–229
　　sequence rearrangements, 227
　ribozyme
　　introns, 76
　　rRNA and miRNA, 77–78
　　types, 75
　secondary structure prediction, 90–91
　spontaneous synthesis
　　Koch's approach, 224–225
　　nonrandom nucleotide
　　　sequences, 222–223
　structural elements, 72–73
　structural property analysis
　　atomic force microscope, 85–86
　　cross-linking and photoaffnity
　　　agents, 81–82
　　cryo-AFM, 86
　　enzyme and chemical probing, 78
　　hydroxyl radical footprinting, 78–80
　　nucleotide analogues, 80–81
　　single-molecule method, 83–85
　　X-ray crystallography and NMR, 82–83
　tertiary structure modeling, 91–93
　vs. protein folding, 73–75

S

Schizosaccharomyces pombe
　accessory subunits, 132
　catalytic subunit, 131–132
Sfold algorithm, 90
Signal-to-noise ratio, 64–65
Silicomimetic molecular computing, 199.
　See also Deoxyribozymes
Single-nucleotide polymorphisms (SNPs)
　drug-resistant mutation, 244
　haplotype information, 243
Site-directed spin labeling (SDSL)
　basic principles
　　dipolar interaction, 152–153
　　Zeeman and hyperfine
　　　interaction, 150–152
　chemical structure, 149
　cw-EPR spectrum, 152
　distance measurements

　　internitroxide distances
　　　interpretation, 168–172
　　nanometer distances, 164–168
　　nitroxides and macromolecules, 163
　　pulsed EPR, 163–168
　　short-range distances, 163–168
　　vs. FRET, 162–163
　electron spin energy level diagram, 152
　EPR measurements, 148–149
　future Directions, 189
　local environment characterization, 149
　nitroxide labeling methods
　　enzymatic methods, 161–162
　　future developments, 162
　　solid-phase chemical synthesis,
　　　153–160
　nitroxide-solvent accessibility, 187–188
　paramagnetic metal ions probing, 189
　paramagnetic relaxation
　　enhancement, 188–189
　polarity effects, 188
　solid-phase chemical synthesis, 154–160
　structural and dynamic information
　　applications, 179–187
　　cw-EPR spectral analysis, 172–173
　　EPR spectral lineshape and nitroxide
　　　dynamics, 173–174
　　nitroxide dynamics
　　　characterization, 174–177
　　nitroxide motion correlation, 177–179
　　UUCG loop, 183
Slowly relaxing local structure (SRLS)
　model, 176
Small nuclear ribonucleoproteins
　(snRNPs), 77
Solid-phase chemical synthesis, 154–160
Spontaneous RNA synthesis
　Koch's approach
　　incubation and colony formation, 225
　　isolation of bacteria, 224
　　nonrandom nucleotide sequences, 222–223
　Qβ replicase, 220
Stochastic Liouville equation (SLE), 175–176

T

Taiman, p160-type steroid hormone
　coactivator, 6
Tau protein, 3

INDEX

Telomere-proximal silencing, 122
Tetraloop-receptor, 73
TLR-type containing GAAA tetraloop receptor, 182–184
TMR. *See* 5-carboxytetramethylrhodamine
trans-activation responsive (TAR), 184
τ_R effect, 179–180
Triplet dynamics, 44–45
Two-photon excitation (TPE), 65

X

Xenopus pol ε, 133
X-ray crystallography, RNA folding and structure, 82–83

Y

Yeast pol ε
  cell cycle progression and replication, 119–120
  DNA repair, 127
  Dpb4/p17, 123
  fidelity, 113–114
  holoenzymes, 106–108
  polymerisation, 111–112
  replication fork, 116–118
  replicative polymerase, 115–116

Z

Zeeman and hyperfine interaction, 150–152